AF835509

Building Integrated Photovoltaic Thermal Systems

For Sustainable Developments

Energy Series

Series Editor:
Julian Hunt FRS, *University College London, London, UK*

Titles in the Series:
1: Hydrogen Energy: Challenges and Prospects
2: Fundamentals of Photovoltaic Modules and its Applications
3: Compound Energy Systems: Optimal Operation Methods
4: Building Integrated Photovoltaic Thermal Systems: For Sustainable Developments

How to obtain future titles on publication:
A standing order plan is available for this series. A standing order will bring delivery of each new volume immediately on publication.

For further information please contact:
Book Sales Department, Royal Society of Chemistry, Thomas Graham House, Science Park, Milton Road, Cambridge, CB4 0WF, UK
Telephone: +44 (0)1223 420066, Fax: +44 (0)1223 420247, Email: books@rsc.org
Visit our website at http://www.rsc.org/Shop/Books/

Building Integrated Photovoltaic Thermal Systems

For Sustainable Developments

Basant Agrawal and G. N. Tiwari
Centre for Energy Studies, Indian Institute of Technology Delhi, New Delhi, India

RSC Publishing

RSC Energy Series No. 4

ISBN: 978-1-84973-090-7
ISSN: 1757-6741

A catalogue record for this book is available from the British Library

© B. Agrawal and G. N. Tiwari 2011

All rights reserved

Apart from fair dealing for the purposes of research for non-commercial purposes or for private study, criticism or review, as permitted under the Copyright, Designs and Patents Act 1988 and the Copyright and Related Rights Regulations 2003, this publication may not be reproduced, stored or transmitted, in any form or by any means, without the prior permission in writing of The Royal Society of Chemistry or the copyright owner, or in the case of reproduction in accordance with the terms of licences issued by the Copyright Licensing Agency in the UK, or in accordance with the terms of the licences issued by the appropriate Reproduction Rights Organization outside the UK. Enquiries concerning reproduction outside the terms stated here should be sent to The Royal Society of Chemistry at the address printed on this page.

The RSC is not responsible for individual opinions expressed in this work.

Published by The Royal Society of Chemistry,
Thomas Graham House, Science Park, Milton Road,
Cambridge CB4 0WF, UK

Registered Charity Number 207890

For further information see our web site at www.rsc.org

Preface

Solar photovoltaic (PV) systems are one of the most promising renewable energy technologies, producing electricity on site directly from solar radiation without harming the environment or depleting materials. The building integrated photovoltaic thermal (BIPVT) system is a technology that merges PV and thermal systems, simultaneously providing both electric and thermal energy. Through this combination, more energy is generated per unit surface area in comparison to the stand-alone photovoltaic system.

This book is intended for specialists, scientists and people involved in research in the disciplines of renewable energy, energy studies, building energy or carbon credits. For the practicing professionals, advanced senior or graduate students with work experience, the book should be used as part of an integrative programme enabling deep linkages to be made and thus better decisions in the professional world. The work is a summary of the knowledge gained by the authors from the experience of years of research and teaching.

The book has been divided into nine chapters. Chapter 1 begins with the fundamental concepts of solar energy and the methodology to determine its availability in terrestrial regions. Chapter 2 reviews the technology for manufacturing silicon and non-silicon photovoltaic solar cells and modules. Chapters 3 and 4 describe human comfort conditions and review passive heating and cooling concepts. Chapter 5 deals with the worldwide installation of photovoltaic systems and a few case studies in this context with regard to BIPVT systems. Chapter 6 deals with thermal modelling, energy and exergy analysis. It shows that the use of BIPVT systems significantly increases electrical performance. Chapters 7 and 8 deal with embodied energy analysis, energy payback periods, carbon mitigation and trading. This concludes that the suggested system has a negligible impact on the environment. Chapter 9 provides the techno-economics of the BIPVT system, showing that it has a faster payback time than any traditional system. It is recommended that beginners read Chapters 6 through 9 in the same chronological order as given in the book. SI

RSC Energy Series No. 4
Building Integrated Photovoltaic Thermal Systems: For Sustainable Developments
By Basant Agrawal and G. N. Tiwari
© B. Agrawal and G. N. Tiwari 2011
Published by the Royal Society of Chemistry, www.rsc.org

units have been used throughout. Appendices and a glossary have also been included at the end of the book.

There is no other single book which covers all the basic and the advanced concepts related to the implementation of solar energy for the passive heating and cooling of buildings. In addition, the present book covers the concepts of modelling and analysis of the BIPVT system, which has not been made elsewhere. All chapters are supplemented with numerous diagrams to understand the concepts better. Anyone who goes through the book cannot miss the enormous work that has gone into preparing the text in the present form. We hope that this book will be a milestone in the widespread deployment of solar designed buildings.

It is our immense pleasure to express our heartfelt gratitude to the Director of IIT Delhi and Head of CES, IIT Delhi for their kind encouragement. We express our appreciation to the reviewers of this book. We wish to thank all the authors that are represented in the references for granting us permission to reproduce their work. We acknowledge with thanks the financial support by the Curriculum Development Cell, IIT Delhi, for preparation of the book. Full credit is due to our publishers, RSC Publishing, Cambridge, UK, for producing the book.

We express our deep gratitude to our respected parents for inspiration and their blessings. Not the least, we thank our patient families for their forbearance during the lengthy process of putting this book together. Comments and suggestions for further improvements of the book can be mailed to the authors at bas_agr@yahoo.co.in or gntiwari@ces.iitd.ernet.in.

Basant Agrawal
G. N. Tiwari

Contents

RSC Energy Series No. 4
Building Integrated Photovoltaic Thermal Systems: For Sustainable Developments
By Basant Agrawal and G. N. Tiwari
© B. Agrawal and G. N. Tiwari 2011
Published by the Royal Society of Chemistry, www.rsc.org

About the Authors

Basant Agrawal obtained his BE and MTech from Maulana Azad National Institute of technology (MANIT), Bhopal. He was awarded a gold medal for securing first position in MTech by Rajiv Gandhi Technical University, Bhopal. He is currently working as Senior Lecturer in the Department of Mechanical Engineering at the Shri G. S. Institute of Technology and Science, Indore, and is on deputation pursuing a PhD degree from the Indian Institute of Technology, Delhi. He has more than 10 years teaching experience. His broad areas of interest include thermodynamics and heat engines, heat transfer, hybrid PV thermal (HPVT) systems and passive building design. He has authored four books with well-known publishers and has published over 30 research papers in various international and national journals and conferences. He is a life member of the Indian Society for Technical Education and the Institution of Engineers (India).

Professor Gopal Nath Tiwari received postgraduate and doctoral degrees from Banaras Hindu University (BHU). Since 1977 he has been actively involved in teaching and research at the Centre for Energy Studies, Indian Institute of Technology, Delhi. His area of research includes solar distillation, water/air heating systems, greenhouse technology, earth–air heat exchangers, passive building design and hybrid PV thermal (HPVT) systems. He has guided about 60 PhD students and published over 400 research papers in well-known refereed journals. He has written 18 books in association with respected publishers. He is a co-recipient of the Hariom Ashram Prerit S. S. Bhatnagar Award in 1982. Dr Tiwari worked at the University of Papua New Guinea between 1987 and

1989 as Energy and Environment Expert and became a European Fellow at the University of Ulster, Belfast, Northern Ireland (UK) in 1993. He is responsible for developing the Solar Energy Park at IIT Delhi and the Energy Laboratory at the University of Papua New Guinea, Port Moresby. He has been invited to speak in many countries, including Italy, Canada, the US, the UK, Australia, Greece, Thailand, Singapore and Taiwan. He has chaired international conferences as an expert in renewable energy, presenting research papers, *etc.* He was appointed as a "Solar Energy Valued Associate Editor" by the publisher Elsevier, Oxford (UK), in 2008 for his exceptional contribution to the quality of solar energy. He was editor of the *International Journal of Agricultural Engineering* for three years (2006–2008). He is also a member of the editorial advisory board for more than a dozen journals, including *Solar Energy*, *International Journal of Photoenergy*, *International Journal of Energy Research*, *The Open Environmental Sciences*, *The Open Renewable Energy Journal* and *Open Fuels & Energy Science Journal.* On 26 March 2008 the government of Uttar Pradesh, India, awarded him the Vigyan Ratna.

Dedication

Our respected teacher and Guruji
Padamshri Professor M. S. Sodha F.N.A.
On his 79th birthday (8 February 2011)

CHAPTER 1

Solar Radiation and its Availability on Earth

1.1 Introduction

The Sun flux entering the terrestrial region is nearly 1.2×10^{17} W, being absorbed by the Earth as shown in Figure 1.1. Out of this, 66.67% energy is available as the sensible amount that can be used for heating. However, we know that the Earth's surface area covered by ocean is 71% (361×10^6 km^2). Also, forest covers nearly 10% of the Earth's surface or nearly 35% of the total land area. Therefore the solar flux available on the plain area of the land is nearly 2.3×10^{16} W, which can be utilized for heating or generating electricity using photovoltaic modules. If only $\frac{1}{5750}$th part of this energy is utilized for power generation through photovoltaic systems, it is sufficient to meet the world's total electricity demand. This chapter introduces solar radiation and methodology to determine its availability on roofs and facades for utilization.

1.2 The Sun

It is considered that about 4.59 billion years ago the rapid collapse of a hydrogen molecular cloud led to the formation of a third-generation T Tauri population I star called the Sun. The surface contains nearly 92% of hydrogen (on volume basis), 7% of helium and trace quantities of other elements such as iron, nickel, oxygen, silicon, sulfur, magnesium, carbon, neon, calcium and chromium. Figure 1.2 shows the structure of the Sun. Here the Sun behaves like a fluid responsible for heat transfer by convection inside the Sun. The Sun rotates on its axis as a gaseous body in which the rotation of the equator takes about 27 days and the rotation of the polar regions take about 30 days. The

RSC Energy Series No. 4
Building Integrated Photovoltaic Thermal Systems: For Sustainable Developments
By Basant Agrawal and G. N. Tiwari
© B. Agrawal and G. N. Tiwari 2011
Published by the Royal Society of Chemistry, www.rsc.org

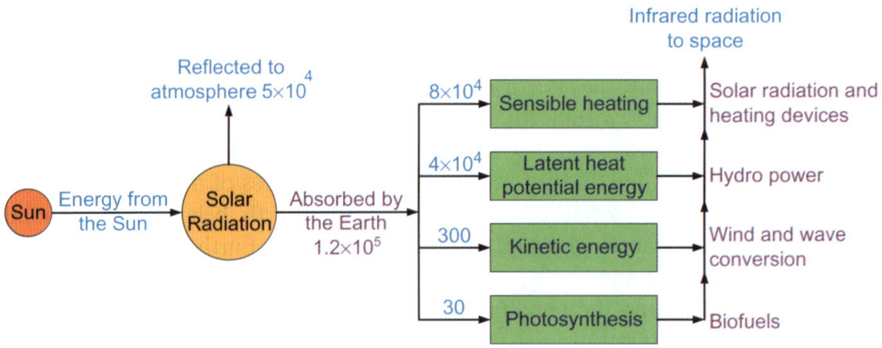

Figure 1.1 Available solar energy for utilization in heating devices such as photovoltaic modules.

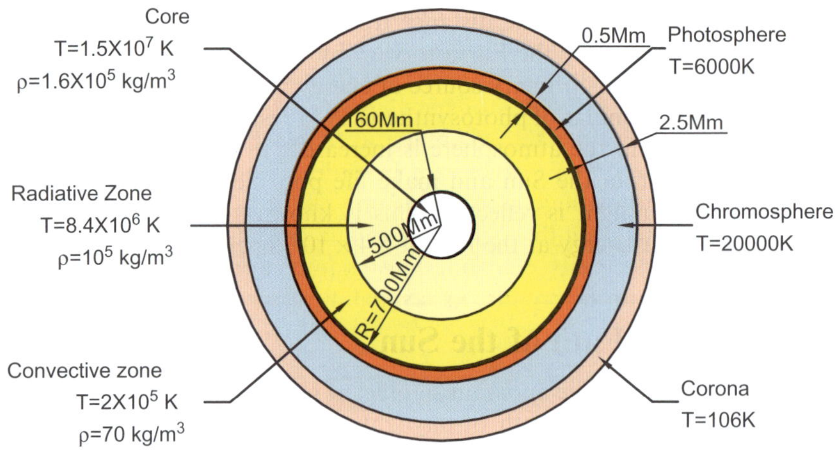

Figure 1.2 Structure of the Sun.

Sun's radius is measured from its centre to the edge of the photosphere, which is nearly 7×10^8 m.

The energy radiated by the Sun is due to continuous fusion reactions occurring simultaneously. The most important is the one in which hydrogen molecules combine to form a helium nucleus:

$$H_2 + H_2 \rightarrow He + 25\,MeV \tag{1.1}$$

The energy is produced in the interior of the solar sphere, which is transferred out to the surface. The average temperature at the surface of the Sun is nearly 5777 Kelvin. The Sun provides energy in the form of sunlight and heat, which are responsible for climatic and weather changes and life on Earth.

1.3 The Earth

The Earth is the only place in the universe where life is known to exist. Scientific evidence indicates that the Earth formed 4.54×10^9 years ago and that life appeared on its surface within 10^9 years. It is composed mostly of 32.1% iron, 30.1% oxygen, 15.1% silicon, 13.9% magnesium, 2.9% sulfur, 1.8% nickel, 1.5% calcium, 1.4% aluminium and the remaining 1.2% as traces of the other elements. Figure 1.3 shows the structure of the Earth. The Earth behaves like a solid responsible for heat transfer through conduction inside the Earth. The average heat flow from the centre of the Earth is 0.04–0.06 W m^{-2}. About 71% of the surface is covered with oceans and the remaining 29% consists of continents and islands.

The Earth's axis of rotation is tilted 23.45° away from the perpendicular to its orbital plane, which helps in producing seasonal variations with a period of one tropical year (365.24 solar days). The Earth's shape is very close to an oblate spheroid with an average diameter of about 12 742 km. The Earth's orbit around the Sun is an elliptical path (Figure 1.4), which causes the Earth's distance from the Sun to vary over a year. It is on average 1.495×10^{11} m from the Earth. The variation in the distance from the Sun causes the amount of solar radiation received by the Earth to vary by 6% annually.

Sunlight is the Earth's primary source of energy. The existence of blue-green algae marks the beginnings of photosynthesis. As a result of photosynthesis, the level of O_2 and O_3 in the atmosphere is increased. It blocks ultraviolet solar radiation coming from the Sun and make life possible on the Earth. Nearly one-third of the sunlight is reflected. This is known as Earth's albedo. The Earth receives solar energy at the rate of 5.4×10^{24} J per year.

1.4 Apparent Path of the Sun

The Sun's path refers to the apparent significant seasonal and hourly positional changes of the Sun (and length of daylight) as the Earth rotates and orbits

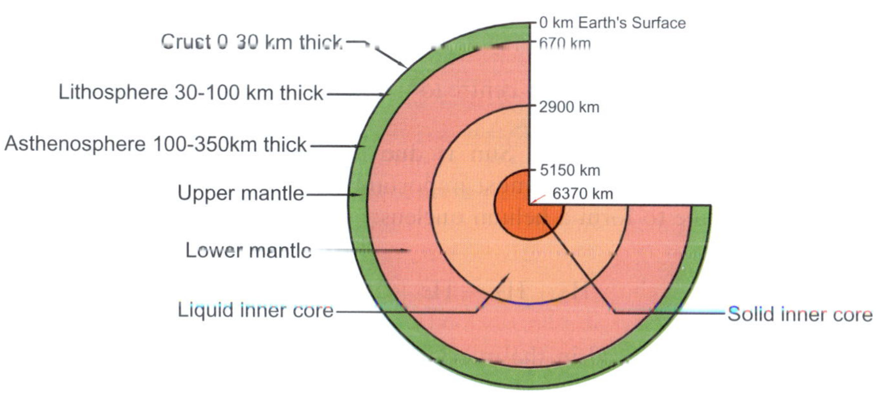

Figure 1.3 Structure of the Earth.

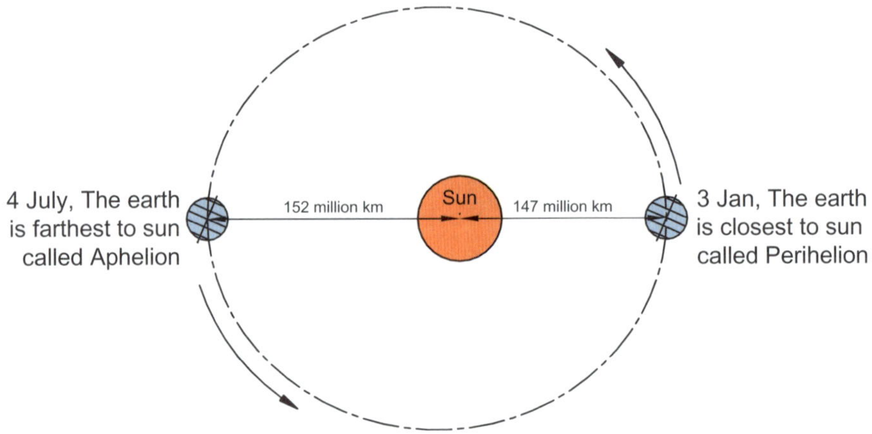

Figure 1.4 The Earth's orbit around the Sun.

around the Sun. The duration of sunshine is determined by the length of time when the Sun is above the horizon and varies throughout the year as the Earth–Sun geometric relationship changes.

Figure 1.5 (*left*) shows the Sun's path for the northern hemisphere occurring during the year. On 21st December, the sunshine duration is the shortest of the year and the Sun traces the lowest path in the southern sky, called the winter solstice. The Sun rises not exactly in the east but south of east and sets south of west. Each day after the winter solstice the Sun begins to rise closer to the east and set closer to the west until it rises exactly in the east and sets exactly in the west. On this day, about 21st March, sunshine lasts for 12 hours and is called the spring equinox.

After the spring equinox, the Sun still continues to follow a higher path through the sky with the sunshine duration growing longer. On 21st June, the sunshine duration is the longest and the Sun traces the highest path through the sky and directly above the Tropic of Cancer, called the summer solstice. This day the Sun rises not exactly in the east but north of east and sets north of west. After the summer solstice the Sun follows a lower path through the sky each day until it reaches the point where the sunshine lasts exactly 12 hours. This day is called the autumn (fall) equinox. After the this equinox, the Sun still continues to follow a lower path through the sky, with the sunshine hours growing shorter, until it reaches its lowest path at the winter solstice.

Figure 1.5 (*right*) shows the Sun's path for the southern hemisphere occurring in the same manner during the year. About 21st June the winter solstice occurs and on 21st December the summer solstice occurs. Note that it is incorrect to say that the summer solstice is the longest day of the year; the fact is that the day is still 24 hours long. However, it is correct to say that the number of sunshine hours is greatest.

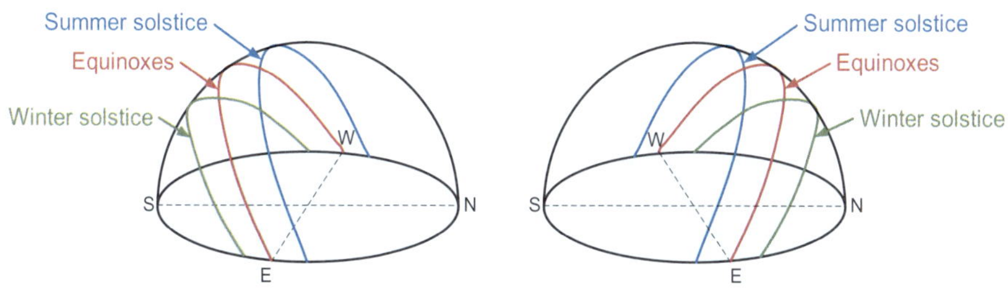

Figure 1.5 Apparent daily path of the Sun in the sky throughout the year: *left*, in the northern hemisphere of the Earth; *right*, in the southern hemisphere of the Earth.

The relative position of the Sun is a major factor in the heat gain of buildings and in the performance of solar energy systems. Accurate location-specific knowledge of the Sun's path and climatic conditions are essential for economic decisions about solar collector area, orientation, landscaping, summer shading and the cost-effective use of solar trackers.

1.5 Solar Radiation on the Earth

The radiation intensity on the surface of the Sun is approximately $6.33 \times 10^7\ \mathrm{W\,m^{-2}}$. The radiation spreads out as the distance squared by the time it travels to the Earth. The radiant energy flux received per second by a surface held normal to the direction of the Sun's rays at the mean Earth–Sun distance, outside the atmosphere, is practically constant throughout the year. This is termed the *solar constant* and its value adopted by the World Radiation Centre is $1367\ \mathrm{W\,m^{-2}}$ with an uncertainty of 1%. As the Earth revolves around the Sun in an elliptic path, the extraterrestrial radiation varies as:

$$I_{ext} = I_{sc}\left[1.0 + 0.033\cos\left(\frac{360 \times n}{365}\right)\right] \tag{1.2}$$

where n is the nth day of the year and $I_{sc} = 1367\ \mathrm{W\,m^{-2}}$. Figure 1.6 shows the annual variation of radiation in the extraterrestrial region.

The Earth is in equilibrium; therefore all gains of incoming energy are approximately equal to all losses of outgoing energy. The total flux of energy entering the Earth's atmosphere is estimated at 174×10^{15} W, which is equal to the product of the solar constant and the area of the Earth's disc ($1.28 \times 10^{14}\ \mathrm{m^2}$) as seen from the Sun. Nearly 30% of the incident solar energy is reflected back into space (20% from clouds, 6% from the atmosphere and 4% from the surface), called the albedo. The remaining 70% of the incident solar energy is absorbed by the Earth and then is re-radiated (64% by the clouds and atmosphere and 6% by the ground). Figure 1.7 shows a schematic

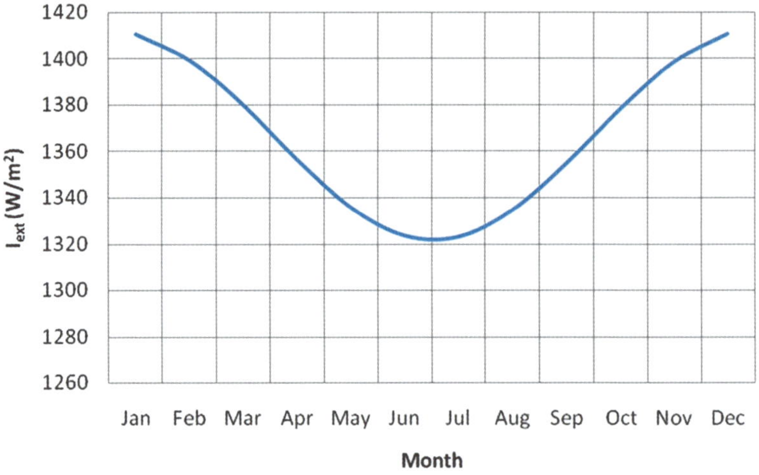

Figure 1.6 Variation of extraterrestrial solar radiation with time of year.

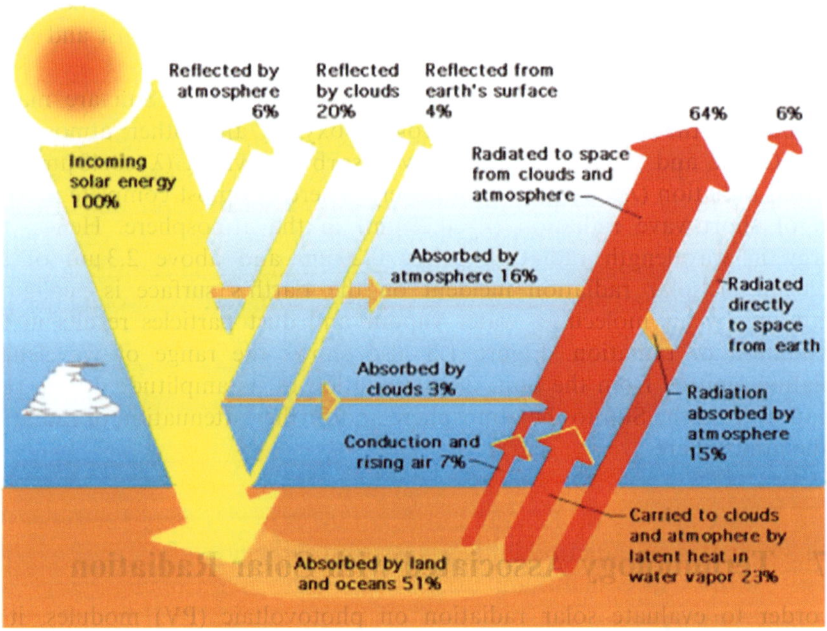

Figure 1.7 Energy equilibrium between outer space and the Earth.

representation of the energy exchanges between the Earth's surface, the Earth's atmosphere and outer space. Emissions of greenhouse gases, and other factors such as land-use changes, modify the balance of energy exchange between outer space and the Earth slightly but significantly.

1.6 Variation of Radiation in Extraterrestrial and Terrestrial Regions

The solar irradiance from a black body, either the Sun or the Earth, is a function of wavelength and is governed by Plank's law of radiation, which is given by:

$$E_{\lambda b} = \frac{C_1}{\lambda^5[\exp\{C_2/(\lambda T)\} - 1]} \tag{1.3}$$

where $E_{\lambda b}$ represents the energy emitted per unit area per unit time per unit wavelength (μm) interval at a given wavelength, $C_1 = 3.74 \times 10^{-16}\,m^2\,W$ and $C_2 = 0.0143879\,mK$.

While passing through the Earth's atmosphere, solar radiation is subjected to the mechanisms of atmospheric absorption and scattering. A fraction of the radiation reaching the Earth's surface is reflected back into the atmosphere and is subjected to this atmospheric phenomenon again. The remainder is absorbed by the Earth's surface. Figure 1.8 shows the position of the terrestrial and extraterrestrial regions. The atmospheric absorption is due to ozone (O_3), oxygen (O_2), nitrogen (N_2), carbon dioxide (CO_2), carbon monoxide (CO), water vapour and scattering from air molecules, dust and water droplets.

The X-rays and extreme ultraviolet radiation from the Sun are mainly absorbed in the ionosphere by nitrogen, oxygen and other atmospheric gases. Ozone and water vapour largely absorb ultraviolet ($\lambda < 0.40\,\mu m$) and infrared radiation ($\lambda > 2.3\,\mu m$), respectively. There is almost complete absorption of short-wave radiation ($\lambda < 0.29\,\mu m$) in the atmosphere. Hence, the energy in wavelength radiation below 0.29 μm and above 2.3 μm of the spectrum of solar radiation incident on the Earth's surface is negligible. Scattering by air molecules, water vapour and dust particles results in the attenuation of radiation. Figure 1.8 also shows the range of wavelength radiation emitted from the Sun, the attenuation of its amplitude during propagation from the Sun to the atmosphere and further attenuation of radiation in the atmosphere.

1.7 Terminology Associated with Solar Radiation

In order to evaluate solar radiation on photovoltaic (PV) modules, it is necessary to understand the following definitions.

1.7.1 Air Mass

This is a ratio indicating the amount of atmosphere that light must pass through before reaching an observer on the ground. The standard spectrum outside the Earth's atmosphere is called AM0 and describes solar

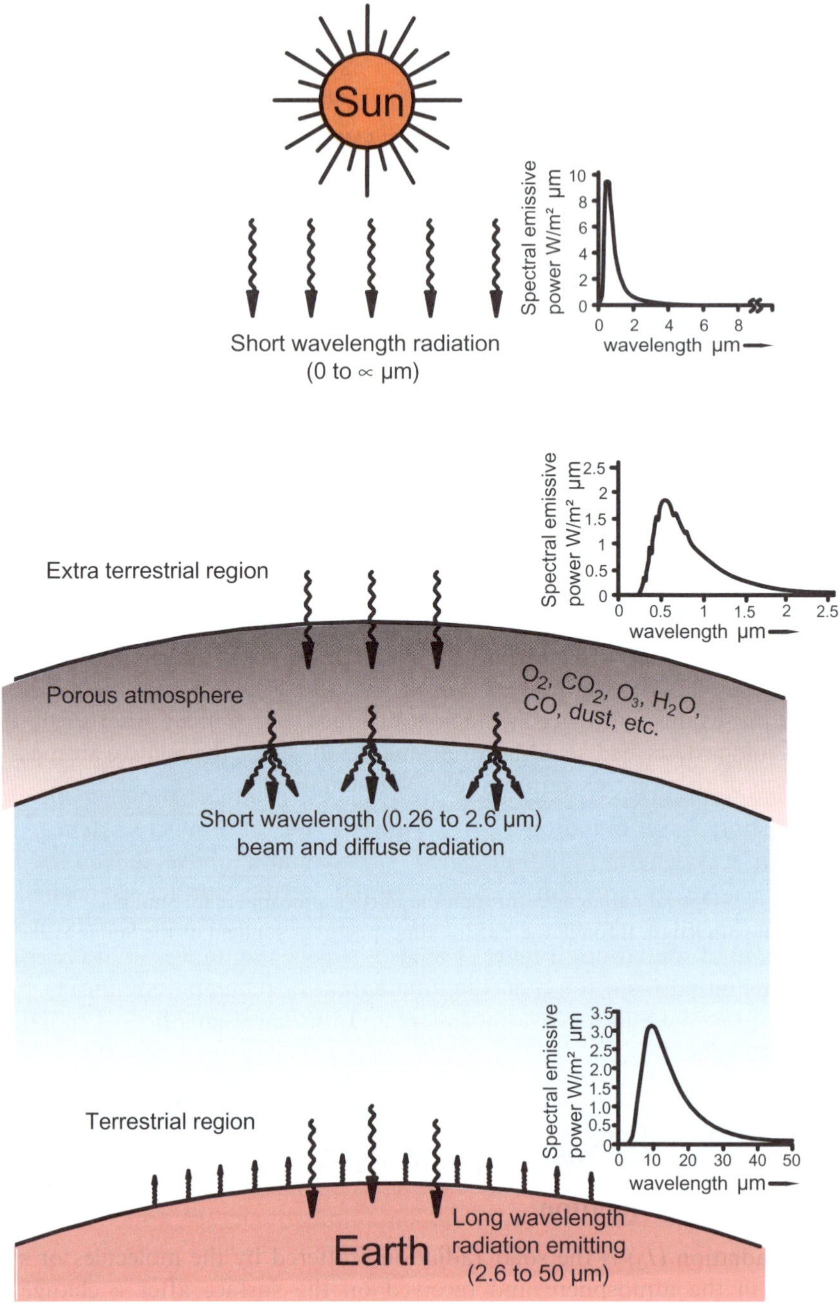

Figure 1.8 Propagation of solar radiation through the atmosphere.

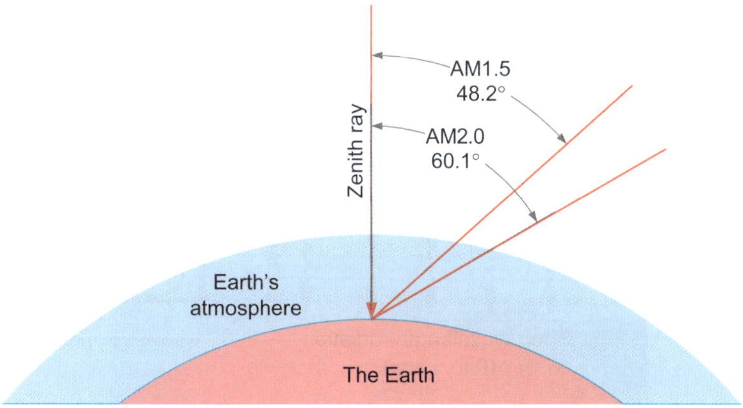

Figure 1.9 Relative air mass.

irradiance in space. The radiation that reaches sea level at high noon in a clear sky is 1000 W m^{-2} and is described as AM1 (or air mass 1) radiation. The Sun is at zenith and sunlight passes through the least amount of atmosphere to reach the ground. The air mass increases as the angle between the sunlight and the zenith increases (see Figure 1.9). When the Sun is 30° above the horizon, sunlight passes through twice as much atmosphere to reach an observer on the ground, and is described as AM2 (or air mass 2.0). Thus, the relative air mass is a function of the solar elevation angle. Mathematically:

$$\text{Air mass} = \frac{\text{Optical path length through the Earth's atmosphere for sunlight}}{\text{Optical path length through the Earth's atmosphere for sunlight, if the Sun is at zenith}}$$

$$\text{Air mass} = \sec\theta_z = \frac{1}{\cos\theta_z} = \frac{1}{\sin\alpha} \tag{1.4}$$

where θ_z is the zenith angle. At noon, $\theta_z = 0$; therefore the air mass = 1.

1.7.2 Diffuse Radiation

Diffuse radiation (I_d) is the solar radiation scattered by the molecules or suspensoids in the atmosphere and received on the surface after a change in direction (see Figure 1.10). It is also called diffuse skylight or sky radiation and is the reason for changes in the colour of the sky. The rate at which this energy falls on a unit horizontal surface per second is called the diffuse radiation. Under a clear sky at noon it is typically 100 W m^{-2}, but under cloudy conditions it may vary from 300 to 600 W m^{-2}.

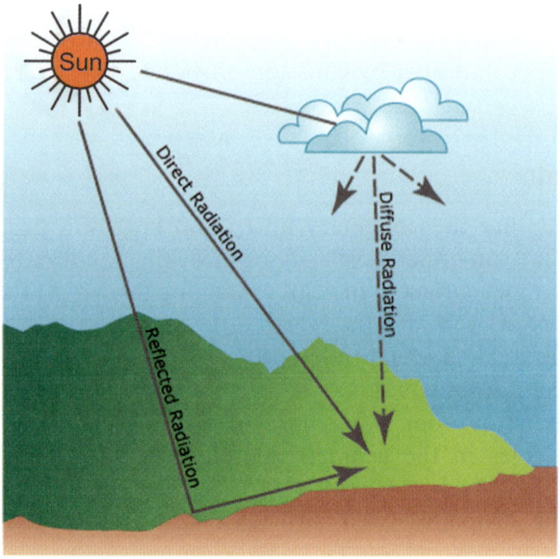

Figure 1.10 Beam, diffuse and reflected radiation.

1.7.3 Beam or Direct Radiation

Beam radiation (I_b) is the solar irradiance measured at a given location on the Earth with a surface element perpendicular to the Sun's rays, excluding diffuse radiation (see Figure 1.10). It is also called direct radiation. Direct insolation is equal to the solar constant minus the atmospheric losses due to absorption and scattering. While the solar constant varies with the Earth–Sun distance and solar cycles, the losses depend on the air mass, cloud cover, moisture content and other impurities.

1.7.4 Total Radiation or Global Radiation

The sum of beam and diffuse radiations falling on a horizontal surface facing upwards is called the total radiation or the global radiation (symbol I).

1.7.5 Insolation

This is a term applied specifically to solar energy irradiation. The symbol H is used for insolation for a day. The symbol I is used for insolation for an hour.

1.7.6 Irradiance, Radiant Exitance and Emissive Power

These are radiometry terms for the power per unit area of electromagnetic radiation at a surface. Irradiance is the rate at which radiant energy is incident on a surface, per unit area. Radiant exitance or radiant emittance is the rate at which radiant energy leaves a surface per unit area by combined emission, reflection and transmission. Emissive power is the rate at which radiant energy leaves a surface per unit area by emission only.

1.7.7 Latitude

Latitude lines (ϕ) are the imaginary horizontal lines shown running east-to-west on maps that run either north or south of the equator. Technically, latitude is an angular measurement in degrees ranging from 0° at the equator to 90° at the poles (90°N or +90° for the North Pole and 90°S or −90° for the South Pole); see Figure 1.11. The latitude is approximately the angle between the zenith and the Sun at an equinox.

Besides the equator, four other lines of latitude are named because they play important roles in the geometrical relationship of the Earth with the Sun, namely the Arctic Circle (66°33′39″N), the Tropic of Cancer (23°26′21″N), the Tropic of Capricorn (23°26′21″S) and the Antarctic Circle (66°33′39″S).

1.7.8 Longitude

Constant longitude is represented by lines running from the north pole to the south pole of the Earth (see Figure 1.12). The line of longitude that passes through the Royal Observatory, Greenwich, in England, establishes the meaning of zero degrees of longitude, or the prime meridian. Any other longitude is identified by the east–west angle, referenced to the center of the Earth as the vertex, between the intersections with the equator of the meridian through the location in question and the prime meridian.

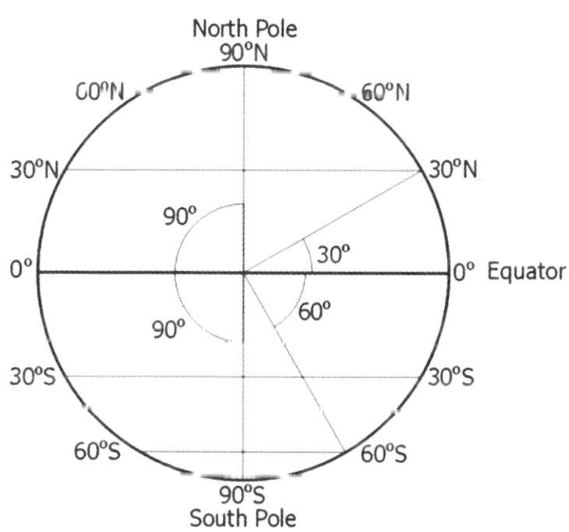

Figure 1.11 Latitude.

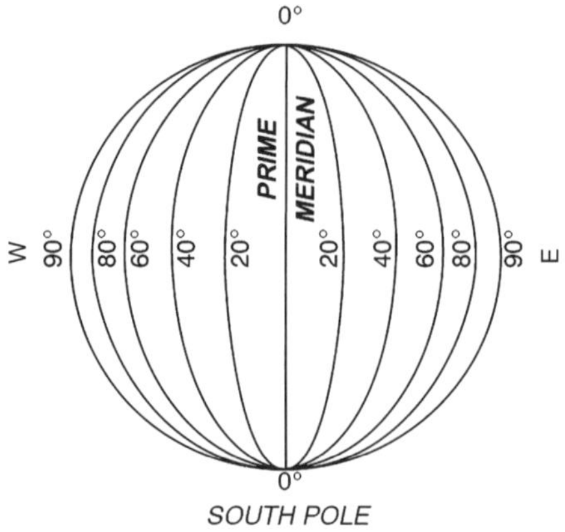

Figure 1.12 Longitude.

1.7.9 Solar Time

Universe time is the time ascertain with respect to the zero meridian. It is the same as Greenwich mean time. Standard time is the time reckoned with respect to the standard meridian of a specific time zone. It is the everyday clock time used within each one hour zone. This kind of time is not very useful in trying to locate the exact position of celestial objects.

Apparent solar time or true solar time is the time measured with respect to the Sun. It is based on the daily apparent motion of the Sun across the sky, with solar noon denoting the time the Sun crosses the local meridian. The difference between the solar time and the standard time is given by:

$$\text{Solar time} - \text{Standard time} = 4(L_{st} - L_{loc}) + E \tag{1.5}$$

where L_{st} is the standard meridian for the local time zone, L_{loc} is the longitude of the location in question (in degrees west) and E is the equation of time (in minutes) which is given by:

$$E = 229.2\begin{pmatrix} 0.000075 + 0.001868\cos B - 0.032077\sin B \\ -0.014615\cos 2B - 0.04089\sin 2B \end{pmatrix} \tag{1.6}$$

where $B = (n-1)\frac{360}{365}$ and n is the day of the year ($1 \leq n \leq 365$).

1.8 Measurement of Solar Radiation on the Earth's Surface

Solar radiation data are becoming increasingly important both for photovoltaic and thermal systems. This applies in activities such as research and development, production quality control, determination of optimum locations, monitoring the efficiency of installed systems and predicting the system output under various sky conditions. PV materials have most of their sensitivity from approximately 0.4 to 1.1 μm, with a peak just beyond the visible range. There is no response to long-wave radiation and little to ultraviolet radiation. Measurements of solar radiation are usually made using thermopile-type radiometers with a flat spectral response. Of course, it is of great importance to set the radiometers in an area where the Sun's rays are not blocked by buildings, trees or poles.

1.8.1 The Normal Incidence Pyrheliometer

This is an instrument used to measure the direct (or beam) component of solar radiation at normal incidence. Figure 1.13 shows a typical normal incidence pyrheliometer (NIP) consisting of a collimating tube having a thermopile sensor at one end while the other end faces the Sun. Solar irradiance entering the collimating tube through a sealed crystal-quartz window is directed onto the black coating plate kept over the thermopile sensor, which absorbs the solar radiation and converts it into heat. The thermopile sensor converts the heat into an equivalent electrical signal. A calibration factor is applied to convert the electrical signal to an equivalent radiant energy flux, measured in watts per square meter. A two-axis tracking mechanism is incorporated to maintain the Sun's disc within the acceptance cone of the instrument. This instrument is sensitive to wavelengths in the band from 0.28 to 3 μm.

It may be noted that the solar intensity thus measured is the radiation from the Sun and from a portion of the sky around the Sun. Since the detectors do not distinguish between forward scattered radiation (which comes from the circumsolar sky) and the beam radiation, the instruments are in effect defining beam radiation. As per the latest ISO 9060 and World

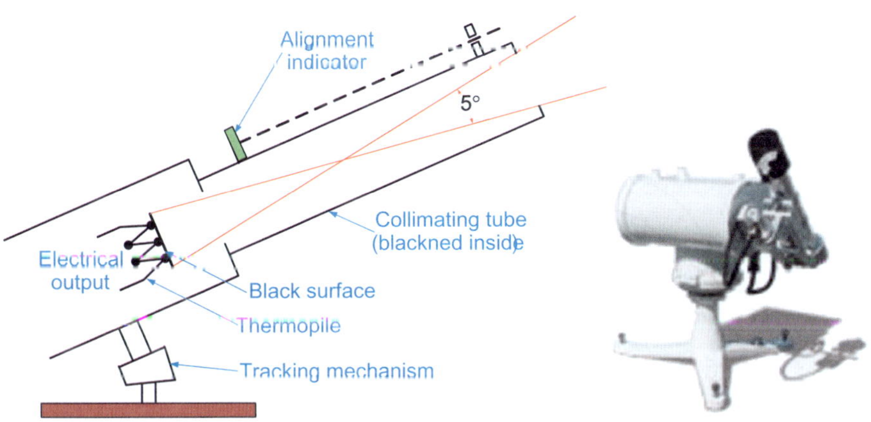

Figure 1.13 A pyrheliometer: *left*, schematic diagram; *right*, photograph.

Meteorological Organization standards, a full opening view angle of precisely 5.0° makes the sensor ideally suited for normal incidence direct solar irradiance measurement.

1.8.2 The Pyranometer

The pyranometer is an instrument used to measure the global (combined beam and diffuse) solar radiation received from the entire hemisphere on a horizontal surface. The principle of working is the same as pyrheliometer except for the fact that a sensitive surface is exposed to the total radiation. Applicable standards are ISO 9060 and 9847 and ASTM E824-94.

Figure 1.14 shows the schematic view and a photograph of a pyranometer. It consists of a thermopile sensor with a black coating plate in order to attain the proper directional and spectral characteristics. The black coating on the thermopile sensor absorbs the solar radiation and converts it into heat. The thermopile sensor senses the heat and generates an equivalent electrical output signal which is proportional to the solar radiation. The sensor has a near-perfect cosine response and can absorb solar radiation in the range of 0.3–50 μm. One or two hemispherical glass domes covering the sensor limit the spectral response from 0.3 to 2.8 μm (cutting off the part above 2.8 μm) while preserving the 180° field of view. The dome also shields the thermopile sensor from wind and rain, thereby reducing the convection currents.

Pyranometers have also been based on PV (solar cell) detectors. Silicon cells are the most common, although cadmium sulfide and selenium cells have been used. Silicon solar cells have the property that their light current is a linear function of the incident solar radiation. They have the disadvantage that their spectral response is not linear, so instrument calibration is a function of the spectral distribution of the incident radiation.

Figure 1.15 shows a pyranometer provided with an occulting disc, which is used to measure the diffuse radiation. This disc, or band, blocks the beam radiation from the surface. It may be noted that the pyranometers are

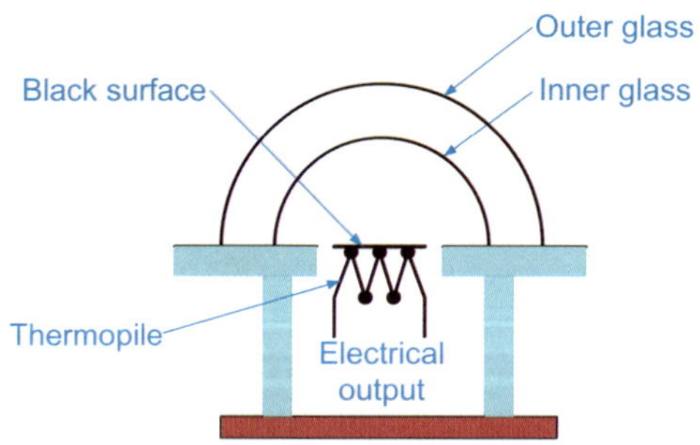

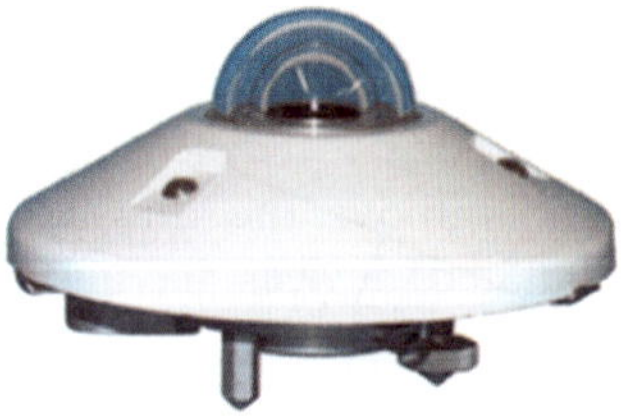

Figure 1.14 A pyranometer: *left*, schematic diagram; *right*, photograph.

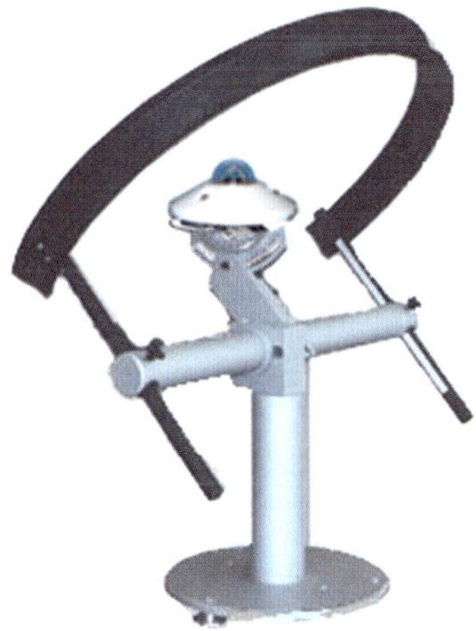

Figure 1.15 A pyranometer with an occulting disc.

calibrated so as to measure the solar radiation on a horizontal surface. Therefore, when tilted, the change in free convection regime within the glass dome may introduce an error in measurement.

1.8.3 The Sunshine Recorder

A sunshine recorder is an instrument used to measure the duration (in hours) of bright sunshine during the course of the day. In 1853, John Francis Campbell developed a burning sunshine recorder. Later, in 1879, Sir George Gabriel Stokes modified it by replacing water-filled glass spheres with solid glass spheres. This device is popularly known as the Campbell–Stokes recorder. Figure 1.16 shows a Campbell–Stokes recorder consisting of a glass sphere of 96 mm diameter. It is designed in a manner such that the rays from the Sun are focused onto a card mounted at the back and is set on a stand. The sphere burns a trace on the card for the bright sunshine threshold of $120\,W\,m^{-2}$. The length of the trace is the direct measure of the duration of bright sunshine. There are sets of grooves for taking three sets of cards, long curved for summer, short curved for winter and straight cards at equinoxes.

Figure 1.16 also shows a specially designed sunshine recorder with a photodiode as the sensing device. It detects the bright sunshine when the beam is above the threshold of $120\,W\,m^{-2}$ with the day length. They are automated devices having much finer time resolution, a far more precise threshold and they eliminate the daily burden of replacing the special card and of manually analyzing the burnt trace to estimate the daily hours of sunshine. These advantages considerably improve the reliability, value and accuracy of this measurement.

Figure 1.16 Sunshine recorders: *left*, Campbell–Stokes; *right*, electronic version.

1.8.4 The World Radiometric Reference

The World Meteorological Organization (WMO) is an intergovernmental organization sponsored by the United Nations through which the United States and more than 160 other nations cooperate in meteorological and weather-related science and engineering. The WMO promotes the development of reference instruments, conducts worldwide and regional field comparisons and evaluations, and recommends test and calibration methods and the corrections applied. WMO sanctions the Physikalisch Meteorologisches Observatorium Davos (PMOD) as its World Radiation Center to establish, maintain and propagate to users the engineering measurement scale for solar radiation measurements called the World Radiometric Reference (WRR). This measurement scale is used by all nations that acquire and report solar radiation measurement data for weather and meteorological purposes, for solar energy and engineering purposes and for space research.

In the past, several radiation references or scales have been used in meteorology, namely the Ångström scale of 1905, the Smithsonian scale of 1913 and the international pyrheliometric scale of 1956 (IPS 1956). The developments in absolute radiometry in recent years have very much reduced the uncertainty of radiation measurements. With the results of many comparisons of 15 individual absolute pyrheliometers of 10 different types, a WRR has been defined. The old scales can be transferred into the WRR using the following factors:

$$\text{WRR} = \frac{1.026}{\text{Ångström scale 1905}} = \frac{0.977}{\text{Smithsonian scale 1913}} = \frac{1.026}{\text{IPS 1956}} \tag{1.7}$$

The WRR is accepted as representing the physical units of total irradiance within 0.3% (99% uncertainty of the measured value).

In order to guarantee the long-term stability of the new reference, a group of at least four absolute pyrheliometers of different design is used as the World Standard Group (WSG). At the time of incorporation into this group, the instruments are given a reduction factor to correct their readings to the WRR. To qualify for membership of this group, a radiometer must fulfill the following specifications:

- Long-term stability must be better than 0.2% of the measured value.
- The 95% uncertainty of the series of measurements with the instrument must lie within the limits of the uncertainty of the WRR.
- The instrument has to have a different design from the other WSG instruments.

To meet the stability criteria, the instruments of the WSG are the subject of an intercomparison at least once a year, and for this reason the WSG is kept at WRC Davos in Switzerland.

1.9 Sun–Earth Geometric Relationship

The geometrical relationships between a plane of any particular orientation relative to the Earth at any time and the incoming beam radiation, *i.e.* position of the Sun relative to that plane, can be described in terms of several angles. Some of these angles and a set of consistent sign conventions are as follows.

1.9.1 The Declination

The declination (δ) is the angle between the line joining the centres of the Sun and the Earth and its projection on the equatorial plane. In other words, it is the angle between the rays of the Sun and the plane of the Earth's equator. The yearly variation of the declination is from $+23.45^\circ$ to -23.45° (see Figure 1.17). It is calculated as:

$$\delta = 23.45 \sin\left[\frac{360}{365}(284 + n)\right] \tag{1.8}$$

where n is the day of the year ($1 \leq n \leq 365$).

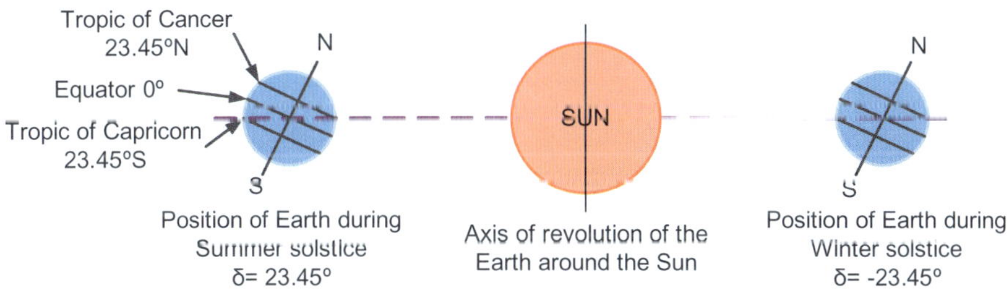

Figure 1.17 Yearly variation of solar declination.

Table 1.1 Recommended average days for months and the value of the day number and declination.

Month of the year	*N for i^{th} day of month*	*For the average day of the month*		
		Date	*Day of the year (n)*	*Declination (δ)*
January	i	17	17	−20.9
February	$31+i$	16	47	−13.0
March	$59+i$	16	75	−2.4
April	$90+i$	15	105	9.4
May	$120+i$	15	135	18.8
June	$151+i$	11	162	23.1
July	$181+i$	17	198	21.2
August	$212+i$	16	228	13.5
September	$243+i$	15	258	2.2
October	$273+i$	15	288	−9.6
November	$304+i$	14	318	−18.9
December	$334+i$	10	344	−23.0

The recommended average days for each month and the value of the day number and declination are given in Table 1.1. A more precise formula for declination is:

$$\delta = \frac{180}{\pi}\left(\begin{array}{r} 0.006918 - 0.399912\cos B + 0.070257\sin B - 0.006758\cos 2B \\ +0.000907\sin 2B - 0.002679\cos 3B + 0.00148\sin 3B \end{array}\right) \tag{1.9}$$

where $B = (n-1)\frac{360}{365}$ and n is the day of the year ($1 \le n \le 365$).

Example 1.1

Calculate the declination on 8th September 2010.

Solution

For the 8th September 2010, the day of the year is:

$$n = 31 + 28 + 31 + 30 + 31 + 30 + 31 + 31 + 8 = 251$$

Using eqn (1.8) we have:

$$\delta = 23.45^\circ \sin\left[\frac{360^\circ}{365}(284 + 251)\right] = 5^\circ.$$

1.9.2 The Hour Angle

The hour angle (ω) is the angle through which the Earth must be rotated to bring the plane of the local meridian directly under the Sun. In other words, it is the angular displacement of the Sun east or west of the local meridian, due to the rotation of the Earth on its axis. Basically, the hour angle expresses the solar time in angular measurement, at 15° per hour. It is zero at the solar noon, negative in the morning and positive in the afternoon for the northern hemisphere and *vice versa* for the southern hemisphere of the Earth. The hour angle can be determined using the expression:

$$\omega = (\text{Solar time} - 12) \times 15^\circ \tag{1.10}$$

where the apparent solar time varies from zero to 24 hours. At noon it is 12 hours.

Example 1.2

Calculate the hour angle at (a) 10 am and (b) 3 pm.

Solution

(a) At 10 am the solar time is 10 hours. From eqn (1.8) the hour angle is:

$$\omega = (10 - 12) \times 15^\circ = -30^\circ$$

(b) At 3 pm the solar time is 12+3=15 hours. From eqn (1.8) the hour angle is:

$$\omega = (15 - 12) \times 15^\circ = 45^\circ$$

1.9.3 The Zenith Angle and the Solar Altitude Angle

The zenith angle (θ_z) is defined as the angle between the Sun's rays and a line perpendicular to the horizontal plane (see Figure 1.18). It varies throughout the day from 0° to 90° with the movement of the Sun. When the Sun is rises or sets, the zenith angle is maximum (closer to 90°) and at the solar noon it is minimum (closer to 0°). It is calculated as:

$$\cos\theta_z = \cos\phi\cos\delta\cos\omega + \sin\delta\sin\phi \tag{1.11}$$

The altitude or solar altitude angle (α) is defined as the angle between the Sun's rays and a horizontal plane (see Figure 1.18). It also varies throughout the day from 0° to 90° with the movement of the Sun, but when the Sun is rising

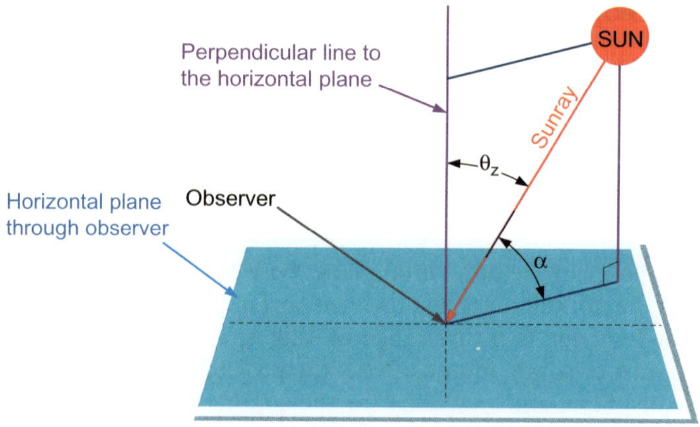

Figure 1.18 Zenith angle and solar altitude angle ($\theta_z + \alpha = 90°$).

or setting the zenith angle is minimum (closer to 0°) and at the solar noon it is maximum (closer to 90°). Thus:

$$\alpha = 90 - \theta_z \tag{1.12}$$

Example 1.3

Calculate the zenith and solar altitude angles at 10 am on 3rd February at a place with a latitude of 30°.

Solution

For the 3rd February, the day of the year $n = 31 + 3 = 34$. From eqn (1.8) the declination is:

$$\delta = 23.45° \sin\left[\frac{360°}{365}(284 + 34)\right] = -16.97°$$

At 10 am the solar time is 10 hours. From eqn (1.8) the hour angle is:

$$\omega = (10 - 12) \times 15° = -30°$$

From eqn (1.11) the zenith angle is:

$$\begin{aligned}\theta_z &= \cos^{-1}(\cos\phi\cos\delta\cos\omega + \sin\delta\sin\phi)\\ &= \cos^{-1}[\cos(30)\cos(-16.97)\cos(-30) + \sin(-16.97)\sin(30)] = 55.15°\end{aligned}$$

From eqn (1.11) the solar altitude angle is:

$$\alpha = 90° - \theta_z = 90° - 55.15° = 34.85°$$

1.9.4 The Slope

The slope (β) is the angle between the inclined surface under consideration and the horizontal plane. It is taken to be positive for a surface sloping towards the south and negative for surfaces sloping towards the north.

1.9.5 The Surface Azimuth Angle

This is the angle (γ) in the horizontal plane, between the line due south and the projection of the normal on to the inclined surface on the horizontal plane (see Figure 1.19). Its value is in the range of 180° to −180°. It is negative if the projection is east of south and positive if the projection is west of south for the northern hemisphere, and *vice versa* for the southern hemisphere.

1.9.6 The Solar Azimuth Angle

The solar azimuth angle (γ_s) is the angle, in a horizontal plane, between the line due south and the projection of beam radiation on the horizontal plane (see Figure 1.20). Its value is in the range of 180° to −180°. It is positive if the projection is east of south and negative if the projection is west of south for the northern hemisphere, and *vice versa* for the southern hemisphere.

The solar azimuth angle can be calculated using the empirical formula as follows:

$$\gamma_s = \text{sign}(\omega)\left|\cos^{-1}\left(\frac{\cos\theta_z \sin\phi - \sin\delta}{\sin\theta_z \cos\phi}\right)\right| \quad (1.13)$$

where sign(ω) represents the sign function and taken as +1 when ω is positive and −1 when ω is negative.

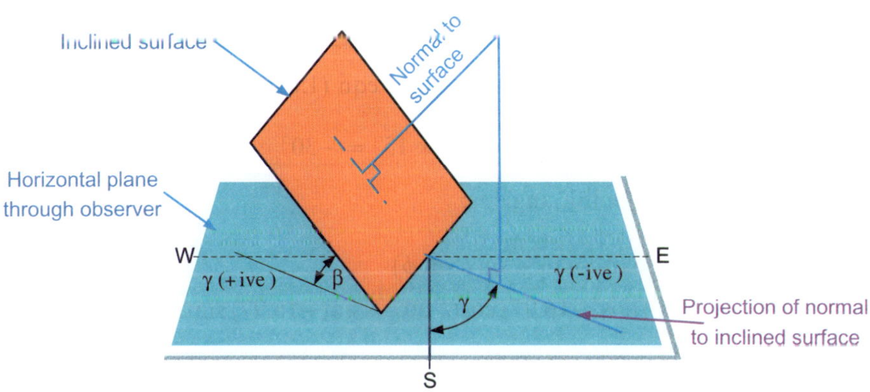

Figure 1.19 Surface azimuth angle, for an observer in the northern hemisphere.

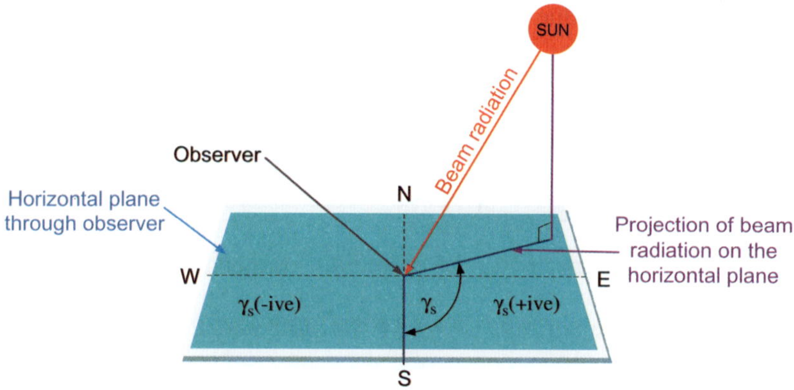

Figure 1.20 Solar azimuth angle, for an observer in the northern hemisphere.

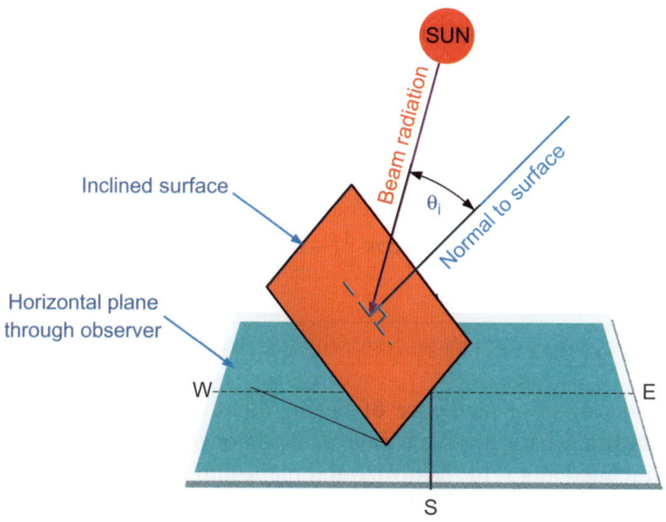

Figure 1.21 The angle of incidence.

1.9.7 The Angle of Incidence

This is the angle (θ_i) between beam radiation on a surface and the normal to that surface (see Figure 1.21). It is calculated as:

$$\cos\theta_i = (\cos\phi\cos\beta + \sin\phi\sin\beta\cos\gamma)\cos\delta\cos\omega \\ + \cos\delta\sin\omega\sin\beta\sin\gamma + \sin\delta(\sin\phi\cos\beta - \cos\phi\sin\beta\cos\gamma) \quad (1.14)$$

For a surface facing due south, $\gamma = 0$:

$$\cos\theta_i = \cos(\phi - \beta)\cos\delta\cos\omega + \sin\delta\sin(\phi - \beta) \tag{1.15}$$

For a vertical plane/wall facing due south, $\gamma = 0$ and $\beta = 90°$:

$$\cos\theta_i = -\sin\delta\cos\phi + \cos\delta\cos\omega\sin\phi \tag{1.16}$$

For a horizontal plane, $\gamma = 0$, $\beta = 0°$ and $\theta_i = \theta_z$; hence:

$$\cos\theta_z = \cos\phi\cos\delta\cos\omega + \sin\delta\sin\phi \tag{1.17}$$

For the sunset hour angle, $\theta_z = 90°$ and $\omega = \omega_s$. Substituting into eqn (1.17) we have:

$$\begin{aligned} 0 &= \cos\phi\cos\delta\cos\omega_s + \sin\delta\sin\phi \\ \Rightarrow \omega_s &= \cos^{-1}(-\tan\phi\tan\delta) \end{aligned} \tag{1.18}$$

The total angle between sunrise and sunset is given by:

$$2\omega_s = 2\cos^{-1}(-\tan\phi\tan\delta) \tag{1.19}$$

As $15° = 1$ hour, the number of sunshine hours is given by:

$$N = \frac{2}{15}\cos^{-1}(-\tan\phi\tan\delta) \tag{1.20}$$

Example 1.4

Calculate the angle of incidence of beam radiation on a surface located in London (latitude 51°) at solar time 1.30 pm on 16th February 2010. The surface is inclined at 45° to the horizontal plane and pointed 30° west of south.

Solution

For the 16th February, the day of the year $n = 31 + 16 = 47$. From eqn (1.8) the declination is:

$$\delta = 23.45° \sin\left[\frac{360°}{365}(284 + 47)\right] = -12.95°$$

At 1.30 pm the solar time is 13.30 hours. From eqn (1.8) the hour angle is:

$$\omega = (13.30 - 12) \times 15° = 22.5°$$

From eqn (1.14) the angle of incidence is:

$$\begin{aligned}\cos\theta_i &= (\cos\phi\cos\beta + \sin\phi\sin\beta\cos\gamma)\cos\delta\cos\omega \\ &\quad + \cos\delta\sin\omega\sin\beta\sin\gamma + \sin\delta(\sin\phi\cos\beta - \cos\phi\sin\beta\cos\gamma) \\ &= (\cos 51\cos 45 + \sin 51\sin 45\cos 30)\cos(-12.95)\cos 22.5 \\ &\quad + \cos\delta\sin 22.5\sin 45\sin 30 + \sin(-12.95) \\ &\quad (\sin 51\cos 45 - \cos 51\sin 45\cos 30)\end{aligned}$$

Thus $\theta_i = \cos^{-1} 0.92423 = 22.45°$.

Example 1.5

Calculate the number of sunshine hours in London (latitude 51°) on (a) 22 December and (b) 22 June 2010.

Solution

(a) For 22nd December, the day of the year $n = 356$. From eqn (1.8) the declination is:

$$\delta = 23.45° \sin\left[\frac{360°}{365}(284 + 356)\right] = -23.44°$$

From eqn (1.20) the number of daylight hours is:

$$\begin{aligned}N &= \frac{2}{15}\cos^{-1}(-\tan\phi\tan\delta) \\ &= \frac{2}{15}\cos^{-1}[-\tan(51)\tan(-23.44)] = 7.68 \text{ hours}\end{aligned}$$

(b) For 22nd June, the day of the year $n = 173$. From eqn (1.8) the declination is:

$$\delta = 23.45° \sin\left[\frac{360°}{365}(284 + 173)\right] = 23.45°$$

From eqn (1.20) the number of sunshine hours is:

$$\begin{aligned}N &= \frac{2}{15}\cos^{-1}(-\tan\phi\tan\delta) \\ &= \frac{2}{15}\cos^{-1}[-\tan(51)\tan(23.45)] = 16.32 \text{ hours}\end{aligned}$$

1.10 Extraterrestrial Radiation on a Horizontal Surface

At any point of time, the solar radiation outside the atmosphere (I_0) incident on a horizontal plane is given as:

$$I_{\text{ext}} = I_{\text{sc}}\left[1.0 + 0.033\cos\left(\frac{360n}{365}\right)\right]\cos\theta_z \tag{1.21}$$

Substituting the value of cos θ_z, we have:

$$I_{\text{ext}} = I_{\text{sc}}\left[1.0 + 0.033\cos\left(\frac{360n}{365}\right)\right](\cos\phi\cos\delta\cos\omega + \sin\delta\sin\phi) \tag{1.22}$$

It is often required to have the integrated daily extraterrestrial radiation on a horizontal surface H_0, for the calculations of daily solar radiation. This can be obtained by integrating the equation over the period from sunrise to sunset. If I_{sc} is in W m^{-2}; and H_0 in J m^{-2}, then:

$$H_0 = \frac{24\times 3600}{\pi}I_{\text{sc}}\left[1 + 0.033\cos\frac{360n}{365}\right]\left[\cos\phi\cos\delta\sin\omega_s + \left(\frac{2\pi\omega_s}{360}\right)\sin\phi\sin\delta\right] \tag{1.23}$$

where ω_s is the sunset hour angle, in degrees.

By integrating the above equation, the extraterrestrial radiation on a horizontal surface for an hour period defined by hour angles ω_1 and ω_2, is:

$$\begin{aligned} I_0 = &\frac{12\times 3600}{\pi}I_{\text{sc}}\left[1 + 0.033\cos\left(\frac{360n}{365}\right)\right] \\ &\times\left[\cos\phi\cos\delta(\sin\omega_2 - \sin\omega_1) + \frac{2\pi(\omega_2-\omega_1)}{360}\sin\phi\sin\delta\right] \end{aligned} \tag{1.24}$$

Example 1.6

Calculate the day's solar radiation on a horizontal surface, in absence of the atmosphere, at latitude 30°N on 30th April 2010.

Solution

For the 30th April, the day of the year $n = 120$. From eqn (1.8) the declination is:

$$\delta = 23.45^\circ\sin\left[\frac{360^\circ}{365}(284+120)\right] = 14.587^\circ$$

From eqn (1.18), $\omega_s = \cos^{-1}(-\tan\phi \tan\delta) = 98.64°$.

From eqn (1.23) the day's solar radiation on a horizontal surface is:

$$H_0 = \frac{24 \times 3600}{\pi} I_{sc}\left[1 + 0.033\cos\frac{360n}{365}\right]\left[\cos\phi\cos\delta\sin\omega_s + \left(\frac{2\pi\omega_s}{360}\right)\sin\phi\sin\delta\right]$$

$$= \frac{24 \times 3600 \times 1367}{\pi}\left[1 + 0.033\cos\frac{360 \times 120}{365}\right]$$

$$\times \left[\begin{array}{l}\cos 30 \cos 14.587 \sin 98.64 \\ +\left(\frac{2\pi \times 98.64}{360}\right)\sin 30 \sin 14.587\end{array}\right] = 37.595 \times 10^6$$

$$\times\, 0.9843 \times 1.0425 = 38.58\,\mathrm{MJ\,m^{-2}}$$

Example 1.7

Calculate the solar radiation on a horizontal surface, in absence of the atmosphere, at latitude 30°N on 30th April 2010, between the hours 9 and 10.

Solution

From eqn (1.10), at 9 hours the hour angle is $\omega_1 = (9 - 12) \times 15° = -45°$, and at 10 hours the hour angle is $\omega_2 = (10 - 12) \times 15° = -30°$.

For the 30th April, the day of the year $n = 120$ and $\delta = 14.587°$. From eqn (1.24) the radiation on a horizontal surface for an hour period is:

$$I_0 = \frac{12 \times 3600}{\pi} I_{sc}\left[1 + 0.033\cos\left(\frac{360n}{365}\right)\right]$$

$$\times \left[\cos\phi\cos\delta(\sin\omega_2 - \sin\omega_1) + \frac{2\pi(\omega_2 - \omega_1)}{360}\sin\phi\sin\delta\right]$$

$$= \frac{12 \times 3600 \times 1367}{\pi}\left[1 + 0.033\cos\left(\frac{360 \times 120}{365}\right)\right]$$

$$\times \left[\begin{array}{l}\cos 30 \cos 14.587\{\sin(-30) - \sin(-45)\} \\ +\frac{2\pi(-30 + 45)}{360}\sin 30 \sin 14.587\end{array}\right]$$

$$= 18.8 \times 10^6 \times 0.9843 \times 0.2065 = 3.82\,\mathrm{MJ\,m^{-2}}$$

1.11 Radiation on an Inclined Surface

The total solar radiation incident on a surface consists of (i) beam radiation, (ii) diffuse radiation and (iii) reflected radiation from the ground and the surroundings. Normally the beam radiation (I_b) and diffuse radiation (I_d) on a horizontal surface are recorded. If the beam and diffuse radiation data are not available, the following expression for beam and diffuse radiation on a horizontal surface can be used (see Figure 1.22).

$$I_b = I_N \cos\theta_z \tag{1.25}$$

$$I_d = \frac{1}{3}(I_{ext} - I_N)\cos\theta_z \tag{1.26}$$

where an expression for I_N (the normal incidence) is given by:

$$I_N = I_{ext} \exp\left(-\frac{T_r}{0.94 + 9.4\cos\theta_z}\right) \tag{1.27}$$

where T_r is called the turbidity factor and varies with the sky conditions and city conditions. Table 1.2 provides tentative values of the turbidity factor T_r for different months, depending upon the region.

After knowing the beam and diffuse radiation on a horizontal surface, to evaluate the total radiation on a surface of arbitrary orientation the Liu and

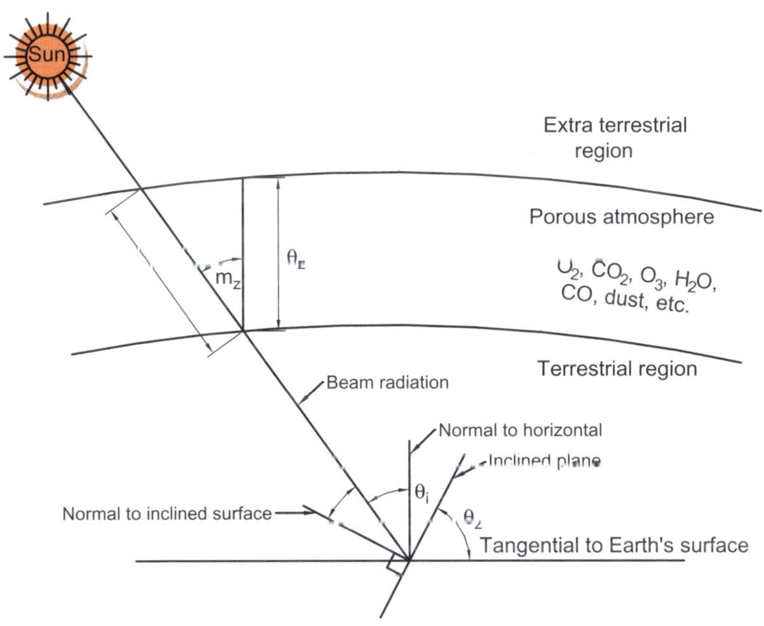

Figure 1.22 Availability of solar radiation.

Table 1.2 Turbidity factor T_r for different months and regions.

Month of the year	*Region*		
	Mountain	*Flat land*	*City*
January	1.8	2.2	3.1
February	1.9	2.2	3.2
March	2.1	2.5	3.5
April	2.2	2.9	3.9
May	2.4	3.2	4.1
June	2.7	3.4	4.2
July	2.7	3.5	4.3
August	2.7	3.3	4.2
September	2.5	2.9	3.9
October	2.1	2.6	3.6
November	1.9	2.3	3.3
December	1.8	2.2	3.1

Jordan formula may be used:

$$I_t = \left(\frac{\cos\theta_i}{\cos\theta_z}\right)I_b + \left(\frac{1+\cos\beta}{2}\right)I_d + \rho\left(\frac{1-\cos\beta}{2}\right)(I_b + I_d) \qquad (1.28)$$

where ρ is the reflection coefficient of the ground. Its value varies from 0.2 to 0.6 for ordinary and snow covered ground, respectively.

Example 1.8

Calculate the hourly direct radiation on (a) a horizontal surface and (b) a surface inclined at 45° to the horizontal facing south, on 15th January at 11 am for a city in a terrestrial region with latitude 30°.

Solution

From Table 1.2, the turbidity factor T_r in the month of January for city conditions is 3.1. On 15th January, the day of the year $n = 15$; the declination $\delta = -21.27°$. The 11 am solar time is 11 hours and the hour angle $\omega = -15°$. From eqn (1.17) the zenith angle is:

$$\theta_z = \cos^{-1}[\cos(30)\cos(-21.27)\cos(-15) + \sin(-21.27)\sin(30)] = 56.26°$$

From eqn (1.27) the normal incidence in the terrestrial region is:

$$I_N = I_{ext}\exp\left(-\frac{T_r}{0.94 + 9.4\cos\theta_z}\right)$$

$$= 1367 \times \exp\left(-\frac{3.1}{0.94 + 9.4\cos 53.26}\right) = 852.37\ \mathrm{W\ m^{-2}}$$

From eqn (1.25) the direct radiation on the horizontal surface is:

$$I_b = I_N \cos\theta_z = 852.37 \times \cos 56.26 = 473.43\ \text{W m}^{-2}$$

From eqn (1.14) the angle of incidence on the inclined surface is:

$$\begin{aligned}\cos\theta_i &= (\cos 30 \cos 45 + \sin 30 \sin 45 \cos 0)\cos(-21.27)\cos(-15) \\ &\quad + \cos(-21.27)\sin(-15)\sin 45 \sin 0 + \sin(-21.27)\begin{pmatrix} \sin 30 \cos 45 \\ -\cos 30 \sin 45 \cos 0 \end{pmatrix} \\ &= 0.9633\end{aligned}$$

The direct radiation on the inclined surface is:

$$I'_b = I_N \cos\theta_i = 852.37 \times 0.9633 = 821.13\ \text{W m}^{-2}$$

Example 1.9

Calculate the total solar radiation received by a surface inclined at 30° to the horizontal facing south at a location with latitude 30°, at 10 am on 3rd February. The measured global and beam radiations were 750 W m^{-2} and 600 W m^{-2}, respectively. Consider the reflection coefficient of the surface as 0.2.

Solution

From Example 1.3, on 3rd February the day of the year $n = 31 + 3 = 34$; the declination $\delta = -16.97°$. At 10 am the solar time is 10 hours; the hour angle $\omega = -30°$. The zenith angle is:

$$\theta_z = \cos^{-1}[\cos(30)\cos(-16.97)\cos(-30) + \sin(-16.97)\sin(30)] = 55.15°$$

From eqn (1.14) the angle of incidence is:

$$\begin{aligned}\cos\theta_i &= (\cos 30 \cos 30 + \sin 30 \sin 30 \cos 0)\cos(-16.97)\cos(-30) \\ &\quad + \cos(-16.97)\sin(-30)\sin 30 \sin 0 + \sin(-16.97) \\ &\quad \times \begin{pmatrix} \sin 30 \cos 30 \\ -\cos 30 \sin 30 \cos 0 \end{pmatrix} \Rightarrow \theta_i = \cos^{-1} 0.8283 = 34.07°\end{aligned}$$

From eqn (1.28) the total solar radiation received on an inclined surface is:

$$\begin{aligned}I_t &= \left(\frac{\cos\theta_i}{\cos\theta_z}\right) I_b + \left(\frac{1+\cos\beta}{2}\right) I_d + \rho\left(\frac{1-\cos\beta}{2}\right)(I_b + I_d) \\ &= \left(\frac{\cos 34.07}{\cos 55.15}\right) 600 + \left(\frac{1+\cos 30}{2}\right)(750 - 600) + 0.2\left(\frac{1-\cos 30}{2}\right) 750 \\ &= 1019.78\ \text{W m}^{-2}\end{aligned}$$

1.12 Estimation of Average Solar Radiation

The regression equation relating monthly average daily radiation ($\bar{H}$) to clear day radiation for a particular location, and average fraction of possible sunshine hours, is given by:

$$\frac{\bar{H}}{\bar{H}_C} = a' + b'\left(\frac{\bar{n}}{\bar{N}}\right) \tag{1.29}$$

where $\bar{H}$ is the monthly average daily total radiation on a horizontal surface, $\bar{H}_C$ is the monthly average clear sky radiation at the location, a' and b' are constants, n is the monthly average daily hours of bright sunshine and $\bar{N}$ is the monthly average of the maximum possible daily hours of bright sunshine (*i.e.* the day length of the average day of the month) (to determine $\bar{N}$, use eqn 1.20).

Owing to difficulty in measuring clear sky radiation, eqn (1.29) has been modified to use extraterrestrial radiation on a horizontal surface as follows:

$$\frac{\bar{H}}{\bar{H}_0} = a + b\left(\frac{\bar{n}}{\bar{N}}\right) \tag{1.30}$$

where $\bar{H}_0$ is the monthly average of the daily extraterrestrial radiation at the location and a and b are modified constants.

1.12.1 Monthly Average of the Daily Total Radiation on a Horizontal Surface

For many applications, one requires knowledge of the monthly average of the daily solar radiation available on a horizontal surface. The correlation for estimating the monthly average daily total radiation on a horizontal surface is given by eqn (1.30) as:

$$\frac{\bar{H}}{\bar{H}_0} = a + b\left(\frac{\bar{n}}{\bar{N}}\right) \tag{1.31}$$

The regression parameters a and b can be determined using the following expressions:

$$a = -0.309 + 0.539\cos\phi - 0.0693E_0 + 0.29\frac{\bar{n}}{\bar{N}} \tag{1.32}$$

$$b = 1.527 - 1.027\cos\phi + 0.0926E_0 - 0.359\frac{\bar{n}}{\bar{N}} \tag{1.33}$$

where E_0 is the elevation of the location above sea level in kilometres and $\frac{\bar{n}}{\bar{N}}$ is the percent possible sunshine.

These coefficients give the global radiation on a horizontal surface with an accuracy of about 10%. The regression coefficients can be computed from latitude and sunshine hours using the parameters as follows:

$$a = -0.110 + 0.235\cos\phi + 0.323\frac{\bar{n}}{\bar{N}} \tag{1.34}$$

$$b = 1.449 - 0.553\cos\phi - 0.694\frac{\bar{n}}{\bar{N}} \tag{1.35}$$

It may be noted that the correlation of eqns (1.34) and (1.35) are valid for the following ranges:

$$0.2 \le \frac{\bar{n}}{\bar{N}} \le 0.6 \text{ for monsoon periods}$$

$$0.4 \le \frac{\bar{n}}{\bar{N}} \le 0.9 \text{ for pre-monsoon and post-monsoon periods}$$

1.12.2 Monthly Average of the Daily Diffuse Radiation on a Horizontal Surface

Investigations relating to the estimation of the monthly average of daily diffuse radiation on a horizontal surface have been less in number than that of global radiation. Mainly two types of correlations have been proposed, which are as follows:

$$\frac{\bar{H}_\mathrm{d}}{\bar{H}} = 1.403 - 1.672\bar{k}_\mathrm{T} \tag{1.36}$$

$$\frac{\bar{H}_\mathrm{d}}{\bar{H}} = 0.931 - 0.814\frac{\bar{n}}{\bar{N}} \tag{1.37}$$

where $\frac{\bar{H}_\mathrm{d}}{\bar{H}}$ is the ratio of monthly average of daily diffuse radiation to the monthly average of daily global radiation; $k_\mathrm{T} = \frac{H}{H_0}$ is the ratio of monthly average of daily global radiation to the monthly average of daily extraterrestrial radiation and $\frac{\bar{n}}{\bar{N}}$ is the number of bright sunshine hours.

1.12.3 Beam and Diffuse Components of Daily Radiation

Studies of the available daily radiation indicate that the average fraction which is diffuse, $\frac{H_\mathrm{d}}{H}$, is a function of $k_\mathrm{T} = \frac{H}{H_0}$, the ratio of monthly average of daily global radiation to the monthly average of daily extraterrestrial radiation. It is given by the following correlation:

$$\frac{H_d}{H} = \begin{cases} 0.99 & \text{for} \quad k_T \leq 0.17 \\ 1.188 - 2.272k_T + 9.473k_T^2 & \text{for} \quad 0.17 < k_T < 0.75 \\ -21.865k_T^3 + 14648k_T^4 & \text{for} \quad 0.17 < k_T < 0.75 \\ -0.54k_T + 0.632 & \text{for} \quad 0.75 < k_T < 0.80 \\ 0.2 & \text{for} \quad k_T \geq 0.80 \end{cases} \tag{1.38}$$

Appendix A2 shows the average annual daily ground solar energy available during clear days on a horizontal surface worldwide.

1.12.4 Beam and Diffuse Components of Hourly Radiation

The fraction of the hourly diffuse radiation on a horizontal plane can be correlated to the hourly clearness index, $k_T = \frac{I}{I_0}$. The ratio of the total radiation to the extraterrestrial radiation for the hour is given by the following correlation:

$$\frac{I_d}{I} = \begin{cases} 1.0 - 0.249k_T & \text{for} \quad k_T < 0.35 \\ 1.557 - 1.84k_T & \text{for} \quad 0.35 < k_T < 0.75 \\ 0.177 & \text{for} \quad k_T > 0.75 \end{cases} \tag{1.39}$$

Once the hourly diffuse radiation (I_d) is known from the above equation, the hourly beam radiation (I_b) can be obtained as:

$$I_b = I - I_d \tag{1.40}$$

Example 1.10

Estimate the global radiation on a horizontal surface at a place with latitude 25°, in the month of May. Assume that the monthly average daily hours of bright sunshine observed is 12.1 hours.

Solution

From Table 1.1, the declination in the month of May is 18.8°. From eqn (1.18):

$$\omega_s = \cos^{-1}(-\tan\phi\tan\delta) = \cos^{-1}(-\tan 25 \tan 18.8) = 99.134°$$

From eqn (1.23) the daily extraterrestrial radiation is:

$$H_0 = \frac{24 \times 3600}{\pi} I_{sc}\left[1 + 0.033\cos\frac{360n}{365}\right]\left[\cos\phi\cos\delta\sin\omega_s + \left(\frac{2\pi\omega_s}{360}\right)\sin\phi\sin\delta\right]$$

$$= \frac{24 \times 3600 \times 1367}{\pi}\left[1 + 0.033\cos\frac{360 \times 135}{365}\right]$$

$$\times \left[\cos 25\cos 18.8\sin 99.134 + \left(\frac{2\pi \times 99.134}{360}\right)\sin 25\sin 18.8\right]$$

$$= 39.78\,\text{MJ m}^{-2}$$

From eqn (1.20) the number of sunshine hours is:

$$\bar{N} = \frac{2}{15}\cos^{-1}(-\tan\phi\tan\delta)$$

$$= \frac{2}{15}\omega_s = \frac{2}{15} \times 99.134 = 13.218 \text{ hours}$$

From eqns (1.34) and (1.35) the values of the constants a and b are:

$$a = -0.110 + 0.235\cos 25 + 0.323 \times \frac{12.1}{13.218} = 0.4$$

$$b = 1.449 - 0.553\cos 25 - 0.694 \times \frac{12.1}{13.218} = 0.338$$

Substituting the values of constants a and b into eqn (1.31), we have:

$$\frac{\bar{H}}{\bar{H}_0} = a + b\left(\frac{\bar{n}}{\bar{N}}\right)$$

$$\Rightarrow \bar{H} = \left[0.4 + 0.338\left(\frac{12.1}{13.218}\right)\right] \times 39.78 = 28.22\,\text{MJ m}^{-2}$$

Example 1.11

Estimate the monthly average of daily diffuse and beam radiations on a horizontal surface at a place with latitude 13°, in the month of May. Consider the average radiation for the place as 22.69 MJ m^{-2}.

Solution

From Table 1.1 the declination in the month of May is 18.8°. From eqn (1.18):

$$\omega_s = \cos^{-1}(-\tan\phi\tan\delta) = \cos^{-1}(-\tan 13\tan 18.8) = 94.5°$$

From eqn (1.23) the monthly average extraterrestrial radiation is:

$$H_0 = \frac{24\times 3600}{\pi} I_{sc}\left[1+0.033\cos\frac{360n}{365}\right]\left[\text{co}\phi\cos\delta\sin\omega_s + \left(\frac{2\pi\omega_s}{360}\right)\sin\phi\sin\delta\right]$$

$$= \frac{24\times 3600\times 1367}{\pi}\left[1+0.033\cos\frac{360\times 135}{365}\right]$$

$$\times\left[\cos 13\cos 18.8\sin 94.5 + \left(\frac{2\pi\times 94.5}{360}\right)\sin 13\sin 18.8\right]$$

$$= 38.18\ \text{MJ m}^{-2}$$

The monthly average clearness index is:

$$\bar{k}_T = \frac{\bar{H}}{\bar{H}_0} = \frac{22.69}{38.18} = 0.594$$

From eqn (1.36) we have:

$$\frac{\bar{H}_d}{\bar{H}} = 1.403 - 1.672\bar{k}_T = 1.403 - 1.672\times 0.594 = 0.41$$

$$\Rightarrow \bar{H}_d = 0.41\times 22.69 = 9.3\ \text{MJ m}^{-2}$$

Thus the beam component of the monthly average radiation is:

$$\bar{H}_b = \bar{H} - \bar{H}_d = 22.69 - 9.3 = 13.39\ \text{MJ m}^{-2}$$

1.13 Heat Transfer through Conduction

The transfer of heat energy occurs owing to a driving force called temperature difference. Heat may be transferred in three distinct modes, namely conduction, convection or radiation. These modes of heat transfer differ profoundly in nature and are governed by different laws.

The phenomenon of heat conduction is a process of the propagation of heat due to a temperature gradient occurring between the particles of either a single body or bodies that are in direct contact with each other. The conduction may take place through solids, liquids or gases. The steady state, one-dimensional heat flow by conduction through a homogeneous material is given by the Fourier's law. According to this, the rate of heat flow ($\dot{Q}$) passing through an isothermal surface of thickness δx with temperature difference between the two surfaces ∂T and area perpendicular to the direction of heat flow (A) is expressed as:

$$\dot{Q} = -KA\frac{\partial T}{\partial x} \tag{1.41}$$

The ratio $\frac{\partial T}{\partial x}$ represents the temperature gradient in the direction of heat flow and K is the proportionality factor usually called the thermal conductivity of the material. Thermal conductivity defines the ability of the substance to conduct heat. The negative sign indicates that the heat flow is in the direction of a negative temperature gradient. The heat flux $\dot{q}$, defined as the rate of heat flow per unit time per unit area of the isothermal surface, is given by:

$$\dot{q} = -K\frac{\partial T}{\partial x} \tag{1.42}$$

1.13.1 Thermal Conductivity

The thermal conductivity of gases, liquids and solids depends on the temperature. Experimental studies have shown that for many materials the dependence of thermal conductivity on temperature can be assumed to be linear as follows:

$$K = K_0[1 + \beta(T - T_0)] \tag{1.43}$$

where K_0 is the thermal conductivity at temperature T_0 and β is the constant for a material.

In general, an increase in temperature causes the conductivity of a gas to increase (positive β) and conductivity of a solid or liquid to decrease (negative β). However, there are some exceptions to this generalization. Thermal conductivity is greatly influenced by the moisture content of the substance. Experimental studies have shown that the thermal conductivity increases substantially with an increase in moisture content. Appendix A3 gives the physical properties of metals and non-metals, such as thermal conductivity, specific heat and density, which may be used to find the heat loss or gain through the system.

1.13.2 Heat Transfer through Parallel Slabs

Consider a composite wall as shown in Figure 1.23, through which heat is transferred from the hot fluid at temperature T_A to the cold fluid at temperature T_B. Assuming a steady state, the heat flux, $\dot{q}$, is given by:

$$\dot{q} = U(T_A - T_B) \tag{1.44}$$

where the overall heat transfer coefficient is:

$$U = \left[\frac{1}{h_i} + \frac{L_1}{K_1} + \frac{L_2}{K_2} + \frac{L_3}{K_3} + \frac{1}{h_o}\right]^{-1} \tag{1.45}$$

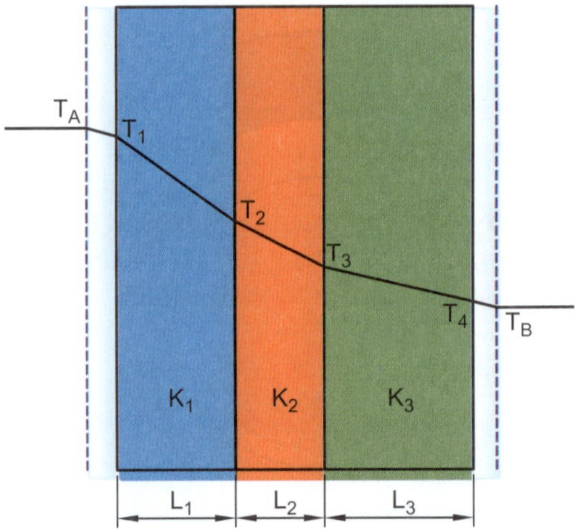

Figure 1.23 One-dimensional heat flow through parallel slabs.

where h_i and h_o are the convective heat transfer coefficients for the inner and outer surfaces of the composite wall, *i.e.* either sides of the wall. L_1, L_2 and L_3 are the thicknesses of the walls as depicted in the figure, and K_1, K_2 and K_3 are the corresponding thermal conductivities.

1.13.3 Heat Transfer through Coaxial Cylinders

Consider a composite coaxial cylinder as shown in Figure 1.24. A hot fluid at temperature T_A flows inside the tube and the exterior surface of the tube is in contact with a cold fluid at temperature T_B. Then, at steady state the heat flux, $\dot{q}$, is given by:

$$\dot{q} = U(T_A - T_B) \tag{1.46}$$

where the overall heat transfer coefficient based on the interior surface is:

$$U = \left[\frac{1}{h_i} + \frac{r_0}{K_1}\ln\left(\frac{r_1}{r_0}\right) + \frac{r_0}{K_2}\ln\left(\frac{r_2}{r_1}\right) + \frac{r_0}{r_2}\frac{1}{h_o}\right]^{-1} \tag{1.47}$$

where h_i and h_o are the convective heat transfer coefficients for the inner and outer surfaces of the composite cylinders, r_0 and r_2 are the inner and outer radii of the composite wall, and $(r_1 - r_0)$ and $(r_2 - r_1)$ are the thicknesses of the two layers of the composite walls having their corresponding thermal conductivities K_1 and K_2, respectively.

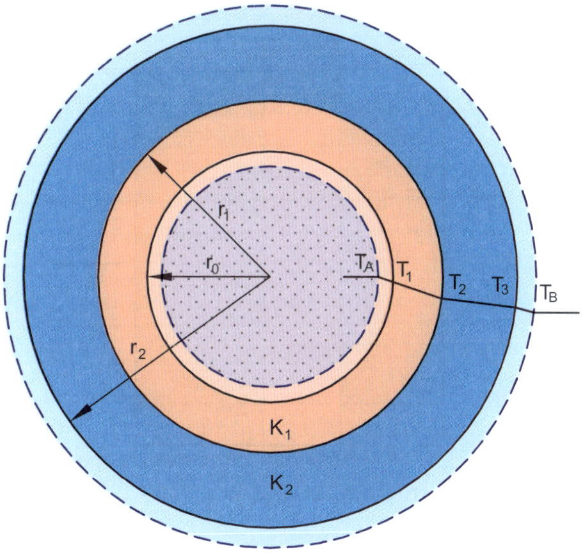

Figure 1.24 Heat flow through coaxial cylinders.

Example 1.12

A wall comprises two layers: an 80 mm inside layer of burnt brick ($K_b = 0.81\,\mathrm{W\,m^{-1}\,K^{-1}}$) and a 50 mm outer layer of flyash brick ($K_f = 0.62\,\mathrm{W\,m^{-1}\,K^{-1}}$). The average values of the convective heat transfer coefficients at the interior and the exterior walls are $12\,\mathrm{W\,m^{-2}\,K^{-1}}$ and $10\,\mathrm{W\,m^{-2}\,K^{-1}}$, respectively. If the inside and outside air temperatures are maintained at 30 °C and 5 °C, respectively, find the rate of heat loss.

Solution

Given that $L_b = 0.08\,\mathrm{m}$, $K_b = 0.81\,\mathrm{W\,m^{-1}\,K^{-1}}$, $L_f = 0.05\,\mathrm{m}$, $K_f = 0.62\,\mathrm{W\,m^{-1}\,K^{-1}}$, $h_i = 12\,\mathrm{W\,m^{-2}\,K^{-1}}$, $h_o = 10\,\mathrm{W\,m^{-2}\,K^{-1}}$, $T_A = 30\,°\mathrm{C}$ and $T_B = 5\,°\mathrm{C}$, then from eqn (1.45) the overall heat transfer coefficient is:

$$U = \left[\frac{1}{h_i} + \frac{L_b}{K_b} + \frac{L_f}{K_f} + \frac{1}{h_o}\right]^{-1} = \left[\frac{1}{12} + \frac{0.08}{0.81} + \frac{0.05}{0.62} + \frac{1}{10}\right]^{-1}$$
$$= 2.76\,\mathrm{W\,m^{-2}\,K^{-1}}$$

and from eqn (1.43) the rate of heat transfer loss is:

$$\dot{q} = U(T_A - T_B) = 2.76 \times (30 - 5) = 68.9\,\mathrm{W\,m^{-2}}$$

Example 1.13

Saturated steam at 110 °C flows inside a copper pipe (thermal conductivity 450 W m^{-1} K^{-1}) having an internal diameter of 10 cm and an external diameter of 12 cm. The surface resistance on the steam side is 12 000 W m^{-2} K^{-1} and that on the outside surface of the pipe is 18 W m^{-2} K^{-1}. Determine the heat loss from the pipe if it is located in space at 25 °C. How would this heat loss be affected if the pipe is lagged with 5 cm thick insulation with thermal conductivity 0.22 W m^{-1} K^{-1}.

Solution

Given that $T_A = 100\,°C$, $K_1 = 450\,W\,m^{-1}\,K^{-1}$, $r_0 = 0.05\,m$, $r_1 = 0.06\,m$, $h_i = 12000\,W\,m^{-2}\,K^{-1}$, $h_o = 18\,W\,m^{-2}\,K^{-1}$, $T_B = 25\,°C$, $r_2 = 0.06 + 0.05 = 0.11\,m$ and $K_2 = 450\,W\,m^{-1}\,K^{-1}$, then when the pipe is not insulated, from eqn (1.47) the overall heat transfer coefficient for the inner surface is given by:

$$U_i = \left[\frac{1}{h_i} + \frac{r_0}{K_1}\ln\left(\frac{r_1}{r_0}\right) + \frac{r_0}{r_1}\frac{1}{h_o}\right]^{-1}$$
$$= \left[\frac{1}{12000} + \frac{0.05}{450}\ln\left(\frac{0.06}{0.05}\right) + \frac{0.05}{0.06} \times \frac{1}{18}\right]^{-1} = 21.55\,W\,m^{-2}\,K^{-1}$$

From eqn (1.46) the heat transfer loss is:

$$Q = U_i A_i (T_A - T_B) = 21.55 \times (2 \times \pi \times 0.05 \times 1) \times (110 - 25)$$
$$= 575.5\,W\,m^{-1}$$

When the pipe is insulated, then from eqn (1.47) the overall heat transfer coefficient for the inner surface is given by:

$$U_i = \left[\frac{1}{h_i} + \frac{r_0}{K_1}\ln\left(\frac{r_1}{r_0}\right) + \frac{r_0}{K_2}\ln\left(\frac{r_2}{r_1}\right) + \frac{r_0}{r_2}\frac{1}{h_o}\right]^{-1}$$
$$= \left[\frac{1}{12000} + \frac{0.05}{450}\ln\left(\frac{0.06}{0.05}\right) + \frac{0.05}{0.22}\ln\left(\frac{0.11}{0.06}\right) + \frac{0.05}{0.11} \times \frac{1}{18}\right]^{-1} = 6.13\,W\,m^{-2}\,K^{-1}$$

From eqn (1.46) the heat transfer loss is:

$$Q = U_i A_i (T_A - T_B) = 6.13 \times (2 \times \pi \times 0.05 \times 1) \times (110 - 25)$$
$$= 163.71\,W\,m^{-1}$$

Thus, the addition of insulation reduces the heat loss from the steam by:

$$\frac{575.5 - 163.71}{575.5} \times 100\% = 71.55\%$$

1.14 Heat Transfer through Convection

Thermal convection is the transfer of heat from one part of a fluid to another part at a lower temperature, by the mixing of fluid particles. Heat is transported simultaneously during the process by convection and by conduction. In most of the solar energy applications, convection is accompanied by conduction.

The temperature difference between the fluid and the contact surface causes a density variation in the fluid, resulting in buoyancy. The fluid motion thus produced is "free convection". However, if the motion of the fluid is caused by forces arising from externally imposed pressure differences and is independent of the temperature difference in the fluid, it is referred to as "forced convection". These two processes may sometimes occur simultaneously. The rate of heat transfer by convection between the fluid and the boundary surface may be evaluated by the formula:

$$\dot{Q} = hA\Delta T \tag{1.48}$$

where h is the local heat transfer coefficient.

The rate of heat flow $\dot{Q}$ at the fluid and body interface is related to the temperature difference between the surface of the body concerned and its surroundings. The heat transfer coefficient h is a complicated function of the fluid flow, thermophysical properties of the fluid and geometric arrangements of the system. However, the appropriate mathematical form of h in a limited domain of applicability is evaluated from empirical equations obtained by correlating experimental results with the method of dimensional analysis.

1.14.1 Dimensionless Heat Convective Parameters

The equations for convective heat transfer contain the following dimensionless terms with dissimilar physical parameters:

1. The Reynolds number is defined as the ratio of inertia force to the viscous force, which indicates the relative importance of inertia and viscous effects in a fluid motion:

$$Re = \frac{\rho V_0 L}{\mu} = \frac{V_0 L}{\nu} \tag{1.49}$$

   where V_0 is the mean fluid velocity in m s^{-1}, L is a characteristic linear dimension, μ is the dynamic viscosity of the fluid in Pa s (or N s m^{-2} or kg m^{-1} s^{-1}), ν is the kinematic viscosity in m^2 s^{-1} and ρ is the density of the fluid in kg m^{-3}.

2. The Nusselt number establishes the relation between the convective film coefficient, the thermal conductivity of the fluid and a significant length parameter of the physical system. The Nusselt number is a convenient measure of the convective heat transfer coefficient:

$$Nu = \frac{hL}{K} \tag{1.50}$$

where h is the convective heat transfer coefficient. It may be interpreted as the ratio of temperature gradient at the surface to an overall reference temperature gradient.

3. The Prandtl number is indicative of the relative ability of the fluid to diffuse momentum and internal energy by molecular mechanisms:

$$Pr = \frac{\mu C_p}{K} \tag{1.51}$$

$$= \frac{\nu}{(K/\rho C_p)} = \frac{\text{kinematic viscosity}}{\text{thermal diffusivity}} \tag{1.52}$$

where K is the thermal conductivity in W m^{-1} K^{-1} and C_p is the specific heat in J kg^{-1} K^{-1}. The Prandtl number is the ratio of the kinematic viscosity to the thermal diffusivity of the fluid. For air and many other gases its value is around 0.7–0.8.

4. The Grashof number indicates the relative strength of the buoyant to viscous forces. It has a role in free convection similar to that played by the Reynolds number in forced convection:

$$Gr = \frac{g\beta\rho^2 L^3 \Delta T}{\mu^2} \tag{1.53}$$

$$= g\beta\rho L^3 \Delta T \frac{\rho V_0 L^2}{(\mu V_0 L)^2} = \text{buoyant force} \times \frac{\text{inertia force}}{(\text{viscous force})^2} \tag{1.54}$$

where g is the acceleration due to Earth's gravity, β is the volumetric thermal expansion coefficient (equal to approximately $1/T$ for ideal fluids, where T is absolute temperature) and $\Delta T = T_s - T_\infty$ is the difference between the surface temperature and the bulk temperature.

5. The Rayleigh number is defined as the product of the Grashof number and the Prandtl number. Mathematically:

$$Ra = Gr \times Pr = \frac{g\beta\rho^2 L^3 C_p \Delta T}{\mu K} \tag{1.55}$$

Hence the Rayleigh number may also be viewed as the ratio of buoyancy forces and (the product of) thermal and momentum diffusivities. When the Rayleigh number is below the critical value for that fluid, heat transfer is primarily in the form of conduction; when it exceeds the critical value, heat transfer is primarily in the form of convection. Appendix A4 gives the thermophysical properties of air and saturated water such as density, specific heat, kinematic viscosity, dynamic viscosity, thermal conductivity, expansion coefficient and Prandtl's number at various temperatures, which can be used to determine the various non-dimensional parameters for a flow.

1.14.2 Free Convection

In the event of free or natural flow, the origin of motion in the considered volume of fluid is due to the heterogeneity of the mass forces acting upon the volume. If this natural flow is not confined within a space by solid boundaries, it is referred to as free convection. It has been found that average free-convection heat transfer coefficients from the horizontal and vertical surfaces can be represented in a functional form for a variety of configurations as follows:

$$Nu = C(Gr \times Pr)^n = CRa^n \tag{1.56}$$

The characteristic dimension to be used in the Nusselt and Grashof numbers depends on the geometry. For a horizontal plate it is the average of the length and breadth of the plate; for a vertical plate it is the height of the plate; and for a horizontal cylinder it is the diameter. Table 1.3 gives the values of the constants C and n to be used in eqn (1.56) for different conditions.

For plates inclined at β to the vertical, multiply the Grashof number by cos β and use the vertical plate constants. Thus, the average free-convection heat transfer coefficients for an inclined surface is represented as:

$$Nu = C(Ra \cos \beta)^n \tag{1.57}$$

Table 1.3 Values of the constants C and n for different geometrical conditions.

Geometry	*Rayleigh number (Gr × Pr)*	*C*	*n*
Vertical planes and cylinders	10^4–10^9	0.59	0.25
	10^9–10^{13}	0.10	0.33
Horizontal cylinder	10^4–10^9	0.53	0.25
	10^{-10}–10^{-2}	0.675	0.058
	10^{-2}–10^2	1.02	0.148
	10^9–10^{13}	0.85	0.188
Upper surface of heated plates or	$2 \times 10^4 - 8 \times 10^6$	0.54	0.25
lower surface of cooled plates	8×10^6–10^{11}	0.14	0.33
Lower surface of heated plates or upper surface of cooled plates	10^5–10^{11}	0.27	0.25
Vertical cylinder of equal height and diameter	10^4–10^6	0.775	0.21

1.14.3 Simplified Free Convection Relations for Air

In the case of air at atmospheric pressure, eqn (1.56) can be simplified as:

$$\begin{aligned} \frac{hL}{K} &= C\left(\frac{g\beta\rho^2 L^3 \Delta T}{\mu^2}\cdot\frac{\mu C_p}{K}\right)^n \\ \Rightarrow h &= CK\left(\frac{g\beta\rho^2 C_p}{\mu K}\right)^n \frac{\Delta T^n}{L^{1-3n}} = C_1\frac{\Delta T^n}{L^{1-3n}} \end{aligned} \tag{1.58}$$

The exponent n is usually $\frac{1}{4}$ for laminar flow and $\frac{1}{3}$ for turbulent flow. Substituting the value of exponent in eqn (1.56) we have:

$$\text{For laminar flow} \quad h = C_1\left(\frac{\Delta T}{L}\right)^{\frac{1}{4}} \tag{1.59}$$

$$\text{For turbulent flow} \quad h = C_1 \Delta T^{\frac{1}{3}} \tag{1.60}$$

where C_1 is a constant and depends on the geometry and flow conditions. Table 1.4 lists the values suggested by McAdams for various geometries, orientations and flow conditions indicated by the magnitude of Rayleigh number (Ra).

1.14.4 Forced Convection

In forced convection, the fluid motion is artificially induced, say with a pump or a fan that forces the fluid flow over the surface. Also considered as forced convection is the flow of the investigated volume of fluid under the action of an

Table 1.4 Simplified free convection equations for air.

Surface and its orientation	*Laminar flow* ($10^4 < Ra < 10^9$)	*Turbulent flow* ($Ra > 10^9$)
Vertical plane or cylinder	$h = 1.42\left(\frac{\Delta T}{L}\right)^{1/4}$	$h = 1.31\Delta T^{1/3}$
Horizontal cylinder	$h = 1.32\left(\frac{\Delta T}{L}\right)^{1/4}$	$h = 1.24\Delta T^{1/3}$
Horizontal plate with heated surface facing upward and cooled downward	$h = 1.32\left(\frac{\Delta T}{L}\right)^{1/4}$	$h = 1.52\Delta T^{1/3}$
Horizontal plate with cooled surface facing upward and heated downward	$h = 0.59\left(\frac{\Delta T}{L}\right)^{1/4}$	

internal homogeneous field of mass forces. The convection heat transfer coefficients can be represented in the functional form as follows:

$$Nu = CRe^{m}Pr^{n}K \tag{1.61}$$

where C and n are constants for a given type of flow and geometry and K is a correction factor (shape factor), added to obtain greater accuracy.

For flow past plates and walls of length L:

$$\text{Laminar flow}: \quad Re < 4 \times 10^5 \text{ and } Pr > 0.6, \; Nu = 0.664 Re_L^{0.5} Pr^{0.33} \tag{1.62}$$

$$\text{Turbulent flow}: \quad Re > 5 \times 10^5, \; Nu = 0.36 Re_L^{0.8} Pr^{0.33} \tag{1.63}$$

$$\text{Transitional flow}: 4 \times 10^5 < Re < 5 \times 10^5, \; Nu = 0.36 Pr^{0.33}(Re_L^{0.8} - A) \tag{1.64}$$

where A is a constant and its value is 18 700 at a transitional Reynolds number of 4×10^5 and 23 100 at a Reynolds number of 5×10^5.

Empirical correlations for the flow past inside pipes and tubes are beyond the scope of the current text.

1.14.5 Combined Free and Forced Convection

A number of practical situations occur where the convective heat transfer is neither forced nor free in nature. The circumstances arise when a fluid is forced over a heated surface at a rather low velocity. Coupled with the forced flow velocity is a convective velocity, which is generated by the buoyant forces resulting from a reduction in fluid density near the heated surface. A large Reynolds number implies a large forced-flow velocity and hence less influence of free convection currents. The larger the value of the Grashof–Prandtl product, the more one would expect free convection effects to prevail. Brown and Gauvin[1,2] developed a correlation for the mixed convection, laminar flow region of flow through horizontal tubes as follows:

$$Nu = 1.75\left(\frac{\mu_b}{\mu_w}\right)^{0.14}\left[Gz + 0.012\,(Gz \times Gr^{1/3})^{4/3}\right]^{1/3} \tag{1.65}$$

where $Gz = Re \times Pr\frac{d}{L}$ is the Graetz number, μ_b is the fluid viscosity at the bulk temperature and μ_w is the fluid viscosity at the wall temperature.

1.14.6 Convective Heat Transfer Due to Wind

The heat transfer from a flat plate exposed to outside winds has been analyzed by several workers. McAdams[3] reported an equation for the convection coefficient as follows:

$$h_c = 5.7 + 3.8V \quad \text{for} \quad 0 \leq V \leq 5\,\text{m}\,\text{s}^{-1} \tag{1.66}$$

where V is the wind speed in m s^{-1}.

The above equation for zero wind speed gives the heat loss by natural convection. It may be mentioned here that the process taking place is not as simple as it appears, as the wind may not always be blowing parallel to the surface. For this reason, Watmuff *et al.*[4] reported that this equation should be:

$$h_c = 2.8 + 3.0V \quad \text{for} \quad 0 \leq V \leq 7\,\text{m}\,\text{s}^{-1} \tag{1.67}$$

Example 1.14

Calculate the rate of heat loss from a plate maintained at 44 °C when exposed to the air at 1 bar and 10 °C under the following conditions: (a) the plate is vertical with 1.8 m height and 2.4 m width; (b) the plate is horizontal with 1.8 m length and 2.4 m breadth.

Solution

The mean film temperature $T_f = \frac{44+10}{2} = 27\,°\text{C} = 300\,\text{K}$. The thermophysical properties of air at 300 K (see Appendix A4.1) are $\rho = 1.177\,\text{kg}\,\text{m}^{-3}$, $\mu = 1.857 \times 10^{-5}\,\text{kg}\,\text{m}^{-1}\,\text{s}^{-1}$, $K = 0.02623\,\text{W}\,\text{m}^{-1}\,\text{K}^{-1}$, $Pr = 0.713$ and $\beta = \frac{1}{T} = \frac{1}{300}\,\text{K}^{-1}$.

(a) Characteristic length $L = 1.8$ m. From eqn (1.53) the Grashof number is:

$$Gr = \frac{g\beta\rho^2 L^3 \Delta T}{\mu^2}$$

$$= \frac{9.81 \times \frac{1}{300} \times (1.177)^2 \times (1.8)^3 \times (44 - 10)}{(1.857 \times 10^{-5})^2} = 2.6 \times 10^{10}$$

From eqn (1.56) and Table 1.3 the correlation for the vertical plate is:

$$Nu = 0.10(Gr \times Pr)^{0.33} = 0.10(2.6 \times 10^{10} \times 0.713)^{0.33} = 264.66$$

From eqn (1.50) the convective heat transfer coefficient is:

$$Nu = \frac{hL}{K} \Rightarrow h = 264.66 \times \frac{0.02623}{1.8} = 3.86\,\text{W}\,\text{m}^{-2}\,\text{K}^{-1}$$

From eqn (1.48) the heat loss by convection from the vertical plate is:

$$\dot{Q}_c = hA\Delta T = 3.86 \times (1.8 \times 2.4) \times (44 - 10) = 567\,\text{W}$$

(b) Characteristic length $L = \frac{1.8+2.4}{2} = 2.1\,\text{m}$. From eqn (1.53) the Grashof number is:

$$Gr = \frac{g\beta\rho^2 L^3 \Delta T}{\mu^2}$$

$$= \frac{9.81 \times \frac{1}{300} \times (1.177)^2 \times (2.1)^3 \times (44 - 10)}{(1.857 \times 10^{-5})^2} = 4.14 \times 10^{10}$$

From eqn (1.56) and Table 1.3 the correlation for the horizontal plate is:

$$Nu = 0.14(Gr \times Pr)^{0.33} = 0.14(4.14 \times 10^{10} \times 0.713)^{0.33} = 432.5$$

From eqn (1.50) the convective heat transfer coefficient is:

$$Nu = \frac{hL}{K} \Rightarrow h = 432.5 \times \frac{0.02623}{2.1} = 5.4\,\text{W m}^{-2}\,\text{K}^{-1}$$

From eqn (1.48) the heat loss by convection from the horizontal plate is:

$$\dot{Q}_c = hA\Delta T = 5.4 \times (1.8 \times 2.4) \times (44 - 10) = 793.5\,\text{W}$$

Example 1.15

Air at 24 °C approaches a 0.9 m long and 0.6 m wide flat plate with an approach velocity 5 m s^{-1}. Determine the total rate of heat transfer from the plate for a plate surface temperature of 130 °C.

Solution

The mean film temperature $T_f = \frac{24+130}{2} = 77\,°\text{C} = 350\,\text{K}$. The thermo-physical properties of air at 350 K (see Appendix A4.1) are $\rho = 1.008\,\text{kg m}^{-3}$, $\mu = 2.09 \times 10^{-5}\,\text{kg m}^{-1}\,\text{s}^{-1}$, $K = 0.02984\,\text{W m}^{-1}\,\text{K}^{-1}$, $Pr = 0.707$ and $\beta = \frac{1}{T} = \frac{1}{350}\,\text{K}^{-1}$.

Characteristic length $L = 0.9$ m; thus from eqn (1.49) the Reynolds number is:

$$Re = \frac{\rho V_0 L}{\mu} = \frac{1.008 \times 5 \times 0.9}{2.09 \times 10^{-5}} = 2.17 \times 10^5$$

From eqn (1.62) the correlation for laminar flow is:

$$Nu = 0.664 Re_L^{0.5} Pr^{0.33} = 0.664 \times (2.17 \times 10^5)^{0.5} \times (0.707)^{0.33} = 275.57$$

From eqn (1.50) the convective heat transfer coefficient is:

$$Nu = \frac{hL}{K} \Rightarrow h = 275.57 \times \frac{0.02984}{0.9} = 9.14\,\mathrm{W\,m^{-2}\,K^{-1}}$$

From eqn (1.48) the heat loss by convection is:

$$\dot{Q}_c = hA\Delta T = 9.14 \times (0.9 \times 0.6) \times (130 - 24) = 523\,\mathrm{W}$$

1.15 Heat Transfer through Radiation

Thermal radiation involves the transfer of heat from a body at a higher temperature to another at a lower temperature by electromagnetic waves (0.1 to 100 μm). The temperature is transmitted in the space in the form of electromagnetic waves. Thermal radiation is in the infrared range and obeys all the rules as that of light, *viz.* it travels in a straight line through a homogenous medium, is converted into heat when it strikes any body which can absorb it, and is reflected and refracted according to the same rules as that of light.

When radiant energy falls on a body, a part of it is reflected, another is absorbed and the rest is transmitted through it. The conservation of energy states that the total sum must be equal to the incident radiation; thus:

$$I_{\mathrm{T}} = I_{\mathrm{r}} + I_{\mathrm{a}} + I_{\mathrm{t}} \tag{1.68}$$

$$\text{or } 1 = \rho' + \alpha' + \tau \tag{1.69}$$

where ρ', α' and τ are the reflectivity, absorptivity and transmissivity of the intercepting body, respectively. The ratio of the energy reflected to that which is incident is called reflectivity. The ratios of the energy absorbed and the energy transmitted to that which is incident are the absorptivity and transmissivity, respectively. Appendix A5 provides the absorptivity of various surfaces for the Sun's rays.

1.15.1 Radiative Heat Transfer Coefficient

The radiant heat exchange between two infinite parallel surfaces at temperatures T_1 and T_2 may be given as:

$$\dot{q}_r = \varepsilon\sigma(T_1^4 - T_2^4) \tag{1.70}$$

The above equation may be rewritten as:

$$\dot{q}_r = h_r(T_1 - T_2) \tag{1.71}$$

$$\text{where } h_r = \varepsilon\sigma \frac{(T_1^4 - T_2^4)}{(T_1 - T_2)} \tag{1.72}$$

for $\bar{T}_1 \approx \bar{T}_2 \approx \bar{T}$:

$$h_r = \varepsilon(4\sigma\bar{T})^3 \tag{1.73}$$

$$\varepsilon = \frac{1}{\varepsilon_1} + \frac{1}{\varepsilon_2} - 1, \quad \text{for two parallel surfaces}$$

$$= \varepsilon \quad \text{for a surface exposed to the atmosphere.}$$

1.15.2 Sky Radiation

In order to evaluate radiation exchange between a body and the sky, a certain equivalent black-body sky temperature is defined. This accounts for the fact that the atmosphere is not at a uniform temperature and that it radiates only in certain wavelength regions. Thus, the net radiation to a surface with emittance ε and temperature T is:

$$\dot{q} = A\varepsilon\sigma(T_{sky}^4 - T^4) \tag{1.74}$$

In order to express the equivalent sky temperature, T_{sky}, in terms of ambient air temperature, various expressions have been given by different people. These relations, although simple to use, are only approximations. Swinbank[5] relates sky temperature to the local air temperature by the relation:

$$T_{sky} = 0.0552T_a^{1.5} \tag{1.75}$$

Another commonly used relation is that given by Whillier:[6]

$$T_{sky} = T_a - 6 \tag{1.76}$$

Example 1.16

A grey surface has an emissivity 0.35. If the surface is opaque, calculate its reflectivity.

Solution

Given that the absorptivity $\alpha' = 0.35$ and for an opaque surface the transmissivity $\tau = 0$. From eqn (1.69) the reflectivity is given as:

$$1 = \rho' + \alpha' + \tau$$

Thus $\rho' = 1 - 0.35 = 0.65$

Example 1.17

A pipe of 60 mm diameter is placed horizontally and covered with 20 mm thick insulation. The insulation has emissivity of 0.92, surface temperature 75 °C and the pipe is placed in the ambient air at 25 °C. Considering heat loss both by radiation and natural convection, calculate the rate of heat loss from a 5 m length of pipe.

Solution

The mean film temperature $T_f = \frac{75+25}{2} = 50\,°C$. The thermophysical properties of air at 50 °C (see Appendix A4.1) is $\rho = 1.092\,kg\,m^{-3}$, $\mu = 1.957 \times 10^{-5}\,kg\,m^{-1}\,s^{-1}$, $K = 0.02781\,W\,m^{-1}\,K^{-1}$, $Pr = 0.708$ and $\beta = \frac{1}{T} = \frac{1}{50+273} = 3.096 \times 10^{-3}\,K^{-1}$. The characteristic length L = diameter of cylinder $= 60 + (20 \times 2) = 100\,mm = 0.1\,m$. From eqn (1.53) the Grashof number is:

$$Gr = \frac{g\beta\rho^2 L^3 \Delta T}{\mu^2}$$
$$= \frac{9.81 \times 3.096 \times 10^{-3} \times (1.092)^2 \times (0.1)^3 \times (75 - 25)}{(1.957 \times 10^{-5})^2} = 4.73 \times 10^6$$

From eqn (1.56) and Table 1.3 the correlation for a horizontal cylinder is:

$$Nu = 0.53(4.73 \times 10^6 \times 0.708)^{0.25} = 22.67$$

From eqn (1.50) the convective heat transfer coefficient can be determined as:

$$Nu = \frac{hL}{K} \Rightarrow h = 25.24 \times \frac{0.02781}{0.1} = 6.3\,W\,m^{-2}\,K^{-1}$$

From eqn (1.48), the heat loss by convection is:

$$\dot{Q}_c = hA\Delta T = 6.30 \times (\pi \times 0.1 \times 5) \times (75 - 25) = 495.2\,W$$

From eqn (1.70), the heat loss by radiation is:

$$\dot{Q}_r = \varepsilon \sigma A(T_1^4 - T_2^4)$$
$$= 0.92 \times (5.67 \times 10^{-8})(\pi \times 0.1 \times 5)(348^4 - 298^4) = 555.5\mathrm{W}$$

Thus the total heat loss is:

$$\dot{Q} = \dot{Q}_c + \dot{Q}_r = 495.2 + 555.5 = 1050.7\,\mathrm{W}$$

References

1. C. K. Brown and W. H. Gauvin, Combined free-and-forced convection. I. Heat transfer in aiding flow, *Can. J. Chem. Eng.*, 1965, **43**, 306–312.
2. C. K. Brown and W. H. Gauvin, Combined free-and-forced convection. II. Heat transfer in opposing flow, *Can. J. Chem. Eng.*, 1965, **43**, 313–318.
3. W. H. McAdams, *Heat Transmission*, McGraw-Hill, New York, 1954.
4. J. H. Watmuff, W. W. S. Charters and D. Proctor, Solar and wind induced external coefficients for solar collectors, *COMPLES*, 1977, **2**, 56.
5. W. C. Swinbank, Long-wave radiation from clear skies, *Q. J. R. Meteorol. Soc.*, 1963, **89**, 339–348.
6. A. Whillier, Design factors influencing solar collectors, In: *Low temperature engineering applications of solar energy*, ed. R. C. Jordan, ASHRAE, New York, 1967, 254–262.

Further Reading

G. N. Tiwari, *Solar Energy Fundamentals, Design, Modelling and Applications*, Narosa, New Delhi, 2002.

G. N. Tiwari and S. Dubey, *Fundamentals of Photovoltaic Modules and their Applications*, RSC, Cambridge, 2010.

J. A. Duffie and W. A. Beckman, *Solar Engineering of Thermal Processes*, Wiley, Hoboken, NJ, 2006.

CHAPTER 2

Photovoltaic Technology and its Development

2.1 Introduction

Solar energy can be directly converted into electricity with the help of a solar cell. Assemblies of cells are used to make solar panels, modules or photovoltaic arrays. The field of technology and research related to the application of solar cells in producing electricity for practical use is called photovoltaics. This chapter discusses the basics and the technological developments in photovoltaics.

2.2 Evolution of the Solar Cell

Photovoltaic generation of power is caused by radiation that separates positive and negative charge carriers in absorbing material. In the presence of an electric field, these charges can produce a current for use in an external circuit. Such fields exist permanently at junctions or inhomogeneities in materials as "built-in" electric fields and provide the required electromagnetic force (emf) for useful power production.

Junction devices are usually known as photovoltaic cells or solar cells, although the term is a misnomer in the sense that it is the current that is produced by the radiation photons and not the voltage. The cell itself provides the source of the emf. It is to be noted that photoelectric devices are electrical current sources driven by a flux of radiation. A majority of photovoltaic cells are silicon semiconductor junction devices. Thus, in order to study photovoltaic cells, an understanding of the basics of semiconductors is required.

RSC Energy Series No. 4
Building Integrated Photovoltaic Thermal Systems: For Sustainable Developments
By Basant Agrawal and G. N. Tiwari
© B. Agrawal and G. N. Tiwari 2011
Published by the Royal Society of Chemistry, www.rsc.org

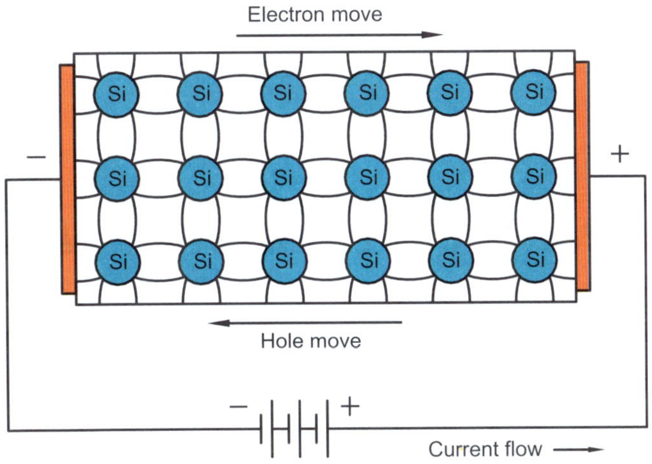

Figure 2.1 Intrinsic-type semiconductor.

2.2.1 Intrinsic Semiconductors

It has been noted that silicon is an insulator at absolute zero temperature. This is because there is insufficient energy available from the lattice to cause a covalent bond to break. Thus, there are no free charges. However, as the temperature of the crystal is raised there is a finite probability that a small concentration of bonds will gain enough energy to break and release an electron to become free as a conduction band electron. At the same time, it leaves behind an empty state in the valence band, which in turn can meander around the lattice as shown in Figure 2.1. This unfilled bond is called a hole and mathematically is treated as a positive charge. The process of free carrier generation is balanced by recombination under thermal equilibrium conditions so that the electron and hole concentrations reach steady-state values. The concentrations of electrons and holes in an intrinsic semiconductor, n_i and p_i, are equal, and may be calculated from the expression:

$$n_i = p_i = (N_c N_v)^{1/2} \exp\left(-\frac{E_g}{2kT}\right) \tag{2.1}$$

where E_g is the forbidden energy gap or bond strength of the semiconductor, k is Boltzmann's constant (1.38×10^{-23} J K^{-1}), and N_c and N_v are the parameters called the effective densities of states in the two bands.

For silicon at 300 K, $N_c = 2.8 \times 10^{19}$ cm^{-3} and $N_v = 1.04 \times 10^{19}$ cm^{-3}. The value of n_i for silicon at 300 K is 1.45×10^{16}, which increases exponentially with temperature. The bonds in silicon become progressively weaker, so it is archetypal of semiconductors with a very high concentration of broken bonds.

The width of the bandgap, $E_c - E_v$, is a very important characteristic of the semiconductor and is usually denoted by E_g. Table 2.1 gives the bandgaps of

Table 2.1 Bandgaps of semiconductors used for solar-cell applications.

Material	*Crystal-line silicon*	*Amor-phous silicon*	*Copper indium selenide, $CuInSe_2$*	*Cadmium telluride, CdTe*	*Gallium arsenide, GaAs*	*Indium phosphide, InP*
Energy gap (eV)	1.11	1.75	1.05	1.49	1.43	1.35

the most important semiconductors for solar-cell applications. Intrinsic semiconductors contain just the right number of electrons to fill the valence band, and the conduction band is therefore empty. Electrons in the full valence band cannot move. For practical purposes, a pure semiconductor is therefore an insulator.

2.2.2 Extrinsic Semiconductors

The utility of an intrinsic semiconductor for electronics is rather limited. However, if appropriate impurities are added, it becomes very versatile indeed. It is now called an extrinsic or impurity semiconductor. Rather than treat extrinsic semiconductors in general, it is simpler to deal first with a specific material and generalize later.

The silicon atom has four valence electrons. When silicon crystallizes into a solid, the valence electrons are shared between adjacent atoms to form covalent bonds. Each silicon atom is surrounded by four covalent bonds to give a crystal with tetrahedral symmetry. At absolute zero temperature, all the valence electrons are tightly held in these bonds; none of them is free to move through the crystal. At higher temperatures, some bonds are broken to release electrons which can move through the lattice. At the same time, an electron can move from a complete bond into an adjacent broken bond to give the effect of a hole moving through the lattice.

2.2.2.1 n-Type Semiconductors

If a silicon atom is replaced by an atom with five valence electrons (phosphorus or arsenic), four of these electrons can fit into the four covalent bonds. However, there is no bond to accommodate the fifth electron. The extra electron tends to drift away from the impurity atom, but it is attracted back by the net positive charge on the impurity ion. So the electrons are held weakly and move around as shown in Figure 2.2 (*left*). Because the electrons are bonded weakly, at higher temperatures the electrons are excited out of these states into the conduction band. Since much less energy is needed for this excitation than for a band-to-band excitation, the process can also occur at temperatures well below where there is significant excitation of electrons from the valence band to the conduction band. Now the solid has free electrons in the conduction band and

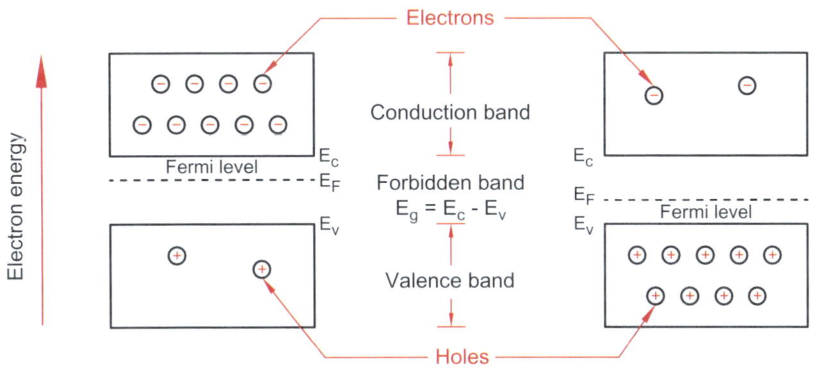

Figure 2.2 Energy band diagram: *left*, n-type semiconductor material; *right*, p-type semiconductor material.

no holes in the valence band. Any electric current is carried by the free electrons which have come from the donor states. Because the current is due to negatively charged carriers, this semiconductor is said to be n-type. The electrons are called the majority carriers and the holes are called minority carriers. At still higher temperatures the material acts as an intrinsic semiconductor.

2.2.2.2 *p-Type Semiconductors*

If a silicon atom is replaced by an atom with three valence electrons (boron or gallium), one of the covalent bonds around the impurity will lack an electron. So there would be a hole in the bond structure as shown in Figure 2.2 (*right*). At higher temperatures, electrons would be excited from the valence band to go into these states, leaving free holes in the valence band. As with the n-type material, this excitation can occur without any appreciable band-to-band excitation. Because the current is due to positive holes in the valence band, the semiconductor is called p-type. At some elevated temperatures the holes are called the majority carriers and the electrons are minority carriers. At still higher temperatures the material acts as an intrinsic semiconductor.

2.2.2.3 *Fermi Level*

The Fermi level is the apparent energy level within the forbidden band gap from which majority carriers (electrons in n-type and holes in p-type) are excited to become charge carriers. For n-type material the Fermi energy level (E_F) is given by:

$$E_F = E_c + kT \ln\left(\frac{N_0}{N_c}\right) \quad (2.2)$$

where E_c is the conduction band energy, k is the Boltzmann constant, N_o is the donor concentration, N_c is the effective density of states in conduction band and T is the fixed temperature. The Fermi energy level of an n-type semiconductor is towards the valence band, as shown in Figure 2.2 (*left*).

For p-type material the Fermi energy level is given by:

$$E_F = E_v - kT \ln\left(\frac{N_a}{N_v}\right) \quad (2.3)$$

where E_v is the valence band energy, N_a is the acceptor ion concentration and N_v is the effective density of states in the valence band. The Fermi energy level of a p-type semiconductor is towards the valence band, as shown in Figure 2.2 (*right*).

2.2.2.4 Light Absorption

The excitation of an electron directly from the valence band to the conduction band is called fundamental absorption. During the passage of light through the crystal its intensity is weakened owing to absorption. Figure 2.3 and shows the plot between the energy and the crystal momentum where the conduction and valence bands are shown by parabolas. In the case of direct band gap semiconductors such as GaAs, GaInP, CdTe and Cu(InGa)Se_2, the minimum energy of the conductance band in relation to the crystal momentum lies directly above the maximum of the valence band, as shown in Figure 2.3 (*left*). When the photon is absorbed, the energy $E = hv$ is the energy difference between the initial and the final conditions of the energy of the crystal. In indirect band gap semiconductors like Si and Ge, the minimum of the conductance band and the maximum of the valence band lie at different crystal momentums, as shown in Figure 2.3 (*right*). The probability of absorption in

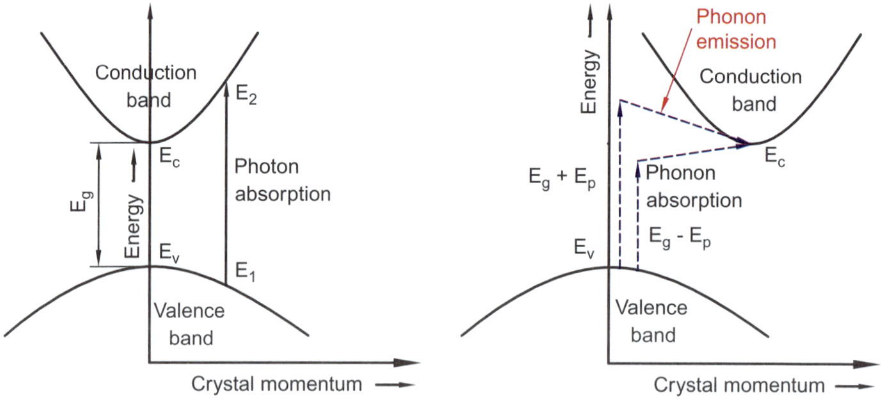

Figure 2.3 Energy of conductance and valence bands as a function of crystal momentum for: *left*, a direct semiconductor; *right*, an indirect semiconductor.

this case is much lower than for a direct semiconductor due to the involvement of two different particles.

As silicon is an indirect semiconductor, it has a low absorption coefficient at photon energies near the band edge. This means that a relatively large thickness of material is necessary to absorb the long wavelength part of the solar spectrum.

2.2.3 p–n Junctions

When n-type and p-type silicon semiconductors are placed in contact with each other as shown in Figure 2.4 (*left*), a p–n junction is formed. The n-type semiconductor has free electrons and the p-type semiconductor has holes. Near the p–n junction the electrons diffuse into the vacant holes in the p-type material, causing a depletion zone as shown in Figure 2.4 (*right*). This depletion zone acts like an insulator, preventing other free electrons in the n-type silicon and holes in the p-type silicon from combining. Thus, a p–n junction creates a diode that allows electricity to flow in one direction and prevents electricity from flowing in the other direction. In addition, a p–n junction leaves a small electrical imbalance inside the crystal. Unfortunately, power cannot be generated from this electrical imbalance.

2.2.3.1 Forward Biased

Figure 2.5 (*left*) shows a p–n diode connected by an external power such that the negative terminal of the battery is connected to the n-type silicon and the positive terminal of the battery is connected to the p-type silicon. As like charges repel each other, so the free electrons and holes are pushed toward the p–n junction. If the voltage pushing the electrons and holes has sufficient strength to overcome the depletion zone, the electrons and holes combine at the junction and current passes through the diode. This arrangement is called forward biased. The direction of current is opposite to the flow of electrons in the outer circuit, as shown in figure.

2.2.3.2 Reversed Biased

Figure 2.5 (*right*) shows a p–n diode connected by an external power such that the negative terminal of the battery is connected to the p-type silicon and the

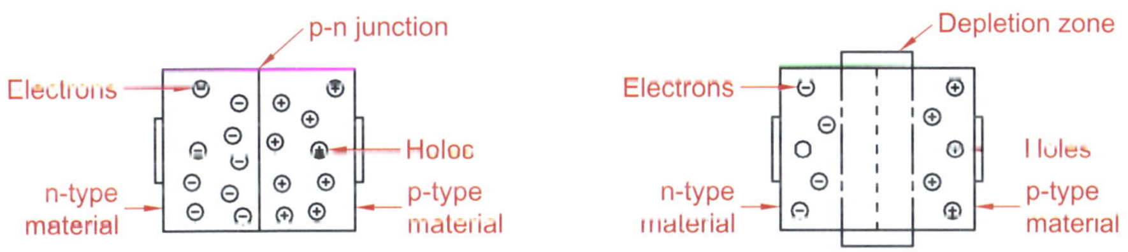

Figure 2.4 *Left*, a p–n junction before migration of electrons and holes; *right*, a depletion zone formed after migration of electrons and holes.

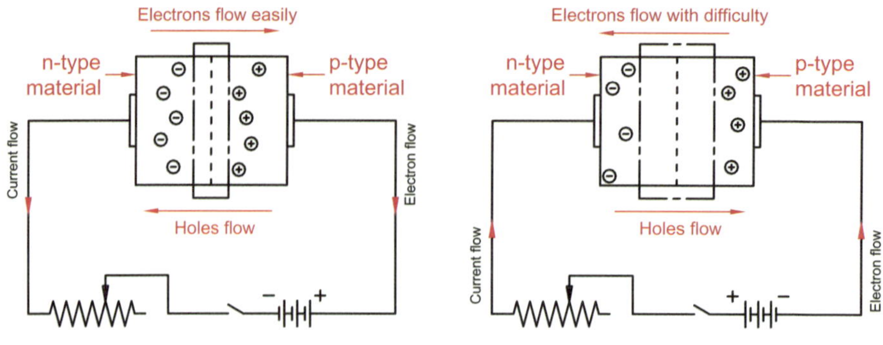

Figure 2.5 *Left*, forward biasing; *right*, reversed biasing.

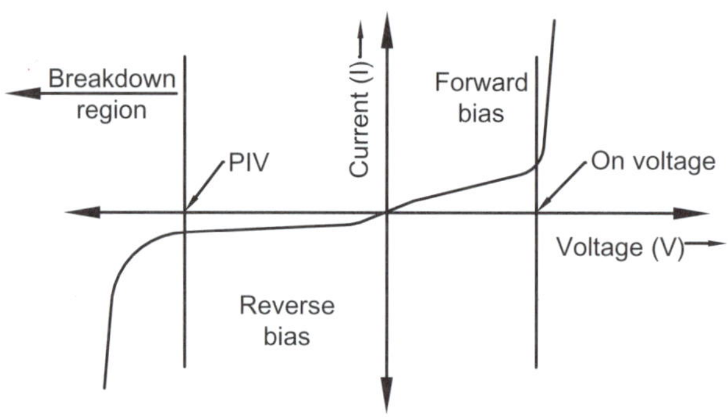

Figure 2.6 *I–V* characteristic curve for a p–n junction diode.

positive terminal of the battery is connected to the n-type silicon. The negative terminal attracts the positive holes in the p-type silicon and the positive terminal of the battery attracts the free electrons in the n-type silicon. All the charge carriers are pulled away from the p–n junction, which essentially creates a larger depletion region and no current flows. Such an arrangement is called reverse biased.

2.2.3.3 I–V Characteristics of a p–n Junction Diode

Figure 2.6 shows the current–voltage variation for the p–n junction diode, generally called the *I–V* characteristic curve. At very large reverse bias, beyond the peak inverse voltage (PIV), a process called reverse breakdown (or avalanching) occurs which causes a large increase in current that usually damages the device permanently. In the second region, at reverse biases more positive than the PIV, only a very small reverse saturation current flows. The third region is a forward but small bias, where only a small forward current is conducted. Finally, as the potential difference is increased above a cut-in voltage or on-voltage, the diode current becomes appreciable and where it can be assumed as a connection with negligible resistance. More precisely, the *I–V* curve is exponential.

With no external bias ($V = 0$), the generation current (I_g) and the reverse recombination current (I_r) are equal:

$$I_g = I_r \tag{2.4}$$

With a forward bias of voltage V, the recombination current becomes an increased forward current:

$$I_r = I_g \exp(qV/kT) \tag{2.5}$$

where q is the charge of the electron (1.602×10^{-19} coulombs or 4.8×10^{-10} esu), k is Boltzmann's constant and T is the working temperature of the cell in degrees kelvin.

The total current (with no illumination) is:

$$I_D = I_r - I_g = I_g[\exp(qV/kT) - 1] \tag{2.6}$$

The above equation is the Shockley equation and can be written as:

$$I_D = I_0[\exp(qV/kT) - 1] \tag{2.7}$$

where I_0 ($= I_g$) is the saturation current under reverse bias, before avalanche breakdown occurs. It is also known as the leakage or diffusion current. For good solar cells, $I_0 \approx 10^{-8}$ A m^{-2}. The value increases with temperature.

2.2.4 Photovoltaic Cells: Generating Electricity

Incident solar radiation can be considered as discrete energy units called photons. The product of the frequency (ν) and wavelength (λ) is the speed of light (c). The energy of a photon is a function of the frequency of the radiation and is given by:

$$E = h\nu = \frac{hc}{\lambda} \tag{2.8}$$

where h is Planck's constant (6.626×10^{-34} J s or 4.136×10^{-15} eV s).

The solar cell is essential a p–n junction with a large surface area. The n-type material is kept thinner to allow light through it and reach the depletion layer. When a photon of light is absorbed by one of these atoms in the n-type silicon it will dislodge an electron, creating a free electron and a hole. The free electron has sufficient energy to jump out of the depletion zone. If a

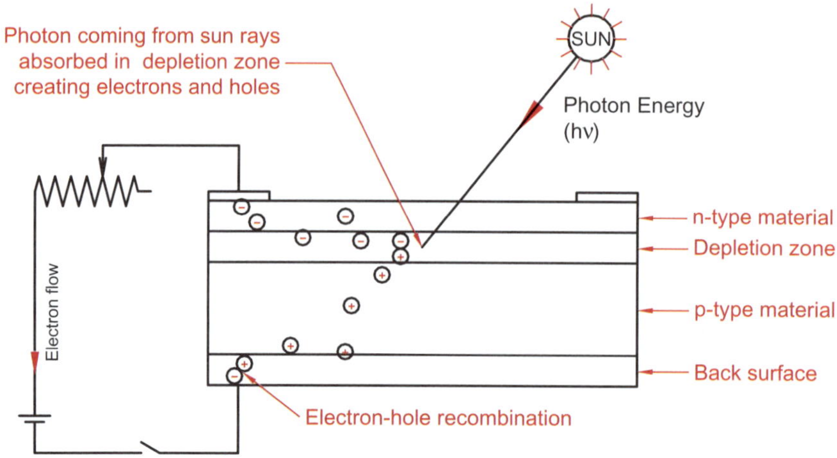

Figure 2.7 Photovoltaic solar cell.

wire is connected from the cathode (n-type silicon) to the anode (p-type silicon) as shown in Figure 2.7, electrons will flow through the wire. The electron travels through the external load, creating a flow of electric current. The hole created by the dislodged electron migrates to the back electrical surface. As the electron enters the p-type silicon from the back electrical contact into the back surface it combines with the hole, restoring the electrical neutrality. Covering the solar cell with thin layer of dielectric material or an antireflecting coating (ARC) minimizes the reflection from the top surface.

The total series resistance of the cell can be expressed as:

$$R_s = R_{cp} + R_{bp} + R_{cn} - R_{bn} \tag{2.9}$$

where R_{cp} is the metal contact to p-type semiconductor resistance, R_{bp} is the bulk p-type resistance, R_{cn} is the contact to n-type semiconductor resistance and R_{bn} is the bulk n-type resistance.

The idealized p–n junction diode current is given as:

$$I_D = I_0\left\{\exp\left[\frac{q(V + IR_s)}{kT}\right] - 1\right\} \tag{2.10}$$

The overall cell current I is determined by subtracting the light-induced current I_L from the diode dark current I_D. Therefore:

$$I = I_D - I_L \Rightarrow I = I_0\left\{\exp\left[\frac{q(V + IR_s)}{kT}\right] - 1\right\} - I_L \tag{2.11}$$

In addition, a shunt path may exist for current flow across the junction due to surface effects or a poor junction region. This alternative path for the current constitutes a shunt resistance R_p across the junction. Then:

$$I = I_L - I_0\left\{\exp\left[\frac{q(V + IR_s)}{AkT}\right] - 1\right\} - \left(\frac{V + IR_s}{R_p}\right) \tag{2.12}$$

where A is the ideality factor or diode quality factor and is usually equal to unity.

Certain parameters to be mentioned in the I–V characteristics of a solar cell are as follows:

1. The *short circuit current* (I_{sc}) is the light generated current, I_L.
2. The *open circuit voltage* (V_{oc}) is obtained by setting $I = 0$ in eqn (2.12):

$$V_{oc} = \frac{kT}{e}\ln\left(\frac{I_L + I_0}{I_0}\right) \approx \frac{kT}{e}\ln\left(\frac{I_L}{I_0}\right) \tag{2.13}$$

   Both I_L and I_0 depend on the structure of the solar cells. No power is generated under short or open circuit conditions.
3. The *power output* is defined as:

$$P_{out} = V_{out} \times I_{out} \tag{2.14}$$

4. The *maximum power*, P_{max}, provided by the device is achieved at a point on the characteristics where the product of current times voltage is a maximum. Thus:

$$P_{max} = V_{max} \times I_{max} \tag{2.15}$$

5. The *fill factor* is given by:

$$FF = \frac{P_{max}}{V_{oc} \times I_{sc}} \tag{2.16}$$

   The fill factor, also known as the curve factor, is a measure of the sharpness of the knee in the I–V curve. It indicates how well a junction was made in the cell and how low the series resistance has been made. It can be lowered by the presence of a series resistance and tends to be higher whenever the open circuit voltage is high. The maximum value in silicon is 0.88. The maximum possible output can be given as:

$$P_{max} = V_{oc} \times I_{sc} \times FF \tag{2.17}$$

An empirical expression for the fill factor is:

$$FF = \frac{V_{oc} - \frac{kT}{q}\ln\left[\frac{qV_{oc}}{kT} + 0.72\right]}{V_{oc} + \frac{kT}{q}} \tag{2.18}$$

6. The solar cell *power conversion efficiency* can be given as:

$$\eta = \frac{P_{max}}{P_{in}} = \frac{V_{oc} \times I_{sc} \times FF}{\text{Incident solar radiation} \times \text{Area of solar cell}} \tag{2.19}$$

Thus, the efficiency of a solar cell is the power P_{max} supplied by the cell at the maximum power point under standard test conditions, divided by the power of the incident solar radiation upon it. The most frequent conditions are a temperature of 25 °C, an irradiance of 1000 W m^{-2} with an air mass of 1.5 (AM1.5).

2.2.4.1 *I–V Characteristics*

A more detailed consideration makes it possible to draw an equivalent circuit of a solar cell in terms of a current generator and a diode, as shown on the left of Figure 2.8. This equivalent circuit has a current–voltage relationship as shown on the right of Figure 2.8.

In solar cell applications this characteristic is usually drawn inverted about the voltage axis, as shown in Figure 2.9. The cell generates no power in short-circuit (when the current I_{sc} is produced) or open-circuit conditions (when the cell generates the voltage V_{oc}). The cell delivers the maximum power P_{max} when operating at a point on the characteristic where the product of the current and

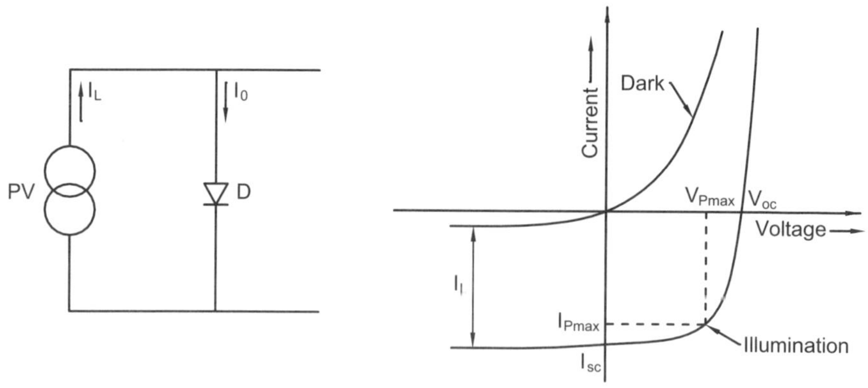

Figure 2.8 A solar cell: *left*, equivalent circuit; *right*, *I–V* characteristics.

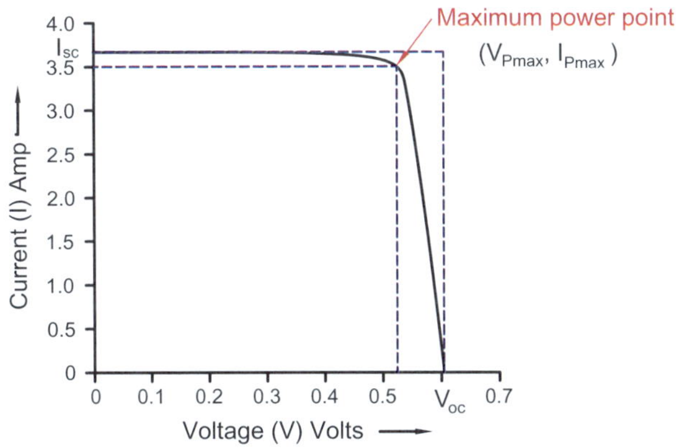

Figure 2.9 Maximum power of a silicon solar cell through the *I–V* curve.

voltage is a maximum. This is shown graphically in Figure 2.9, where the position of the maximum power point represents the largest area of the rectangle.

The optimum load resistance $R_L(P_{max}) = R_{P\max}$ is connected if the PV generator is able to deliver maximum power:

$$P_{max} = V_{P\max} \times I_{P\max} \tag{2.20}$$

$$\Rightarrow R_{P\max} = \frac{V_{P\max}}{I_{P\max}} \tag{2.21}$$

The efficiency is defined as:

$$\eta = \frac{P}{\phi} \tag{2.22}$$

where $P = V \times I$ is the power delivered by the PV generator, $\phi = I_T \times A$ is the solar radiation falling on the PV generator, I_T is the solar intensity and A is the surface area irradiated.

2.2.5 The Limits to Cell Efficiency

Photovoltaic cells are limited in efficiency by many losses. Some of these are avoidable but others are intrinsic to the system and may be described as follows:

1. The electric current leaves the top surface by a web of metal contacts arranged to reduce series resistance losses at the surface. These contacts have a finite area and thus cover part of the active surface and block 5–15% of the incident solar radiation.
2. Without special precautions, the reflectance from semiconductors is high (about 40% of the incident solar radiation). However, this may be reduced to 3% or less by the use of thin film surfaces.

3. The output of cells is limited by several factors. There is a minimum energy level of photons that can cause the creation of a hole–electron pair. For silicon, the maximum wavelength is 1.15 μm. Radiation at higher wavelengths does not produce hole–electron pairs but heats the cell. Each photon wavelength does not produce hole–electron pairs but heats the cell. Each photon causes the creation of a single hole–electron pair and the energy of photons in excess of that required to create hole–electron pairs is also converted into heat. From these considerations alone, the maximum theoretical efficiency of silicon cells is 23%.
4. The net effect of the spectral characteristics of the input solar radiation is that 30–40% of the incident photon energy may not be utilized for the generation of electron–hole pairs in silicon solar cells. The excess energy of the photons ($h\nu - E_g$) appears as heat.
5. The quantum efficiency is defined as the percentage of photons that are converted to electrical current when the cell is operated under short-circuit conditions. The external quantum efficiency is the fraction of incident photons that are converted to electrical current, while the internal quantum efficiency is the fraction of absorbed photons that are converted to electrical current. Since some wavelengths are absorbed more effectively than others in most semiconductors, spectral measurements of quantum efficiency can yield information about which parts of a particular solar cell design are most in need of improvement.
6. The collection efficiency is defined as the proportion of radiation-generated electron–hole pairs that produce current in the external circuit. The solar cell with a collection efficiency factor of 0.7 has an overall efficiency of 10%. Increasing the collection efficiency to about 0.9 would produce an overall efficiency greater than 20%.
7. Each absorbed photon produces electron–hole pairs with an electric potential difference of E_g/q (1.1 V in Si). However, only a part (V_B) of this potential is available for the emf of an external circuit. The voltage factor F_v is equal to qV_B/E_g. The missing emf occurs because in the open circuit the Fermi level across the junction equates at the dopant n and p levels and not at the displaced conduction to valence band levels. Increased dopant concentration increases F_v. The loss due to the voltage factor is about 20% of the insolation.
8. In practice, the cell characteristics do not follow eqn (2.11) and are better represented as:

$$I = I_0\left\{\exp\left[\frac{q(V + IR_s)}{AkT}\right] - 1\right\} - I_L \tag{2.23}$$

The factor A results from increased recombination in the junction and tends to change V_{oc} and I_0. So the optimum output would occur if $A = 1$.

Within the cell, recombination is lessened if:

(a) The diffusion paths are long (50–100 μm in Si); this requires long minority carrier lifetimes.

(b) The junction is near the top surface (within 0.15 μm).

(c) The material has few defects other than the dopant.

9. The solar cell *I–V* characteristic is strongly influenced by the p–n diode characteristics. Thus as the solar cell (Figure 2.10) output is raised towards V_{oc}, the diode becomes increasingly forward biased, so increasing the internal recombination current I_r across the junction. This necessary behaviour is treated as a fundamental loss in the system. The loss due to the curve factor is about 4% of the insolation.
10. Cell parameters are inherent electrical losses in a solar cell. The equivalent series and shunt resistance as shown in Figure 2.10 can represent these losses. Incorporating parasitic series and shunt resistances with real solar cells is given by:

$$I = I'_{sc} - I_0\left\{\exp\left[\frac{q(V + IR_s)}{AkT}\right] - 1\right\} - \frac{(V + IR_s)}{R_{sh}} \tag{2.24}$$

The effects of these parasitic resistances on the *I–V* characteristic are shown in Figure 2.11. The series resistance arises mainly due to the thin diffused layer on a silicon solar cell. The ohmic contacts on the two opposite surfaces and the bulk silicon resistivity also contribute to the series resistance of a cell. The shunt resistance is the result of the leakage current across the junction and the edge of the solar cell. Both these contribute to the ohmic losses in the devices. These adverse effects can be minimized by a good engineering design of the solar cell parameters to minimize the series resistance and maximize the shunt resistance.

11. Temperature effects. An increase in the intrinsic carrier concentration increases the dark saturation current and results in a decrease in the open-circuit voltage. The band gap decreases with temperature and its

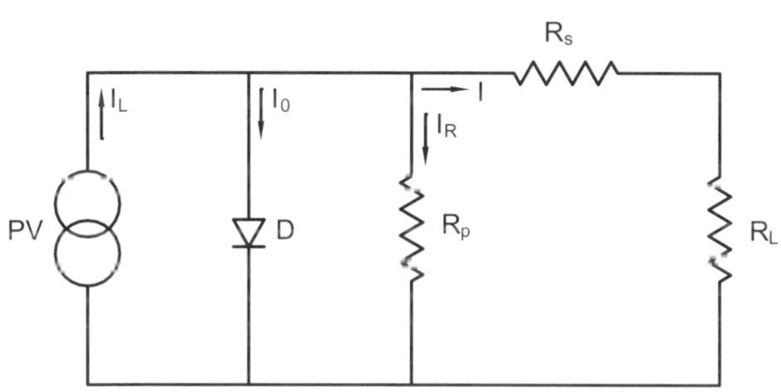

Figure 2.10 Equivalent circuit of a solar cell.

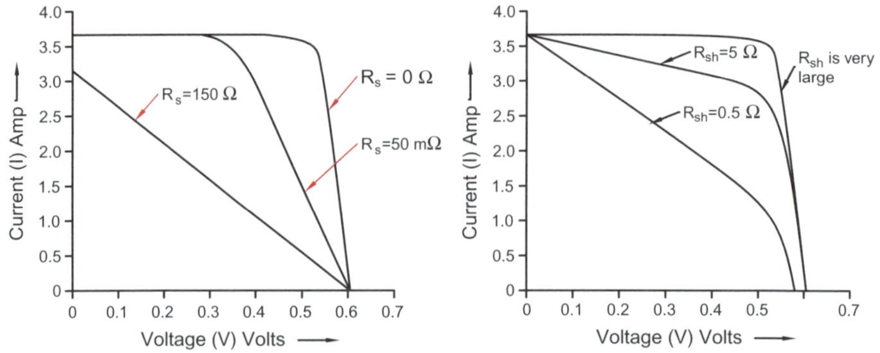

Figure 2.11 *Left*, effect of series resistance on the *I–V* curve; *right*, effect of parallel resistance on the *I–V* curve.

temperature dependency can be given by:

$$E_g(T) = E_g(0) - \frac{\alpha T^2}{T + \beta} \tag{2.25}$$

where α and β are constants specific to each semiconductor.
The short-circuit current is relatively unaffected by temperature under typical operating conditions. Thus, the temperature dependency of the open-circuit voltage for silicon at 300 K can be given by:

$$V_{oc}(T) = \frac{1}{q} E_g(0) - \frac{kT}{q} \ln\left(\frac{BT^{\xi}}{I_{sc}}\right) \tag{2.26}$$

Thus, V_{oc} varies roughly linearly and inversely with temperature and an extrapolation of V_{oc} to $T = 0$ is approximately the band gap because $\lim_{T \to 0}[T \ln T] = 0$.

12. Dependence of efficiency on radiation. Consider that the short-circuit voltage, I_{sc}, is proportional to the level of radiation. Replace V_{oc} in eqn (2.19) for efficiency:

$$\eta = \frac{I_{sc} \times \dfrac{kT}{q} \ln\left(\dfrac{I_{sc}}{I_0}\right) \times FF}{\text{Incident solar radiation} \times \text{Area of solar cell}} \tag{2.27}$$

The temperature dependency of the fill factor can be disregarded here. After some manipulation we find that for the relative change in efficiency:

$$\frac{\Delta\eta}{\eta} = \frac{\ln(n)}{\ln\left(\dfrac{I_{sc}}{I_0}\right)} \tag{2.28}$$

where n is the air mass factor of light radiation ($n = 1$ for one Sun). Introducing V_{oc} we find that the change in relative efficiency is given by:

$$\frac{\Delta\eta}{\eta} = \frac{kT}{q} \times \frac{\ln(n)}{V_{oc}} \tag{2.29}$$

As the value of $\frac{kT}{q \times V_{oc}}$ for silicon is nearly 0.04, the efficiency of a silicon solar cell reduces by approximately 3% (relative) at half the radiation intensity.

13. Dependence of efficiency on temperature. Consider constant radiation and therefore a constant short-circuit voltage. The fill factor is also assumed to be constant. Thus, the relative change in efficiency is then equal to the relative change in open-circuit voltage. If we differentiate the equation of the open-circuit voltage for temperature, then disregarding the small change in band gap with temperature we have:

$$\frac{dV_{oc}}{dT} = \frac{1}{T}\left(V_{oc} - \frac{E_g}{q}\right) \tag{2.30}$$

As a rule of thumb we find that for silicon:

$$\frac{dV_{oc}}{dT} = \frac{1}{T}(V_{oc} - 1.1) \tag{2.31}$$

Now taking V_{oc} as 0.6 V, at room temperature the relative change in the open-circuit voltage is nearly (–)1.7 mV per C. Thus efficiency drops approximately 0.3% per degree rise in temperature.

2.3 Historical Developments

Edmond Bequerel[1] is usually credited as being the first using sunlight to demonstrate the photovoltaic effect between two platinum electrodes coated with either silver chloride or silver bromide inserted into an acidic solution (Figure 2.12). In 1876, Adams and Day[2] demonstrated a photoconductive effect in a selenium bar merely by the action of light (Figure 2.13).

The first true solar cell was constructed by Charles Fritts[3] in 1883 using junctions formed by coating the semiconductor selenium with an ultrathin, nearly transparent, layer of gold (Figure 2.14). These devices were very inefficient, transforming less than 1% of the absorbed light into electrical energy. By the 1930s,[4] both the selenium cell and the copper oxide cell were being employed in light-sensitive devices, such as photometers for use in photography (Figure 2.15). These early solar cells, however, still had energy conversion efficiencies of less than 1%. In 1941, Russel Ohl[5] described the first semiconductor p–n junction solar cell (Figure 2.16). Most photovoltaic systems produced to date are based on silicon p–n junctions

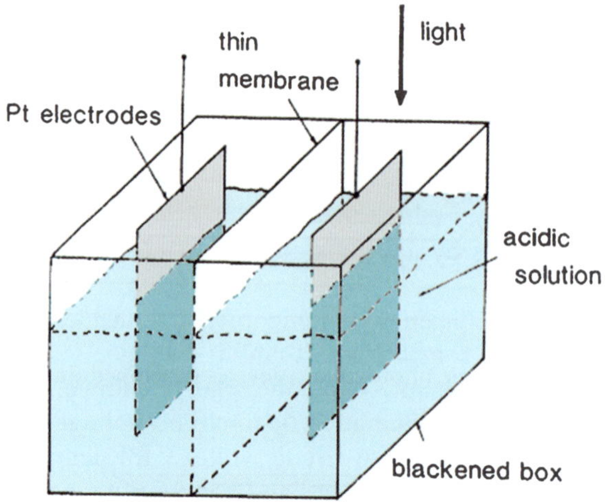

Figure 2.12 Becquerel apparatus.

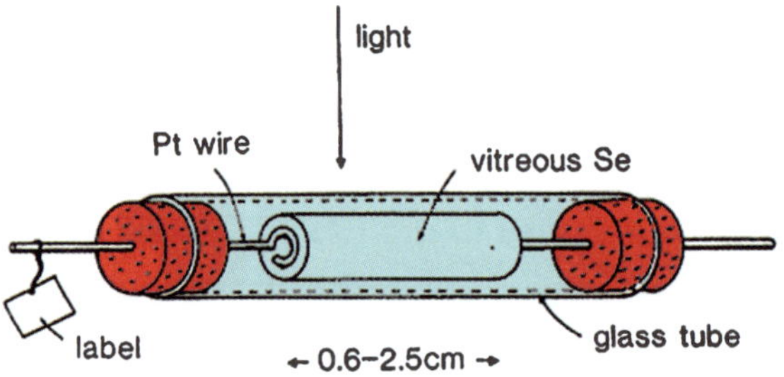

Figure 2.13 Adams and Day apparatus.

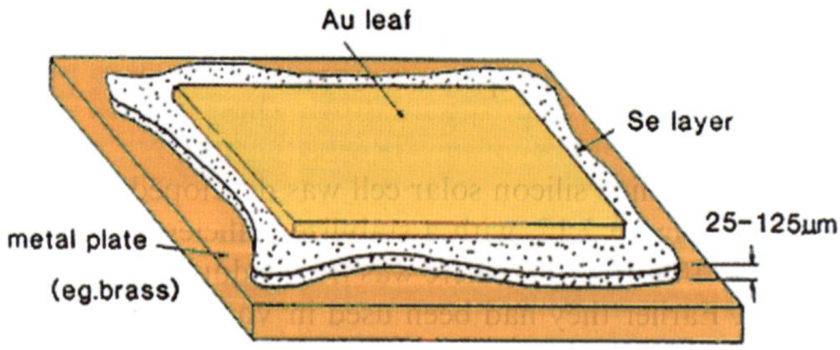

Figure 2.14 Thin-film selenium demonstrated by Fritts in 1883.

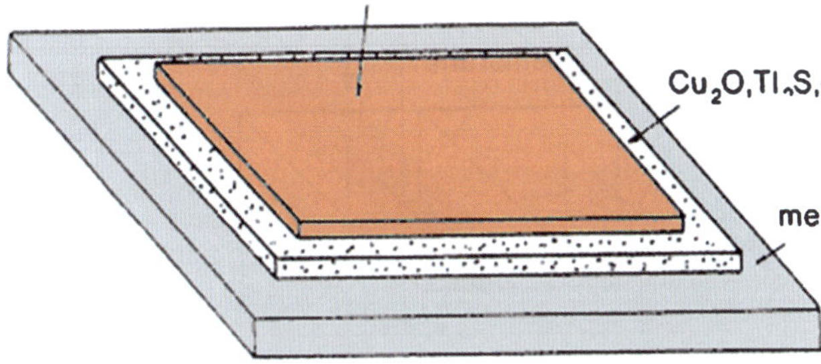

Figure 2.15 Structure of the photovoltaic devices developed during the 1930s.

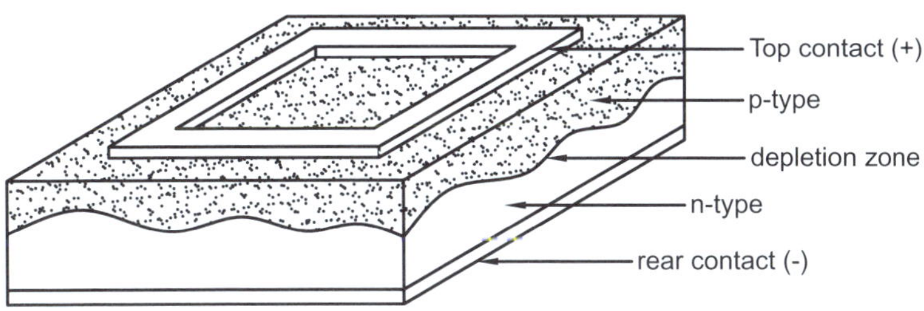

Figure 2.16 Silicon solar cell reported in 1941 by Ohl.

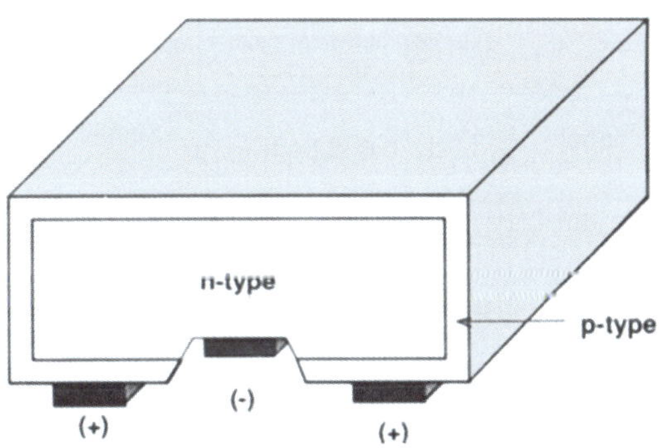

Figure 2.17 The first silicon cell, reported in 1954.

The first monocrystalline silicon solar cell was developed in 1954 by Chapin *et al.*,[6] as shown in Figure 2.17, with a stabilized efficiency of 6%, which was rapidly increased to 10%. In 1958, there was tremendous activity for using solar cells in spacecraft. Earlier they had been used in Vanguard-I to power a small radio transmitter, but soon their application was extended for generating power for newly developed spacecraft such as Explorer-III, Vanguard-II, Sputnik-3,

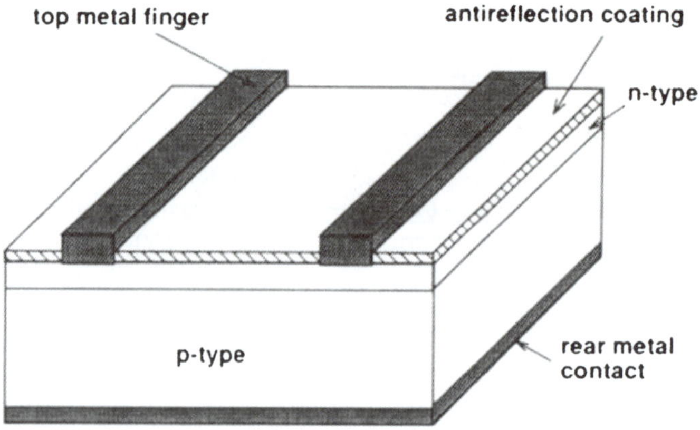

Figure 2.18 Silicon cell design developed for use in space in the early 1960s.

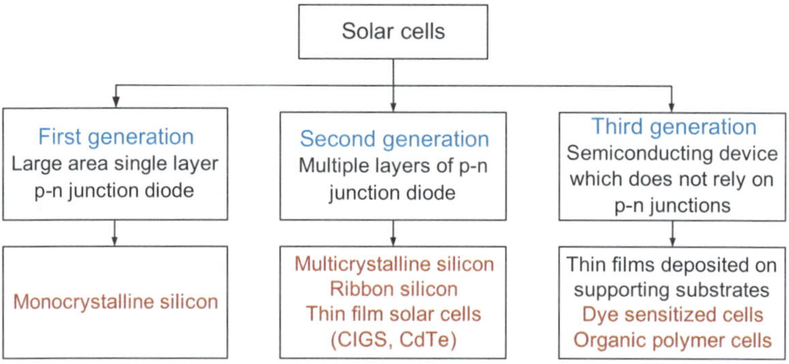

Figure 2.19 The three generations of development in solar cells.

Explorer-VI and Explorer-VII. In the early 1960s[7] the cell shown in Figure 2.18 was developed, with an energy conversion efficiency of about 14% under terrestrial sunlight.

2.4 Technology Generation

The development of solar cells is classified into three technology generations, as shown in Figure 2.19. Detailed descriptions of each generation follow in the next sections.

2.4.1 First-generation Technology

First-generation technology consisted of a single-junction solar cell based on silicon wafers, including single-crystal (c-Si) and multi-crystalline silicon (mc-Si). In 2007, first-generation solar cells accounted for more than 85% of commercial production, the majority of which were based on screen printing-based devices. Generally, cells were sold as a module containing 36 cells in series. Each cell gives a maximum output of about 0.6 V in sunlight, resulting in a maximum output voltage of a module as 20 V. This is sufficient to fully charge a 12 V lead–acid battery.

BP solar, Sanyo, Sunpower and others are producing commercial single-crystal silicon wafers using modern technologies with efficiencies in the range of 19–22%. It is well known that 50% of the cost of these cells is the cost of the silicon wafers with thicknesses of 200–250 μm. Although research is continuing to reduce wafer thickness, they are still too expensive for truly competitive commercial production. It is likely that the cost reduction trend will reach its limit before the first-generation technology reaches full cost competitiveness.

2.4.2 Second-generation Technology

The second-generation technology includes amorphous silicon (a-Si), poly-crystalline silicon (p-Si), copper indium gallium selenide [$CuIn(Ga)Se_2$ or CIGS] and cadmium telluride (CdTe).These are aimed at reducing the cost by eliminating the silicon wafers but maintain the efficiency of the first-generation photovoltaic systems. The technology uses only 1–10 μm of active material and absorbs the solar spectrum much more efficiently. New manufacturing techniques such as vapour deposition and electroplating are advantageous as they reduce the need for processing solar cells at high temperatures. In 2007, CdTe solar cell production represented 4.7% of total market share. Nanosolar commercialized its CIGS in 2007 with a production capacity of 425 MW and represented 0.5% of the total market share.

Although expansion of second-generation technology is slower than expected, it has the potential to reduce the cost of photovoltaic systems in large-scale production. It seems that the potential of second-generation solar cells is mostly likely to be realized by silicon-based thin film devices.

2.4.3 Third-generation Technology

Third-generation photovoltaic technology includes GaInP/GaAs/Ge (gallium indium phosphide/gallium arsenide/germanium) triple-junction solar cells. These cells are expensive for terrestrial applications and are preferred for space satellites. Basically the technology aims to maintain the low production costs of second-generation technology but enhance the poor electrical performance. Current research shows that the emergence of third-generation technologies has a conversion efficiency of 30–60%. It is concluded that to achieve high efficiency there is a need to modify the incident spectrum and use excess thermal generation to enhance carrier collection.

2.5 Silicon Solar Cell Materials and Technology

2.5.1 Production of Silicon

Nearly 25% of the Earth's crust contains silicon, mainly in the form of silica (silicon oxide, SiO_2) and silicates. Metallurgical grade silicon is produced by heating silica with carbon up to 2000 °C in an electric arc furnace:

$$SiO_2 + 2C \rightarrow Si + 2CO \tag{2.32}$$

The molten silicon goes to the bottom of the furnace, from which it is removed periodically. Oxygen and chlorine are blown through the molten silicon to reduce the metallic impurities such as aluminium, calcium and manganese. To be useful for semiconductor materials the metallurgical grade silicon is purified further using the Siemens process:

$$Si(solid) + 3HCl(gas) \xrightarrow{300\,^{\circ}C} SiHCl_3(gas) + H_2(gas) \tag{2.33}$$

The gaseous products are sent through a condenser, which removes the volatile products. The condensed liquid is then put through a series of fractional distillation columns whose final output is semiconductor-grade trichlorosilane of very high purity. The products trichlorosilane and hydrogen are then fed to the decomposition reactor where following reverse reaction occurs:

$$SiHCl_3(gas) + H_2(gas) \xrightarrow{1100\,^{\circ}C} Si(solid) + 3HCl(gas) \tag{2.34}$$

The silicon is deposited on the heated rods by a chemical vapour deposition process to increase their diameter (see Figure 2.20). The rods grow with a fine-grain polycrystalline silicon microstructure. Finally the rods are broken into smaller chunks, which may be used for ingot production of good crystalline quality. All the processes require a considerable amount of heat energy.

2.5.2 Bulk Monocrystalline Silicon Ingot Production

After purification, silicon ingots are produced using one of the ways described in the following sections.

2.5.2.1 Czochralski Method

The highly purified polycrystalline silicon is melted using induction heating under inert gases, in a quartz crucible. Dopant impurity atoms such as boron or phosphorus are added to the molten intrinsic silicon in precise amounts in order to change it into an n-type or p-type extrinsic semiconductor. A seed crystal

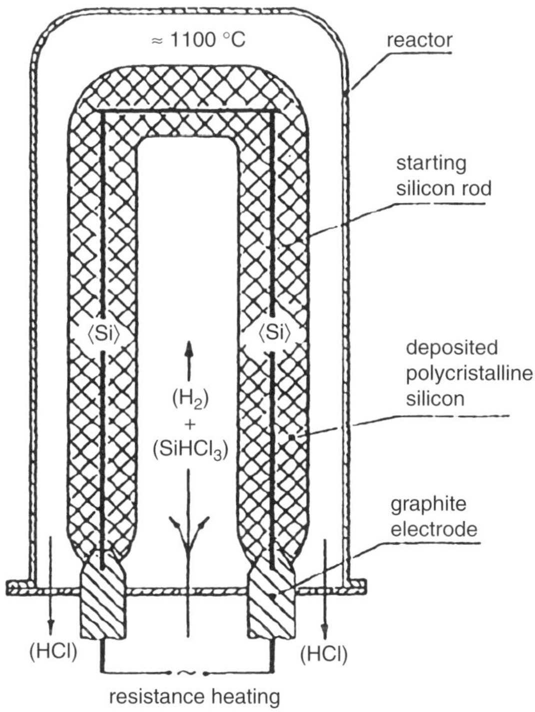

Figure 2.20 Equipment for the industrial preparation of polysilicon.

is dipped into the molten silicon and pulled out at a constant rate of 0.5–1.2 mm min^{-1} (Figure 2.21). While being pulled, the crystal is rotated slowly to ensure even angular distribution of temperature and dopant distribution at the growth front. The ingots are pulled in a defined ⟨100⟩ orientation. The monocrystalline silicon produced has a diameter up to 400 mm and a length of 2 m. Owing to the molten silicon, the reaction occurs at the walls of the quartz crucible, introducing nearly 10^{18} oxygen atoms cm^{-3}, which has a beneficial effect.

The main disadvantage of the Czochralski method is that it produces the ingot with a circular cross-section. In order to use both the crystal and the module area in the best manner, the ingots are usually cut into a pseudo-square cross-section before they are cut into wafers. Also a part of the ingot is cut from the top and the bottom to have smooth surfaces for wafer production.

2.5.2.2 *Float-zone Technique*

Figure 2.22 shows the float-zone technique for monocrystalline silicon ingot production. In this method, a radiofrequency field is used to produce a local melted zone. The highly purified polycrystalline silicon rod is passed at the rate of 2–3 mm min^{-1} along the melted zone in a vacuum or inert gas environment. The

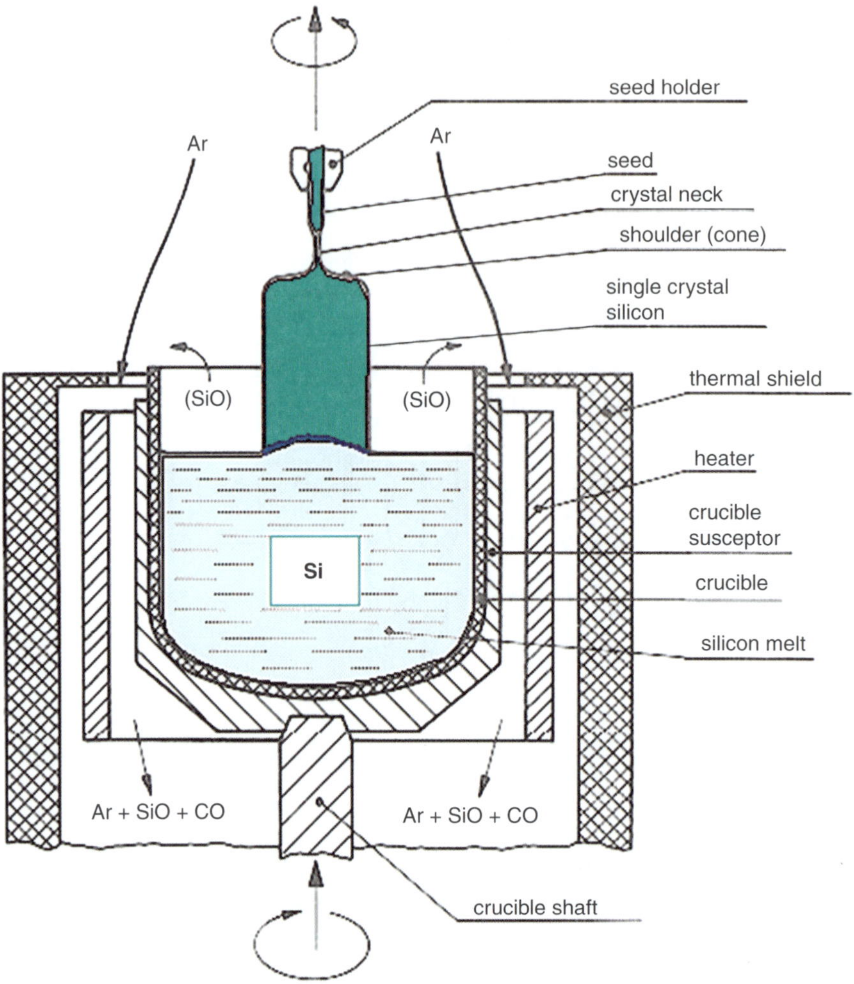

Figure 2.21 Czochralski technique for monocrystalline silicon ingot production.[8]

molten silicon is held in place both by surface tension and by the levitation effect of the radiofrequency field. Thus, the polycrystalline silicon is converted into monocrystalline silicon. There is no direct physical contact between the silicon and the surroundings, except for the inert gas. The concentrations of impurities, such as carbon and oxygen, are extremely low. Impurities such as nitrogen, which are thought to bring about an improvement in mechanical strength of the wafers, are now being intentionally added during the growth stages. The product monocrystals are free from volume defects, planar defects or line defects.

The float-zone technique for production of monocrystal ingots is somewhat faster than the Czochralski method, but is too expensive for regular solar cell production.

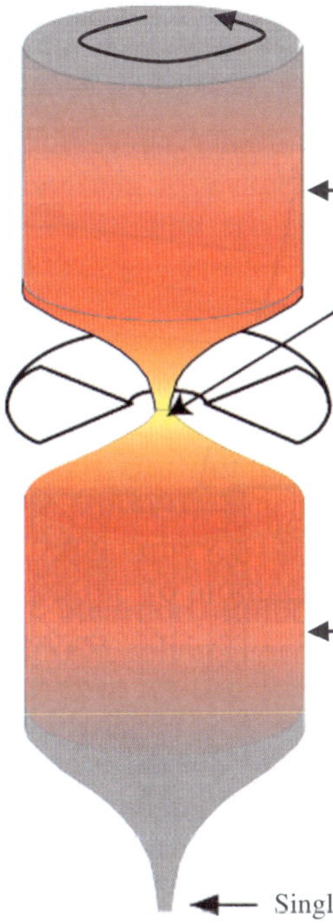

Figure 2.22 Float-zone technique for monocrystalline silicon ingot production.[9]

Figure 2.23 A tricrystalline silicon ingot.

2.5.2.3 Tricrystalline Silicon

Figure 2.23 shows a tricrystalline silicon ingot, which consists of three mutually tiled monocrystalline silicon grains. The crystal compound has a ⟨110⟩ surface orientation in all the grains. All boundaries are perpendicular to the ⟨110⟩ wafer plane and meet at the ingot centre, forming a characteristic tristar. The tricrystalline silicon production rate is faster, the ingots have mechanical stability and are fully compatible with Czochralski monocrystalline silicon. The ingots of tricrystalline silicon having diameters of 100–150 mm and lengths up to 700 mm are grown in a standard commercial crystal puller. The use of stronger wafers with improved light-trapping properties could allow thinner cells to be fabricated, thereby increasing the number of wafers produced per ingot by 60% and markedly reducing cell costs.

2.5.3 Bulk Multicrystalline Silicon Ingot Production

Considerable attention has been directed towards reducing the cost of the silicon wafer, usually grown by the Czochralski technique, since this accounts for about 40% of the cost of a standard silicon module. The most successful approach has been the simplification of the ingot growth processes by using cruder directional solidification or "casting" approaches to produce multicrystalline ingots. The first multicrystalline silicon cells developed specifically for the terrestrial market were reported in 1976. These multicrystalline approaches involve basically a reversion to the earlier ingot-forming approaches for crystal rectifiers, techniques pre-dating the microelectronics explosion.

Multicrystalline silicon ingot is generally produced by either the Bridgman process or the block-casting process. In the Bridgman process, both melting and crystallization processes take place in the same crucible, whereas in the block-casting process the melting and crystallization processes takes place in a different crucible.

2.5.3.1 The Bridgman Process

In the case of the Bridgman process, a silicon nitride-coated quartz crucible is usually employed for melting the silicon raw material and subsequent solidification of the multicrystalline ingot. The silicon nitride coating thereby serves as an anti-sticking layer, preventing the adhesion of the silicon ingot to the quartz crucible walls that, owing to the volume expansion during crystallization of the silicon material, would inevitably lead to a destruction of both the silicon ingot and the crucible. Common crystallization speeds used are in a range of about 0.2 mm min^{-1}.

2.5.3.2 Block-casting Process

Figure 2.24 shows the block-casting process, where the melting of the silicon raw material is performed in a quartz crucible without a coating. For crystallization the melted silicon is then poured into another crucible having a silicon nitride coating. The second crucible is designed in such a manner that heat is rejected only from the bottom surface. This helps in achieving considerably higher crystallization speeds.

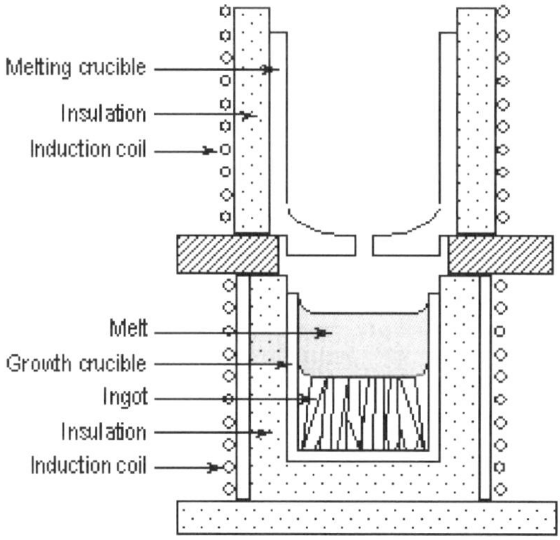

Figure 2.24 Block-casting process for multicrystalline silicon ingot production.

2.5.3.3 Advantages and Disadvantages of Producing Multicrystalline Silicon Ingots

Multicrystalline ingots as large as 690 mm × 690 mm in cross section and weighing as much as 240 kg are grown in total cycle times of 56 hours. The resultant throughput is 4.3 kg h^{-1}. Either induction heating or resistance heating can be used. The energy consumption for directional solidification or casting is in the range of 8–15 kWh kg^{-1}.

The main advantage of multicrystalline silicon is its manufacturing cost, which is lower than that of monocrystalline silicon. Another advantage of multicrystalline silicon is that the blocks can be manufactured into square solar cells, yielding a better utilization of the module area in comparison to the mostly round or pseudo-square monocrystalline wafers.

Owing to their multicrystalline nature, the wafers are generally more brittle than monocrystalline wafers. This demands the use of about 10–20% higher thickness for the multicrystalline solar cell. The efficiencies of multicrystalline silicon solar cells are lower due to recombination-active impurity atoms and extended defects such as grain boundaries and dislocations. With the introduction of modern hydrogen-passivation steps by silicon nitride layer deposition, the efficiencies of industrial multicrystalline silicon solar cells are boosted to 14–15%.

2.5.3.4 The Heat Exchanger Method: an Emerging Technology

The heat exchanger method (HEM), shown in Figure 2.25, is one of the new casting processes for the production of crystalline silicon and was first announced by Crystal Systems in the US. The polysilicon is melted in a specially prepared silica crucible. Then heat extraction is carried out from the

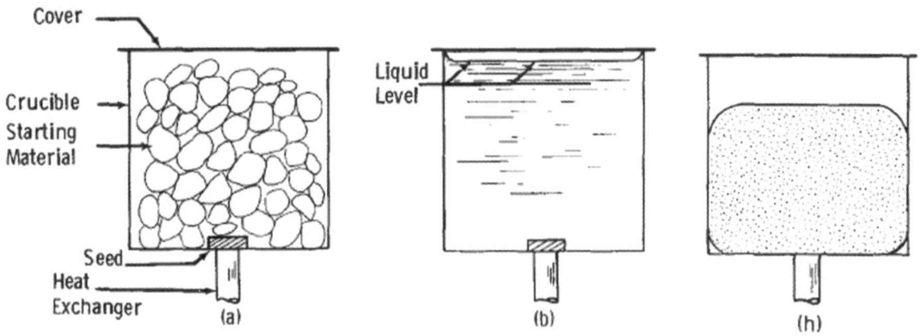

Figure 2.25 The heat exchanger method for multicrystalline silicon ingot production.

bottom of the crucible using a cold finger, which is gradually cooled by a helium gas heat exchanger. The molten silicon freezes in a very controlled manner so that a nearly perfect cubical single-crystal ingot is formed.

To date, ingots of size 340 mm × 340 mm × 170 mm have been produced. Because of the efficient packing possible with rectangular cells, the overall module efficiency may be over 14%. Impure meltstock can be used in the HEM process because a stable interface is present which rejects impurities to the last material to freeze at the crucible wall. The use of a low-cost meltstock could reduce significantly the cost of silicon wafers.

2.5.4 Silicon Wafers

In order to produce silicon cells it is necessary to slice the monocrystalline or multicrystalline silicon into thin wafers. In the case of monocrystalline silicon, first the ends of the ingot are cut to remove its crown and tail. Then a centreless grinding process followed by a light chemical polishing etch is used to smooth the outer surface and to form a perfect cylinder. The cylindrical ingot is then used to produce wafers. The multicrystalline material is nearly parallelepiped in shape and has its outer surface rough. To make it perfect parallelepiped, a thin layer is removed from all six surfaces by a light chemical polish. The perfect parallelepiped ingot is then used to produce wafers.

2.5.4.1 Traditional Wafer Slicing

At the beginning of the photovoltaic industry the ingot was cut by an inner diameter saw. Here a stainless steel circular blade with an annular internal cutting edge impregnated with diamond is used to slice the silicon wafers. The blade is held at its outer periphery in a high-speed rotating chuck. A coolant is also used during the cutting operation to dissipate the excessive amount of heat generated. The rotating cutting blade moves down on the ingot as it cuts. The cutting rate varies between 4 and 6 cm min^{-1}. Thus, to obtain a single slice from the ingot takes more than 2 min. The life span of the saw is limited to about 3000 cuts, which adds further to the cost of the final photovoltaic product.

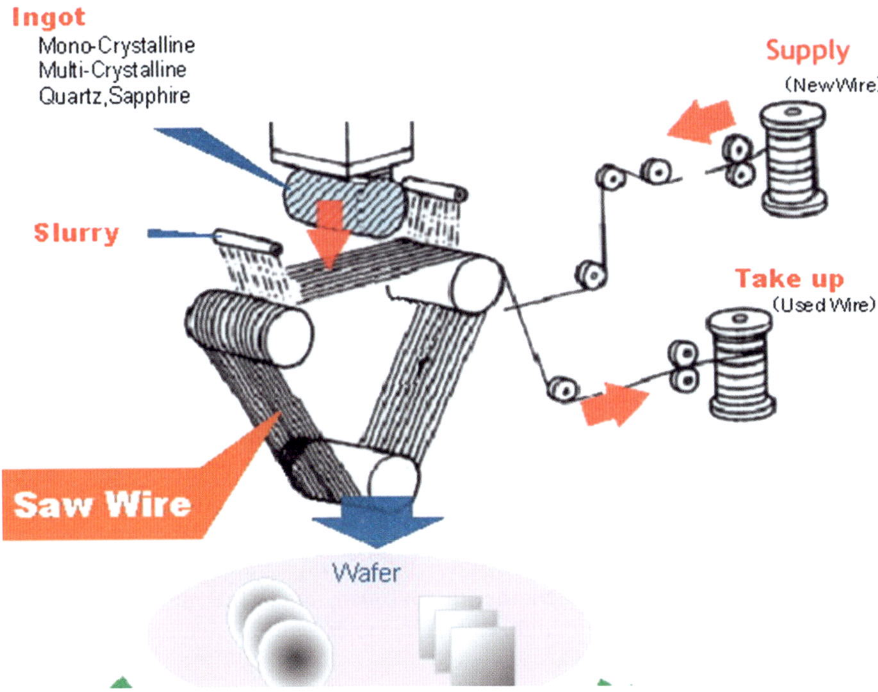

Figure 2.26 Multi-wire sawing for wafer production.[10]

2.5.4.2 *Multi-wire Saw*

The multi-wire saw depicted in Figure 2.26 is the most successful technique developed to increase the rate of slicing and reduce the material wastage. Here a parallel group of thin stainless steel wires carrying along an appropriate abrasive slurry moves either in one direction or oscillates back and forth. This results in slicing of a number of wafers at a time. Silicon carbide and diamond are the most commonly used abrasives. As wires with a diameter of less than 0.18 mm are used, the material loss is very low.

The process of slicing ingots into wafers entails one of the major cost factors. Some attempts have been reported to reuse the waste slurry and the waste ingot material.

2.5.5 Silicon Ribbon and Foil Production

Crystalline silicon in the form of a vertically grown ribbon has long been recognized as an excellent candidate for reaching the goal of low-cost photovoltaic materials. The main aim of this type of manufacturing process is to eliminate the associated cost of silicon wastage caused by the sawing process and also has high potential savings in terms of silicon material utilization. In 1970s there was tremendous activity in the field of silicon ribbon growth technologies for photovoltaic applications, some of which have now reached maturity with the start up of manufacturing on a MW scale. There are three main categories of ribbon technology: dendritic web, string ribbon and edge-defined film-fed growth. Essentially all three technologies begin with a vat of molten silicon.

2.5.5.1 Dendritic Web Growth

Figure 2.27 shows dendritic web growth, introduced by Dermatis and Faust in 1963 for silicon ribbon growth. Growth is initiated from a thin wire-like dendrite seed, which is brought into contact with the molten surface of the silicon. Initially the seed grows laterally to form a button at the melt surface. When the seed is withdrawn, needle-like dendrites propagate from each end of the button into the melt. The web portion of the crystal is formed by the solidification of the liquid film supported by the button and the boundary dendrites. As the crystal grows, it gradually widens until it reaches some relatively steady-state width determined by the thermal conditions in the melt. Web ribbons are currently grown at about 1.5–2 cm min^{-1} pulling rates, with a width of 5 cm, a thickness of 100 μm and in lengths up to 100 m. Since the material is nearly single crystalline, relatively high cell efficiencies can be achieved. The best reported value is 17.3% for a 4 cm^2 cell.

2.5.5.2 Edge-defined Film-fed Growth

The edge-defined film-fed growth (EFG) process, shown in Figure 2.28, was developed by Tyco laboratories in 1965 to grow crystal sapphire (aluminium oxide) ribbons. The EFG process involves the growth of ribbons of crystalline

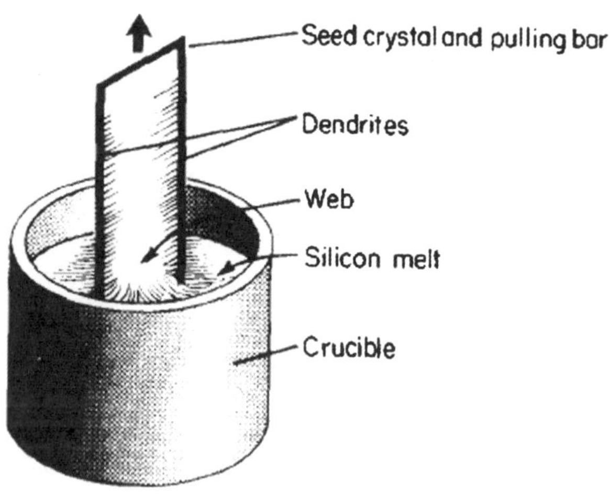

Figure 2.27 Dendritic web development technology.

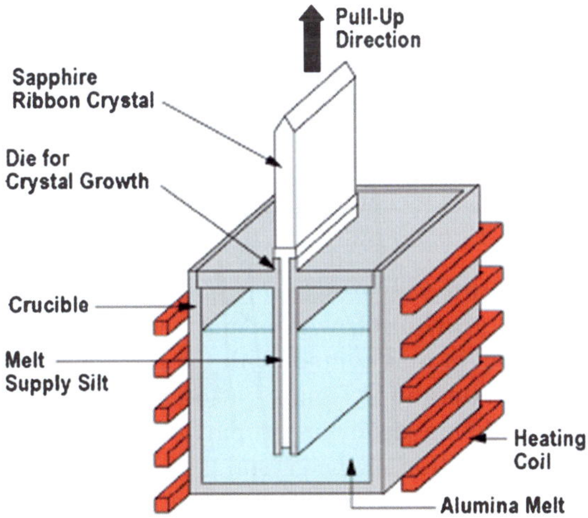

Figure 2.28 The edge-defined film-fed growth process using ribbon die.[11]

silicon directly from the melt by using a die which is immersed partly into the liquid. Molten silicon rises into the die by capillary action and then can be pulled as a crystalline ribbon as the silicon solidifies. In 1974, Tyco Laboratories and Mobil Oil formed Mobile Tyco and were the first to grow silicon solar cells using the EFG process. Multiple ribbons have been successfully grown with thicknesses of 100 μm and a face width of 10 cm at pull rates $0.6\,m^2\,h^{-1}$. The highest cell efficiency obtained is 16.3%, although production efficiencies are less than 14%.

The growth of crystals from the tips of capillary shaping dies was introduced for sapphire growth using molybdenum dies in 1971, and was first applied to silicon tubes using graphite shaping dies by Ciszek in 1975. A graphite crucible and graphite shaping dies are used with induction heating. Liquid silicon rises by capillarity up a narrow channel in the shaping die and spreads across the die's top surface, which defines the base of the meniscus from which the shaped crystal solidifies. Octagonal tubes with 100 mm wide flat faces are now used for production of PV substrates. Pulling rates are comparable to those used in web growth, but the 800 mm effective width increases the throughput to about $20\,m^2\,d^{-1}$. After growth, rectangular 100 mm wide "wafers" are laser-cut from the tube faces. They provide 275 μm thick multicrystalline substrates with longitudinal grains that routinely make 14% efficient solar cells.

2.5.5.3 String Ribbon Growth

Figure 2.29 shows the string ribbon (STR) technique, where ribbon growth takes place directly from a pool of molten silicon without a die. Rather than dendritic web growth the position of the ribbon edges in STR is maintained by

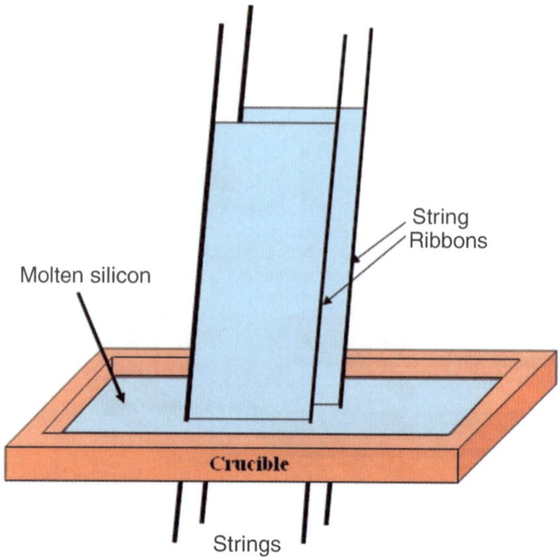

Figure 2.29 String ribbon growth technology.[12]

two strings fed through holes in the bottom of the crucible. The strings are drawn upward out of the melt to support the meniscus and the ribbon, and their pull rate determines the growth speed of the ribbon. The thickness of the ribbon is controlled by surface tension, heat loss from the sheet and pull rate. An important difference of the STR process from WEB growth is that the constraints of maintaining propagating dendrites and a supercooled melt are eliminated, and this relaxes the high degree of temperature control required in the WEB furnace. Careful adjustment of the growth parameters can allow very thin ribbons, down to 5 μm, to be grown.

2.5.5.4 Ribbon Growth on Substrate

Somewhat related to above ribbon approaches are other sheet approaches, which produce silicon films on substrates from which they are subsequently detached. Ribbon growth on substrate (RGS) foil development, shown in Figure 2.30, was initiated in 1984 at Bayer, but now is continuing at the Energy Research Centre of the Netherlands under a multinational cooperative agreement with the new owner, Deutsche Solar. The basic idea of the RGS process is to obtain a large, wedge-shaped crystallization front within a shaping die which sits on a moving substrate.

A substrate having its temperature below the silicon melting temperature is moved underneath a casting frame filled with liquid silicon. Thus heat is extracted from the molten silicon, forcing a crystallization process of silicon from the substrate into the silicon melt. During this process the substrate is moved underneath the casting frame and crystal growth is stopped at the

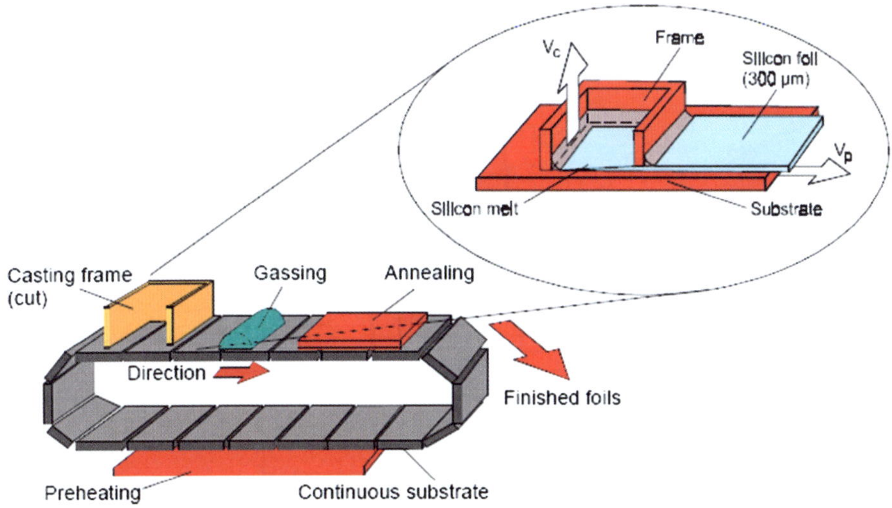

Figure 2.30 Ribbon growth on substrate technology.[13]

moment the substrate leaves the casting frame. Thus crystal growth direction and silicon wafer production direction are perpendicular to each other, which allows the independent control of both. Therefore relatively slow crystal growth can be combined with high substrate transport speed and thus high production volume. After the casting frame, the wafers and the substrates are cooled. During this process the wafer and the substrate separate, forced by their different thermal expansion coefficients. This allows the substrate to be re-used after the wafer has been extracted. The most developed version of this technology is the VEST technology developed by Mitsuishi. A promising efficiency above 16% has been obtained from these substrates that are only 60–70 μm in thickness, but are still self-supporting.

2.5.5.5 Silicon Sheets from Powder

The development of the silicon sheets from powder (SSP) process started in the late 1980s and was considered as a very attractive approach for low-cost photovoltaic material. The method was originally based on a two-step melting process of silicon powder, which is poured onto temporary carrier plates. This resulted in grains of several mm width and some cm length and conversion efficiencies of up to 13% were achieved.[14] Meanwhile, activities on SSP material as the active solar cell material were stopped in favour of SSP pre-sheets as substrate material for subsequent deposition of a thin silicon layer. In this process, only a surface melting of the thin silicon powder layer is applied for mechanical reasons, followed either by direct epitaxial growth of a highly pure active silicon layer or by an insulating and protecting silicon dioxide intermediate layer. This second approach needs a subsequent recrystallization

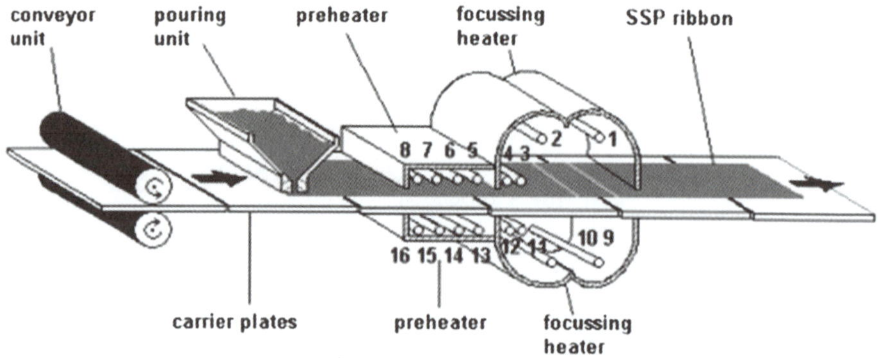

Figure 2.31 Newly designed silicon sheets from the powder technique.[15]

Figure 2.32 Solar cells: *left*, circular shaped monocrystalline; *centre*, pseudosquare shaped monocrystalline; *right*, multicrystalline/polycrystalline.

step since the thin active silicon layer on the silicon dioxide layer is very fine-grained after deposition. The advantage of this approach is that very low-quality and, thus, very inexpensive silicon-like metallurgical grade silicon can be used for the SSP pre-sheet. Figure 2.31 shows a schematic diagram of newly designed silicon sheets from powder equipment.

2.5.6 Crystalline Silicon Solar Cell Manufacturing Process

Figure 2.32 and shows circular and pseudosquare shaped monocrystalline solar cells, as well as the multicrystalline (polycrystalline) silicon solar cell.

The manufacture of solar cells involves the following processes.

1. Silicon wafers. The monocrystalline wafers used are cut into a pseudosquare shape, whereas multicrystalline wafers are cut into square shapes. Doping is p-type (boron) to a resistivity of around 1 Ω cm.
2. Etching. The sawing operation damages the cut surface, which may lead to fracture of the wafer during processing. So about 10 μm are etched off from the face using alkaline or acid solution.

3. Texturing. Texturing is done to create a microscopic pyramid on the surface. Thus, the light ray finds the textured surface as tilted with respect to the normal. This enhances the light absorption capacity. After texturing, to neutralize the alkaline remains the wafers are etched with acid.
4. Diffusion. Junctions are realized by thermal diffusion. Diffusion sources are phosphorus for the emitter junction and boron for the so-called back surface field, which repels carriers from the high recombination back contact. The doping elements are introduced as liquid or gaseous compounds.
5. Junction isolation. The n-type region at the wafer edges may interconnect the front and back contacts. Dry etching with plasma at low temperature is done to remove this region. An alternative method is to use lasers for cutting the wafer edges.
6. Antireflection coating. Titanium dioxide is often used for creating the antireflection coating, using atmospheric pressure chemical vapour deposition.
7. Electrical contacts. For the front contact a paste containing silver power, an organic binder and sintered glass is employed. For the back contact a paste containing aluminium, silver and organic compound is used. Generally, screen printing is used to stick the paste, followed by a sintering operation.
8. Testing. For acceptability the finished cells are tested under standard conditions and the *I–V* curve is plotted.

2.5.7 Thin-film Silicon Cell Approach

Enhanced optical absorption accompanying light-trapping can help lower the cell thickness required for efficient generation of the photocurrent to a few microns. Initially a texturing process was introduced to reduce surface reflectance for broadband illumination. Soon it was realized that texturing also increases the optical path of light transmitted into a silicon wafer. This made texturing and etching a standard process for fabricating silicon solar cells. Since the natural formation of texture on a silicon wafer is made possible, it has great potential to reduce the cost to a great extent by obviating the need for a separate layer to enhance the light-trapping. One of the drawbacks of texturing is that it increases the surface area, thereby increasing the dark current and reducing V_{oc}.

Silicon–metal interfaces and edge leakage leads to surface recombination in a solar cell. Many approaches are used to minimize the effective surface recombination at each surface. A prudent approach to minimize the silicon–metal contact area (both at the front and back) is through appropriate grid design. This consideration has led to the design of point-contact and buried-contact cells. This feature will also minimize the shadow loss for the incident light.

Surface passivation of unmetallized regions can be further improved by oxidation. It has been shown that such passivation can reduce the surface-recombination velocity to about 100 cm s^{-1}. In wafer-based cells, passivation schemes are embodied in various configurations called passivated emitter rear locally diffused (PERL) and passivated emitter rear totally diffused (PERT) methods. Another approach to produce effective surface passivation can be the use of low-temperature plasma-enhanced chemical vapour deposition (PECVD) nitride. It is now well known that silicon nitrides or oxynitrides produce a positive charge at the silicon–metal interface that results in excellent passivation characteristics for p-type silicon.

Another cell that illustrates the high-performance capability of thin cells was fabricated on a crystalline silicon film made by zone-melting recrystallization (ZMR). The current techniques for thin-film silicon solar cell fabrication are diverse and use monocrystalline, large-grain multicrystalline or fine-grain microcrystalline silicon films. Although this distinction appears to be related to grain size, in reality it separates the technologies related to the growth of the film itself. The monocrystalline and multicrystalline silicon films require temperatures higher than 800 °C, the use of a siliocon substrate, some type of epitaxial growth and separation from the substrate. The fine-grain films of microcrystalline material are deposited on a low-cost substrate, typically at a temperature lower than 600 °C. Thus, the approaches currently used in thin-film silicon solar cell fabrication can be categorized on the basis of processing temperature and substrates used for depositing the thin film. These approaches include using a monocrystalline wafer for deposition of a thin silicon layer, which is subsequently separated from it (or removing a thin layer from a single-crystal wafer and transferring it to another substrate), depositing thin films on a multicrystalline silicon wafer and using a non-silicon substrate.

The use of microcrystalline silicon seems to have a large potential because of being a true thin-film technology that can use some established thin-film processes. Furthermore, it can also derive benefits from ongoing research and development of amorphous silicon solar cells.

2.5.8 Transfer Technologies of Monocrystalline Thin Silicon Films onto Glass

The approaches based on transfer technologies of thin monocrystalline silicon films are the latest and very exciting developments within the crystalline silicon thin-film community. The basic idea is to detach a semi-processed thin monocrystalline silicon layer from the float-zone (FZ) wafer and to transfer it onto a glass substrate.[16] The advantages are that the superior material quality enables very high efficiencies even from less than 40 μm thick silicon layers, that inexpensive glass can be used as substrate and, finally, that the FZ wafer can be reused for thin-film formation and detachment several times. In the meantime, some promising results have been obtained but up to now no wafer has been used several times for layer formation. After some pre-studies with semi-processed FZ wafers which were attached via ethylene–vinyl acetate (EVA) foil to a glass substrate and thinned wet-chemically from the back side, there are currently five main approaches under development.

1. The first real silicon-transfer approach was pursued by Brendel with the so-called Ψ-process. A textured (100)-oriented silicon wafer with a porous silicon film serves as an epitaxial seed for the growth of an approximately 10 μm thick monocrystalline silicon film. The epitaxial growth serves to uniformally coat the surface of the anodically etched porous silicon, resulting in a waffle-like structure enclosed by (111) planes. Simulations predicted efficiencies of up to 19% under ideal technological conditions.
2. The epi-lift process, which was developed at the Australian National University, is based on the formation of (111)-oriented crystal planes during near-equilibrium growth of silicon using liquid phase epitaxy. An oxidized, (100)-oriented silicon wafer with oxide-free seed lines oriented close to two orthogonal (110) directions on the wafer surface serves as a substrate for epitaxial silicon growth by liquid phase epitaxy. This approach aims at the fabrication of a solar cell from the monocrystalline silicon net and to detach it from the wafer using a suitable etch.
3. Mitsubishi Electric have developed the so-called via-hole etching for the separation of the thin-film structure where a silicon layer is chemical vapour deposited and recrystallized on a silicon dioxide-covered monosilicon substrate. After the realization of this structure, via-holes are etched with 100 μm diameter and 1.5 mm distance to each other. The silicon layer is subsequently detached from the wafer by HF etching of the silicon dioxide intermediate layer through the via-holes. Solar cells were realized with this approach having a thickness of 77 μm, an area of nearly 100 cm^2 and efficiencies of up to 16%. This is a remarkable success, especially due to the fact that the back contacts were screen printed on this fragile structure.
4. Sony have pursued the epitaxial growth of monocrystalline silicon film on the thermally annealed crystalline surface on top of a highly porous buried silicon level. The silicon layer for the solar cell can be detached from the wafer due to the mechanically fragile separation layer underneath. Using a 12 μm thick epitaxial film that was transferred on a plastic foil, a conversion efficiency of 12.5% on 4 cm^2 was achieved.
5. The method from the Institute of Physical Electronics at Kaunas University of Technology in Lithuania also relies on epitaxial growth of the active silicon layer on a so-called quasi-monocrystalline silicon film. Again, a buried porous layer enables the separation of the solar cell from the wafer and the transfer of the processed cell to a foreign superstrate. Owing to an excellent light-trapping scheme, efficiencies of up to 14% were obtained with this approach. Furthermore, it could be shown that the starting wafer can be used several times as a seeding wafer for porous layer formation and subsequent epi-growth.

2.6 Concentrator Photovoltaic Systems

A concentrator photovoltaic (CPV) system consists of basically an optical collector and a solar cell receiver. The optical collector may be a mirror, a lens or a combination of both. The cell is designed for the irradiance level which will be received by the focus from the collector, and must be thermally bonded to the heat sink in order to remove the energy not converted into electricity. The collector, receiver and heat sink as a whole is commonly known as the concentrator module. Figure 2.33 shows some of the commonly used collectors generally employed to concentrate the light.

2.7 Amorphous Silicon Solar Cells

The first publications on amorphous silicon (a-Si) relevant for solar cell fabrication appeared after the late 1960s.[16] Only five years later, the first consumer products appeared on the market. However, it took quite some time until the basic properties of the material were understood. More than 20 years have already passed since the first solar cell from amorphous silicon was reported by Carlson in 1976. The high expectancy of this material was curbed by the relatively low efficiency obtained and by the initial light-induced degradation for this kind of solar cell (the so-called Staebler–Wronski effect). Today, a-Si has its fixed place in consumer applications, mainly for indoor use. After understanding and partly solving the problems of light-induced degradation, a-Si has begun to enter the power market. Stabilized cell efficiencies reach 13%. Module efficiencies are in the 6–8% range. The visual appearance of thin-film modules makes them attractive for façade applications. An analysis by Suntola in 1992 predicted a stabilized module efficiency of 8% by the turn of the century. This prediction appeared pessimistic at that time; however, this performance is still difficult to achieve on a production level. The present developments appear to sustain the contribution of this technology to large-scale production.

The fundamental photodiode inside an amorphous silicon-based solar cell has three layers deposited in either the p-i-n or the n-i-p sequence. The three layers are a thin p-type layer ($\sim$15 nm), a thicker undoped intrinsic layer ($\sim$160 nm) and a thin n-type layer ($\sim$15 nm). Sunlight enters the photodiode as a stream of photons that pass through the p-type layer, which is a nearly transparent window layer. The solar photons are mostly absorbed in the much thicker intrinsic layer. Each photon that is absorbed will generate one electron and one hole photocarrier.[17] The photocarriers are swept away by the built-in electric field to the n-type and p-type layers, generating the electricity.

The advantages of amorphous silicon solar cells is that they absorb sunlight very efficiently, although the total thickness of the absorbing layers is less than 1 μm. Consequently, these layers need to be supported on a much thicker substrate. Two different designs of amorphous silicon are available, depending upon the type of substrate used: transparent and oblique. Photovoltaic manufacturers such as BP Solar, Energy Photovoltaics, Intersolar, Kaneka, Sanyo and Sharp use commercial float glass as a substrate. Canon and United Solar use stainless steel foil as a substrate, while Fuji Electric and Sanyo use plastic substrates.

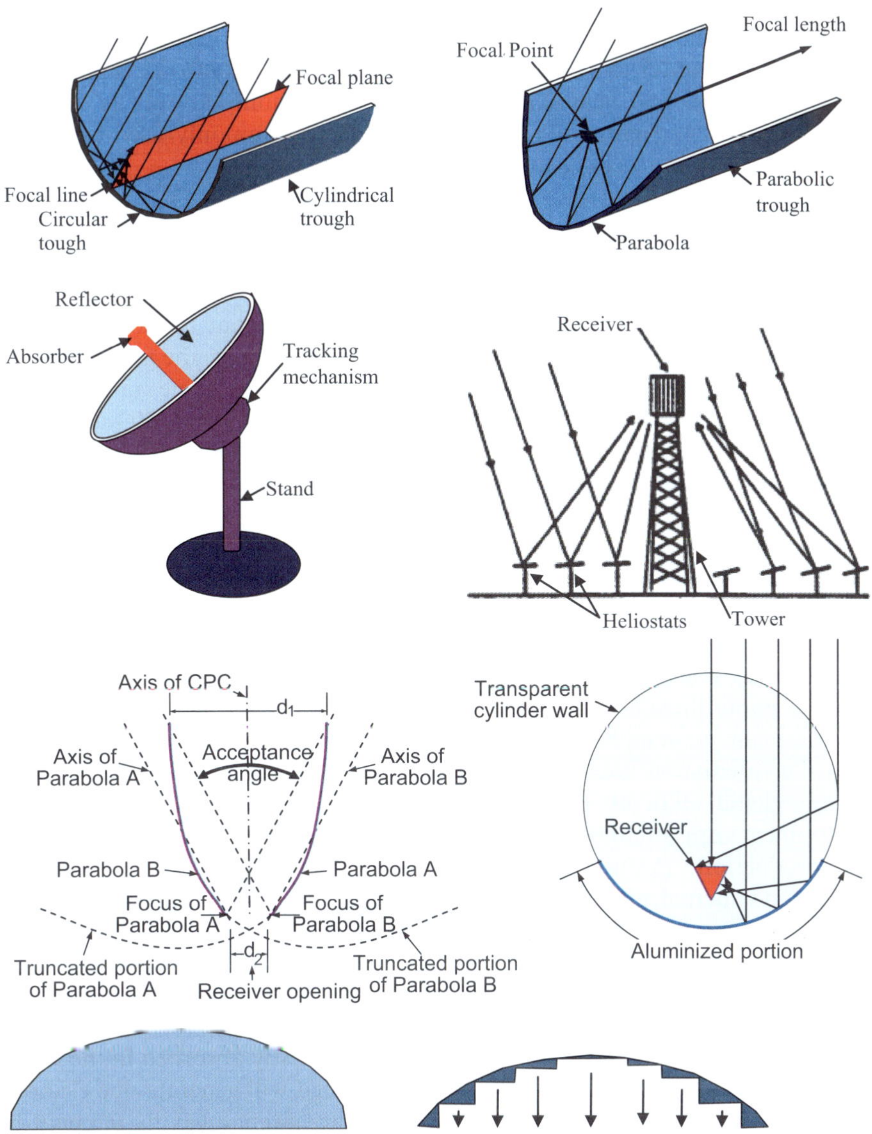

Figure 2.33 CPV systems. *Top row*: *left*, cylindrical trough; *right*, parabolic trough. *Second row*: *left*, parabolic dish; *right*, central receiver concentrator. *Third row*: *left*, compound parabolic concentrator; *right*, cylindrical tubular collector. *Bottom row*: *left*, plain lens; *right*, curved Fresnel lens.

In the superstrate design, shown in Figure 2.34 (*left*), sunlight first enters through the transparent substrate, which is usually glass or a transparent plastic. The insulating substrate needs a conducting layer, which is typically a transparent conductive oxide (TCO) such as tin (IV) oxide (SnO_2). The

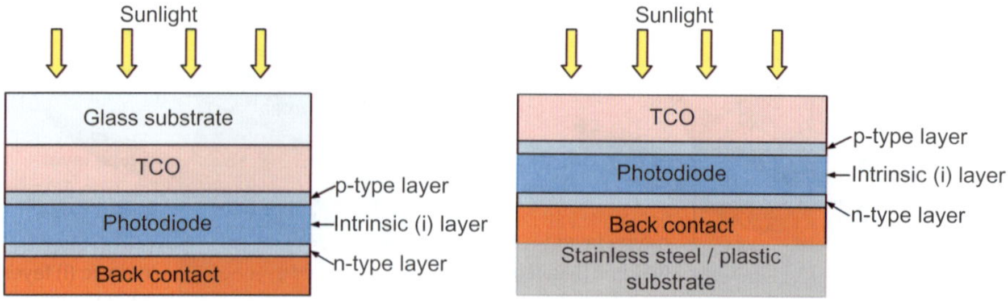

Figure 2.34 Cross-section of hydrogenated amorphous silicon (a-Si:H) solar cell: *left*, superstrate design on glass substrate; *right*, substrate design on stainless steel substrate.

amorphous silicon photodiode layers are then deposited onto the TCO, starting with a p-type window layer. Finally, a back reflector is deposited onto the photodiode, which acts as an electrode to the n-type photodiode layer. The superstrate design is suited to building-integrated solar cells in which a glass substrate can be used as an architectural element. In the substrate design, shown in Figure 2.34 (*right*), sunlight enters the photodiode before it reaches the substrate. Starting with the substrate, the cell is fabricated almost in the reverse order compared to the superstrate design. First is the back reflector, then the photodiode layer as an n-i-p type, and finally a TCO layer to act as a window layer for the photodiode and electrode on the top.

The conversion efficiency of the relatively simple, amorphous silicon p-i-n photodiode structure just described can be significantly improved by depositing two or three such photodiodes, one on top of another, to create a multijunction device. A tandem device using a combination of two p-i-n diodes is shown in Figure 2.35 (*left*). Here the bottom cell is based upon an amorphous silicon–germanium alloy instead of a-Si:H. United Solar Corporation produces an a-Si:H multijunction device using the n-i-p configuration as shown in Figure 2.35 (*right*). The configuration follows stainless steel foil/textured silver/zinc oxide/n-i_3-p/n-i_2-p/n-i_1-p/ITO/EVA/fluoro polymer, where both i_2 and i_3 are a-SiGe:H alloy and i_1 is an a-Si:H alloy.

A great variety of deposition methods have been used to fabricate amorphous silicon-based alloy materials and devices, but among these only the radiofrequency and direct current glow discharge techniques are used by industries. Some of the newer techniques, such as electron cyclotron resonance, plasma-enhanced chemical vapour deposition (PECVD) and hot wire deposition, have produced materials with interesting physical properties, such as lower defect density, greater diffusion length and lower hydrogen concentration. Hydrogen plasma is injected into the silane/methane mixture, by very high hydrogen dilution of the silane/methane mixture, by a high radiofrequency PECVD process or even by a combination of hydrogen dilution and high frequency. To decrease the concentration of impurities such as oxygen and nitrogen, an ultra-high vacuum reaction chamber method has been developed.

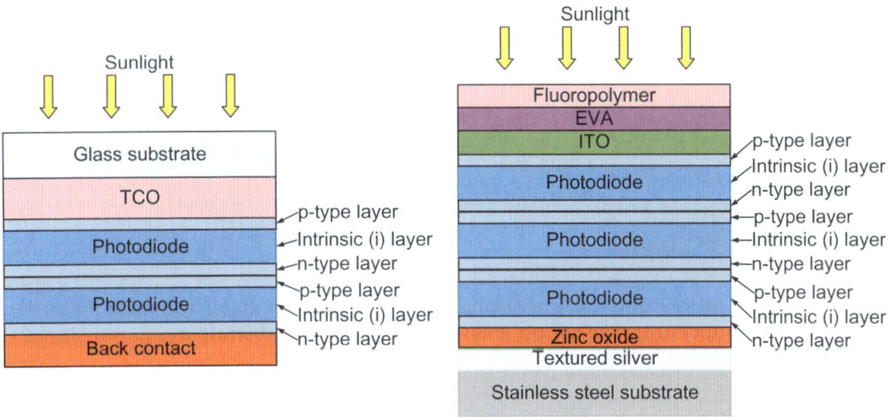

Figure 2.35 Cross section of hydrogenated amorphous silicon (a-Si:H) solar cells: *left*, tandem device with two junctions; *right*, multijunction device with three junctions.

Since the absorption coefficient of light rises rapidly with the photon energy, the topmost layer of a tandem cell acts as a low-pass optical filter. This effect shows that a 0.5 μm layer of a-Si:H absorbs photons with energies larger than 1.9 eV and passes photons with smaller energies. The wasted lower energy photons can be efficiently harvested by amorphous silicon–germanium, which has a much larger optical absorption coefficient below 1.9 eV than does a-Si:H, and hence a lower threshold energy. Overall, the advantages of the multijunction design are sufficiently compelling that they usually overcome the additional complexity and cost of the deposition facility. Both tandem and triple-junction devices are being manufactured today.

One of the most intriguing and actively researched facets of amorphous silicon solar cells is that the single-junction cell loses about 30% of its initial efficiency after about 1000 h, whereas the triple-junction module loses about 15% of its initial efficiency. The behaviour is mostly due to the Staebler–Wronski effect, which is the light-induced change in the a-Si:H and related materials used in the cell. Amorphous silicon modules reach a steady state after about 1000 h of steady illumination.

The solar conversion efficiency of production modules is generally lower than the efficiency of the research and development scale, small-area solar cells since production processes are more restrained by cost reductions. The differences in efficiency are mostly from the TCO performance, semiconductor material quality, deposition uniformity, encapsulation loss, bus bar shadow loss and electrical loss, and small shunts. The amorphous silicon modules when used outdoors under actual working conditions (the light intensity always fluctuates because of clouds and also by the incident angle of the sunlight; the average intensity during day time is less than 50 mW cm^{-2} and the module temperature is usually higher than 25 °C in moderate weather regions) generate electric power steadily compared with crystalline silicon modules.

2.8 Copper Indium Gallium Selenide Solar Cells

A very challenging technology is based on the ternary compound semiconductors $CuInSe_2$, $CuGaSe_2$, $CuInS_2$ and their multinary alloy $Cu(In,Ga)(S,Se)_2$ (hereafter CIGS). The first results of single-crystal work on $CuInSe_2$ (CIS) were extremely promising, but the complexity of the material looked complicated as a thin-film technology. Pioneering work of Kazmerski,[18] however, showed immediate success. It became evident that CIS process technology is very flexible with respect to the process conditions. Establishing a well-controlled system for multisource co-evaporation by Boeing made the CIS cell soon the front-runner with respect to thin-film solar cell efficiencies. In the mid-1980s, ARCO Solar developed a fabrication technology that is more adapted to current thin-film processing, namely sputtering of metal films with a subsequent selenization step. In later developments the addition of Ga and S helped to increase the efficiency.

Copper indium gallium (di)selenide is a I–III–VI_2 compound semiconductor material composed of copper, indium, gallium and selenium. The material is a solid solution of copper indium selenide and copper gallium selenide, with a chemical formula of $CuIn_xGa_{(1-x)}Se_2$, where the value of x can vary from 1 (pure copper indium selenide) to 0 (pure copper gallium selenide). It is a tetrahedrally bonded semiconductor, with the chalcopyrite crystal structure, and a bandgap varying continuously with x from about 1.0 eV (for copper indium selenide) to about 1.7 eV (for copper gallium selenide). It is used as a light absorber material for thin-film solar cells.

The basic structure of a CIGS heterojunction $Cu(In,Ga)Se_2$ thin-film solar cell is shown in Figure 2.36. The most common substrate is soda-lime glass of 1–3 mm thickness. This is coated on one side with molybdenum ($\sim$1 μm thick) that serves as a metal back contact. The $Cu(In.Ga)Se_2$ ($\sim$2 μm thick) is deposited on top of the molybdenum back-electrode as the photovoltaic absorber material. The heterojunction is then completed by chemical bath deposition (CBD) of CdS ($\sim$0.05 μm thick) and by the sputter deposition of a

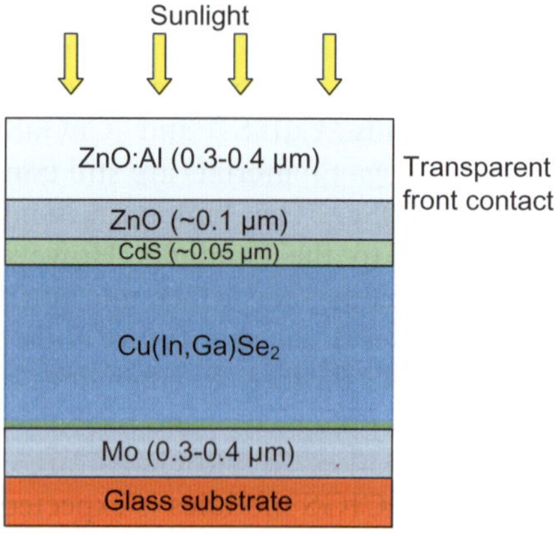

Figure 2.36 Cross-section of CIGS solar cell structure.

Table 2.2 Performance of CIGS cells and modules.[16]

Process	*Laboratory cell efficiency (%)*	*Module efficiency (%), area (cm^2)*	*Laboratory or company status*
Selenization of precursor	>16	12.1, $1 \times 4\,ft^2$	Siemens, pilot production
Metal films	–	14.7, 18	–
Co-evaporation/ sequential evaporation	18.8	–	NREL
	17.2	13.9, 90	IPE
	16.2	12.7, 800	ZSW, laboratory production; Wurth Solar, pilot production
	–	9.6, 135	EPV
	11.5	5.6, 240	Global Solar
	–	16.8, 19	Angstrom Solar Center
Non-vacuum processes	>11	8, 74	ISET, Unisun

nominally undoped ZnO (~0.05 μm thick) as an intrinsic layer and then a heavily doped ZnO layer. As ZnO has a band-gap energy of 3.2 eV, it is transparent for the main part of the solar spectrum and acts as the window layer of the solar cell. Either a current-collecting grid or a monolithic series interconnection completes the module.

Several companies worldwide are in the process of commercial development of CIGS modules. Efficiencies approaching 19% have been reported for laboratory-scale devices. This result has been obtained by this material system by empirical optimization of process parameters. Owing to the high flexibility of designing compounds with defined properties in this material system (*e.g.* bandgap grading), even more improvements can be expected in the near future. Table 2.2 summarizes the best values of solar cell performance obtained by this material system.

A further challenge is to realize high-voltage devices on the basis of the ternary chalcopyrite compounds $CuInS_2$ and $CuGaSe_2$. The efficiencies achieved with these wide energy gap materials are still considerably lower than those of the low-gap $Cu(In,Ga)Se_2$. This limitation is mostly due to the low open-circuit voltage compared to the bandgap. However, improved understanding of the materials, especially the surface and junction properties, may help to further develop the devices in the near future and open new perspectives of ternary thin-film photovoltaic devices. Further developments are directed towards replacement of the CdS buffer layer and probably the reduction of the film thickness or even replacement of In and Ga as rare elements.

The energy payback time for CIGS modules is expected by Alsema and van Engelenburg as about four years, compared to three years for crystalline-silicon modules. Production-cost analyses result in a range of projected manufacturing costs. There are predictions of 1.5–2 $ per W_p for first-generation CIGS plants with a few MW_p yearly capacity and projected costs of 0.4–0.6 $ per W_p for large-volume manufacturing.

Despite the level of effort on developing manufacturing processes, there remains a large discrepancy in efficiency between the laboratory-scale solar cells and mini-modules, and the best full-scale modules. In part, this is due to the necessity for developing completely new processes and equipment for the large-area, high-throughput deposition needed for manufacturing thin-film photovoltaics. This is compounded by the lack of a comprehensive scientific base for $Cu(InGa)Se_2$ materials and devices. This lack of a science base has been perhaps the biggest hindrance to the maturation of CIGS solar cells.

There are also environmental concerns for the hazards during the operation of CIGS modules, with one potential risk being the leaching of critical materials into rainwater. This only happens if a module is broken or crushed, so the normally well-encapsulated active layers are exposed. An experimental study of the emissions of toxic elements into rainwater from crushed CIGS modules and into soil exposed to water concluded that no acute danger to humans or the environment is likely to occur. The main hazard during the active life of the CIGS modules is related to fire accidents. A fire in a commercial-size system could result in harmful concentrations up to 300 m downwind of the fire if most of the CIGS materials are released. With release of 10% of the CIGS materials, concentrations were not harmful even under worst-possible meteorological conditions. The study concluded that there are no immediate risks to the public from fires in sites with CIGS modules.

2.9 Cadmium Sulfide/Cadmium Telluride Solar Cells

Research on cadmium telluride (CdTe) dates back to the 1950s, because it was quickly identified as having a band gap (about 1.5 eV) almost perfectly matched to the distribution of photons in the solar spectrum in terms of optimal conversion to electricity. A simple heterojunction design evolved in which p-type CdTe was matched with n-type cadmium sulfide (CdS). The cell was completed by adding top and bottom contacts. Early leaders in CdS/CdTe cell efficiencies included GE in the 1960s. By 1981, Kodak used close-spaced sublimation (CSS) and made the first 10% cells and first multi-cell devices (12 cells, 8% efficiency, 30 cm^2). Matsushita started with screen printing but shifted in the 1990s to CSS.

An important step forward occurred when cells were being scaled-up in size to make larger area products called modules. These products require higher currents than small cells and it was found that an additional layer, called a transparent conductive oxide (TCO), could facilitate the movement of current across the top of the cell (instead of a metal grid). One such TCO, tin oxide, was already being applied to glass for other uses (thermally reflective windows). Made more conductive for PV systems, tin oxide became and remains the norm in CdTe PV modules.

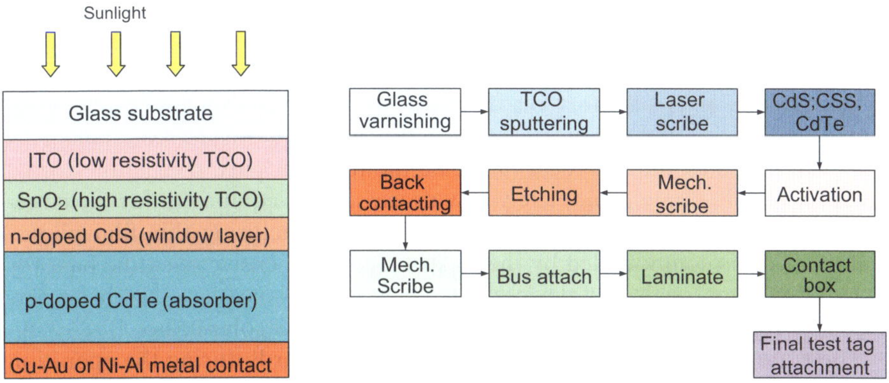

Figure 2.37 CdS/CdTe solar cell structure: *left*, cross-section; *right*, production stage.

In 1992, Chu *et al.*[19] made significant contributions to moving the efficiency of CdTe cells to above 15%, a critical level of success in terms of potential commercial competitiveness. This was done when they added an intervening or buffer layer to the TCO/CdS/CdTe stack and then thinned the CdS to allow more light through. Today, there are a number of start-ups in CdTe, namely Q-Cells' Calyxo (Germany), GE's PrimeStar Solar (Golden, CO, USA), Arendi (Italy) and Abound Solar (Fort Collins, CO, USA). Including Antec, their total production represents less than 70 MW per year.

Figure 2.37 shows the cross-section and production stage, respectively, of high-efficiency CdTe solar cells with transparent glass as the substrate. Usually the cheapest soda-lime glass or window-pane glass is used. For higher temperatures, borosilicate glass is used as the substrate, which can be heated to temperatures above 600 °C without softening. The opaque substrate with the TCO/CdS/CdTe deposited has been much less successful because of poor CdS/CdTe junction quality and poor ohmic CdTe contact. The glass substrate is coated with a transparent conducting oxide (TCO, such as SnO_2), indium tin oxide (In_2O_3:Sn, called ITO) or cadmium stannate (Cd_2SnO_4), which serves as the front contact. To obtain high current density in the completed cell the CdS layer needs to be very thin. The CdS layer is prepared by evaporation, chemical bath deposition (CBD) or closed space sublimation (CSS). It helps to allow a high fraction of the photons with energy above its band gap to reach the CdTe absorber and hence produce a high photocurrent.

The deposition of a second highly resistive transparent (HRT) oxide layer between the TCO and CdS improves both junction quality and uniformity. Materials used for the resistive layer include SnO_2, In_2O_3, Ga_2O_3 and Zn_2SnO_4. Most CdTe cells utilize n-type CdS for the window layer immediately adjacent to the CdTe. CdTe films can be prepared by a variety of techniques, such as atomic layer epitaxy, electrodeposition (ED), electroless deposition (ELD), spraying, close spaced sublimation (CSS), chemical vapour deposition (CVD), hot wall evaporation, screen printing, electron beam evaporation, laser ablation, thermal evaporation, molecular beam epitaxy, metal organic chemical vapour deposition and sputtering. Among these techniques, the most popular methods are ED and CSS. The commercial CdTe cells are manufactured by Matsushita in Japan, First Solar in Ohio, BP Solar in California and Antec Solar in Germany, of $60 \times 120\,cm^2$ module size. CdTe solar cells of efficiency about 16.5% ($V_{oc} = 845\,mV$, $I_{sc} = 25.9\,mA\,cm^{-2}$, $FF = 75.51\%$, area $= 1.032\,cm^2$) on the glass substrate have been fabricated.

Health and environmental risks researchers from the US Department of Energy at the Brookhaven National Laboratory[20] have found that large-scale use of CdS/CdTe photovoltaic modules does not present any risks to health and the environment and recycling the modules at the end of their useful life completely resolves any environmental concerns. During their operation these modules do not produce any pollutants and, furthermore, by displacing fossil fuels they offer great environmental benefits. CdS/CdTe PV modules appear to be more environmentally friendly than all other current uses of cadmium.

2.10 Dye-sensitized Cells

Dye-sensitized solar cell (DSC) technology, also called Grätzel cells, were first introduced in 1991 by Michael Grätzel, a chemist at the Swiss Federal Institute of Technology, Lausanne, and his colleagues. Nano-crystalline dye-sensitized solar cells are based on the mechanism of a fast regenerative photoelectrochemical process. The main difference of this type of solar cells compared to conventional cells is that the functional element which is responsible for light absorption (the dye) is separated from the charge carrier transport itself. In the case of the n-type semiconductor TiO_2 (bandgap 3.2 eV), this results in a working cycle starting with the dye excitation by an absorbed photon at the TiO_2/electrolyte interface and an electron injection into the TiO_2. The injected electrons may migrate to the front electrode (a transparent TCO glass) and can be extracted as an external current. The dye is subsequently reduced by a redox electrolyte, based on an organic solvent and the redox couple iodide/triiodide. The redox electrolyte also accomplishes the charge transport between the counter electrode (also a TCO glass) and the dye molecules. For a low-resistant electron transfer, the counter electrode is covered with some platinum, which acts as a catalyst for the redox reaction.

It could be shown that only dye molecules directly attached to the semiconductor surface are able to efficiently inject charge carriers into the semiconductor, with a quantum yield of more than 90%. As the overall light absorption of a dye monolayer is only small, this limits the photocurrent efficiency with respect to the incident light to a value well below 1%. This mechanism could be evaded by the preparation of titanium dioxide electrodes with a nanoporous morphology, resulting in a roughness factor of about 1000. After the announcement of surprisingly high efficiencies by O'Regan and Grätzel in the early 1990s, this type of solar cell is under reinforced development, aiming at large-area and low-cost solar cells. After the first experimental success, it took nearly a decade until the first quantitative models were established which link the material parameters of the constituents with the electrical performance of the whole cell, such as the *I–V* characteristic and the spectral response.

The major advantage of the concept of dye sensitization is the fact that the conduction mechanism is based on a majority carrier transport, as opposed to the minority carrier transport of conventional inorganic cells. This means that bulk or surface recombination of the charge carriers in the TiO_2 semiconductor cannot happen. Thus, impure starting materials and simple cell processing without any clean-room steps are permitted, yet resulting in promising conversion efficiencies of 7–11% and the hope of a low-cost device for photoelectrochemical solar energy conversion. On the other hand, impure materials can result in a strongly reduced lifetime of the cells.

The most important issue of the dye-sensitized cells is their stability over time and the temperature range which occurs under outdoor conditions. Although it could be shown that intrinsic degradation can considerably be reduced, the behaviour of the liquid electrolyte under extreme conditions is still unknown. For a successful commercialization of these cells the encapsulation/sealing, the coloration and the electrolyte filling have to be transferred into fully automated lines, including the final closure of the filling openings. Therefore, a significant effort has been undertaken in order to replace the liquid electrolyte by a gel electrolyte, a solid-state electrolyte or a p-conducting polymer material. Using p-type conducting materials, the best efficiencies obtained so far are in the 1% range. Recently, high efficiencies above 7% were announced by Toshiba using a gel electrolyte.

In terms of a possible integration of these cells into electronic devices, the necessary sintering step of the nanometer-sized TiO_2 at temperatures of over 400 °C on the transparent conductive oxide (TCO) glass might be a certain drawback. Owing to this thermal budget, the glass electrode is the only solution which limits the shape of the cells to a flat design (see Figure 2.38).

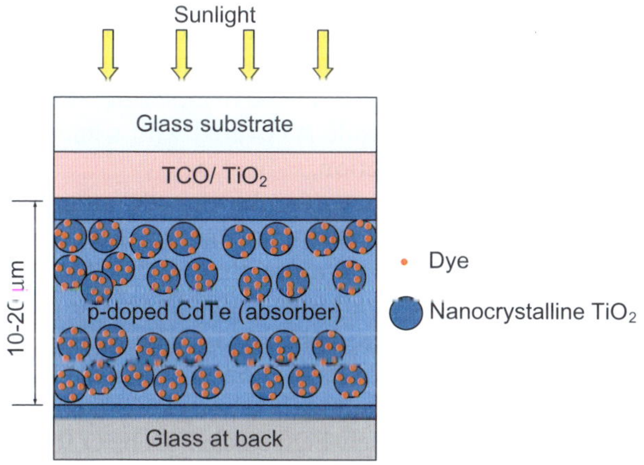

Figure 2.38 Cross-section of a dye-sensitized solar cell.

2.11 Organic Solar Cells

Beside dye-sensitized solar cells, which may be considered as organic/inorganic hybrid cells, other types of organic solar cells are currently becoming of broader interest. These cells can be divided roughly into molecular and polymer organic solar cells, or into flat-layer systems and bulk heterojunctions.

Organic materials such as, for example, conjugated polymers, dyes or molecular organic glasses, can show p- or n-type semiconducting properties. Extremely high optical absorption coefficients are possible with these materials, which offer the possibility for the production of very thin solar cells [far below 1 μm, as shown in Figure 2.39 (*left*)], and therefore only very small amounts of materials are needed. The variability of organic compounds is nearly infinite. Beside this, the large interest in these materials results from technological aspects, including the expected ease of large-scale manufacturing at low-temperature processes and very low costs. The upscaling of organic solar cells into large-area devices, always a big challenge with inorganic solar cells, has already been demonstrated to be straightforward.

The energetic pay-back time of organic solar cells is expected to be very short. Considering the fact that light-emitting films of plastic materials have been realized, there is also a realistic chance to achieve efficient photovoltaic conversion in such materials because this is just the reverse process. Organic solar cells offer the hope of being very inexpensive. Quite a variety of materials, compositions and concepts are being investigated, which reflects the possibilities in terms of device concepts, materials use and materials design. In spite of the many fundamental questions that still exist, these perspectives, and the fact that exploration has only just begun, are causing a large growing interest in the development of such solar cells.

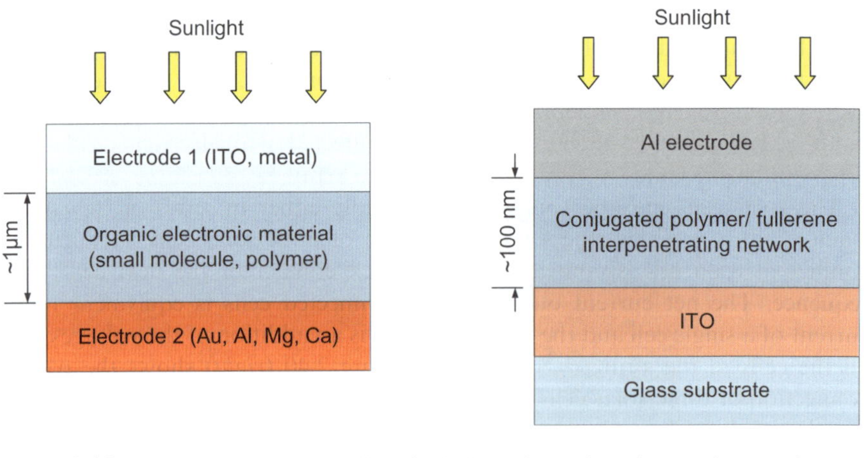

Figure 2.39 Organic solar cells: *left*, a single layer heterojunction; *right*, a conjugated polymer/fullerene "plastic solar cell".

The width of the charge generation layer for organic solar cells is much smaller than in inorganic cells. In order to overcome this obstacle, the concept of interpenetrating networks with a modified fullerene (*i.e.* C_{60}) particles was successfully introduced[21] [see Figure 2.39 (*right*)]. Fullerenes have been proven to be very efficient electron acceptors for photoexcited conjugated polymers; the quantum efficiency for charge separation is near unity. Mixtures of fullerene derivates (*e.g.* phenyl C_{61}-butyric acid methyl ester (PCBM)) and conjugated polymers (*e.g.* poly (2-methoxy-5-(3′,7′-dimethyloctyloxy)-*p*-phenylenevinylene (MDMO-PPV)) are spin-coated or doctor-bladed on suitable substrates, such as ITO-coated polymer foils or glasses. Interpenetrating networks can also be produced by co-evaporation of fullerenes and molecular dyes, *e.g.* zinc phthalocyanine. Another concept is the use of stratified layers made of donor–acceptor blends, avoiding the problem of continuous and separate pathways for the two types of charge carriers, which in interpenetrating networks might exist between anode and cathode.

Only modest solar conversion efficiencies of up to 1% were reached until 1999. Efficiencies increased rapidly within this year: with molecular flat-layer systems based on molecular organic single crystals made of iodine- or bromine-doped pentacene, efficiencies of up to 3.3% under AM1.5 illumination have been reported at Lucent Technologies. Nearly the same value was reported with improved bulk heterojunctions (interpenetrating network) of conjugated polymers and fullerene derivatives.

Before these cells become practical, which at the moment still looks far away, the efficiency will have to be increased further. Also, long-term stability and protection against environmental influences are significant challenges.

2.12 Photovoltaic Modules and Arrays for Crystalline Silicon Solar Cells

In order to provide useful power for any application, the individual solar cells must be connected together electrically. The system should also be protected from the adverse effects of the operating environment. This electrically connected and fully environmental protected unit is usually termed a photovoltaic module. The module is then used alone or connected in an electrical circuit with other similar modules to form a photovoltaic array.

The solar cells may be connected electrically either in series or in parallel. Figure 2.40 (*left*) shows the series connection of three solar cells. Here the top contact of each cell is connected to the back contact of the next cell in the sequence. The net current output of the connected cells is equivalent to the current of a single cell and the voltage output is an addition of the voltages from all the cells in the string (*i.e.* $3 \times V_{cell}$). Figure 2.40 (*right*) shows the parallel connection of three solar cells. Here the top contacts of each cell are connected together and similarly the bottom contacts of each cell are connected together. The current from the cell group is equivalent to the addition of the current from each cell (*i.e.* $3 \times I_{cell}$) and the voltage remains equivalent to that of a single cell.

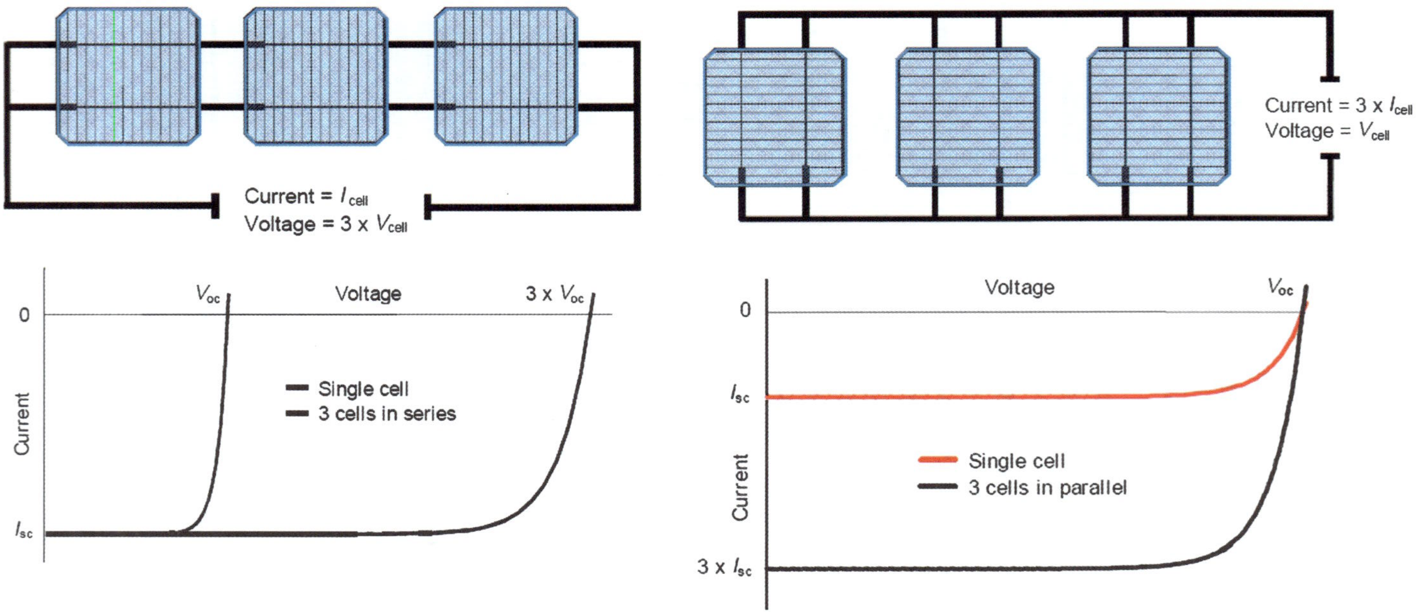

Figure 2.40 Solar cells: *left*, connected in series; *right*, connected in parallel.

In the early 1980s when the development of module fabrication techniques for crystalline silicon cells reached maturity, the majority of the modules required to charge the batteries were for stand-alone applications. Thus, the electrical output was required to be appropriate for battery charging under a range of sunlight conditions and this was found to be most readily achieved by the series connection of 36 crystalline silicon cells. The series connection of these cells produces an open-circuit voltage of around 18 V and a maximum power point voltage of around 14–15 V. This provides a voltage above the 12 V required for battery charging over a wide range of sunlight conditions.

The most common shape of a monocrystalline silicon cell is pseudosquare, where the cell is cut from a circular wafer and is square apart from the cut-off corners. The solar cells are generally arranged in four rows and with spacing between cells. This spacing is required to prevent electrical shorting between cells. The module area is around $0.3\,m^2$, making it suitable for ease in transportation and installation. Because it is necessary to have the cells physically separated, the module area is always larger than the sum of the cell areas and therefore the module efficiency is always lower than the cell efficiency. The amount of reduction due to area effects depends on the configuration of the module and is defined by the packing density. Typically, a crystalline silicon module will have a packing density in the range 80–90% and so if solar cells are typically 14% efficient, the module efficiency would be around 12%. The performance of the module is also a function of its operating temperature and hence the rated efficiency is quoted at a standard temperature of 25 °C. The amount of the change depends on the cell type and structure, with crystalline silicon cells typically losing about 0.4–0.5% of their output per degree Celsius rise.

In modern crystalline silicon modules, the front surface is composed of toughened glass to provide physical strength and with a low iron content to allow transmission of short wavelengths in the solar spectrum. The rear of the module can be made from a number of materials. One of the most common is Tedlar [a poly(vinyl fluoride) film]. If a level of transparency is required, then it is possible to use either a translucent Tedlar sheet or, more commonly, a second sheet of glass. The glass glass structure is popular for architectural applications, especially for incorporation into a glazed façade or roof.

Figure 2.41 shows that the glass–Tedlar module is fabricated by a lamination technique. The electrically connected cells are sandwiched between two sheets of encapsulant, for example EVA (ethylene–vinyl acetate) and positioned on the glass sheet, which will form the front surface of the module. The rear plastic sheet is then added and the whole structure is placed in the laminator. Air is removed and then reintroduced above a flexible sealing membrane above the module to provide pressure. The module is heated and the encapsulant melts and surrounds the cells. Additional encapsulant material is included at the module perimeter to ensure complete sealing of the module edges.

The glass–glass construction is more time and labour intensive, since the removal of air must be accomplished without the aid of lamination. Both film and liquid encapsulants can be used. In the case of a liquid encapsulant, this is

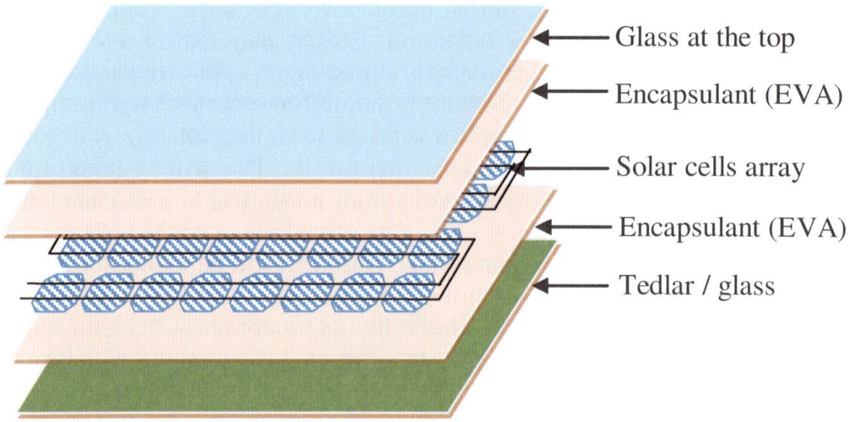

Figure 2.41 Monocrystalline photovoltaic module lamination.

poured between the glass sheets after the module has been sealed on three edges. The connected cells must be fixed in place before this procedure is undertaken. In the thin film module, the glass substrate on which the cell is deposited is often used as the front surface of the finished module. Lamination is then carried out in the same way as for crystalline modules, although only a single layer of encapsulant is required. Lower temperatures are often used to avoid damage to the cells. Particular care must be taken with edge sealing since all thin-film cells are badly affected by the ingress of moisture. In the manufacturing process, a clear gap must be left around the edge of the cell area for proper sealing of the module.

The electrical connections to the module are made via a junction box, usually fixed to the rear of the module, or by flying leads. These typically exit the module through the rear Tedlar sheet. In the case of glass–glass modules, the leads may exit through one edge of the module to avoid drilling holes in the glass sheet. The points at which the electrical connections are brought out of the module are sealed to prevent moisture ingress. The electrical connection is also made in two places on each cell, generally called double tabbing, to allow for any problems with thermal expansion and other stresses during manufacture or operation.

The operating temperature is also influenced by the exterior materials of the module, with glass–glass structures usually running at a higher temperature than the glass–Tedlar module under similar conditions. A module with a white Tedlar backing will reject more heat than one with a black Tedlar backing, so allowing it to operate at higher efficiency.

The optimum array orientation will depend on the latitude of the site, prevailing weather conditions and the loads to be met. It is generally accepted that, for low latitudes, the maximum annual output is obtained when the array tilt angle is roughly equal to the latitude angle and the array faces due south (in the northern hemisphere) or due north (for the southern hemisphere). Although this condition will give the maximum output over the year, there can be considerable variation in output with season. This is particularly true in high-latitude locations, where the day length varies significantly between summer and winter. Therefore, if a constant or reasonably constant load is to be met or, particularly, if the winter load is higher than the summer load, then the best tilt angle may be higher in order to boost winter output. Two-axis tracking, where the array follows both the daily east–west and north–south movement of the Sun, could provide an increase of about 20%.

The output of any cell or module which is shaded will be reduced according to the reduction of light intensity falling on it. However, if this shaded cell or module is electrically connected to other cells and modules which are unshaded, their performance may also be reduced since this is essentially a mismatch situation.

In the case of the thin-film module, the same design principle was adopted when battery charging was required. This was accomplished by the series connection of the cells during fabrication. Since the voltage from the amorphous silicon cell is higher than that from a crystalline silicon device, fewer series-connected cells are required to maintain sufficient voltage to charge the battery. However, the cells must be of larger area in order to reach similar current levels.

References

1. A. E. Becquerel, Recherches sur les effets de la radiation chimique de la lumiere solaire au moyen des courants electriques, *C. R. Acad. Sci.*, 1839, **9**, 145–149.
2. W. G. Adams and R. E. Day, The action of light on selenium, *Proc. R. Soc. (London)*, 1877, **A25**, 113.
3. C. E. Fritts, On a new form of selenium photocell, *Proc. Am. Assoc. Adv. Sci.*, 1883, **33**, 97; also *Am. J. Sci.*, 1883, **26**, 465.
4. L. Bergmann, Uber eine neue selen- sperrschicht photozelle, *Phys. Z.*, 1931, **32**, 286.
5. R. S. Ohl, Light-sensitive electric device, *US Pat.* 2 402 622, 1941; R. S. Ohl, Light-sensitive device including silicon, *US Pat.* 2 443 542, 1941.
6. D. M. Chapin, C. S. Fuller and G. L. Pearson, A new silicon p-n junction photocell for converting solar radiation into electrical power, *J. Appl. Phys.*, 1954, **25**, 676–677.
7. J. Mandelkom, C. McAfee, I. Kesperis, I. Schwartz and W. Pharo, Fabrication and characteristics of phosphorus-diffused silicon solar cells, *J. Electrochem. Soc.*, 1962, **109**, 313–318.
8. The science and art of Si crystal growth, http://www.tf.uni-kiel.de/matwis/amat/elmat_en/kap_6/advanced/t6_1_2.html.
9. Schematic of float-zone wafer growth, http://pvcdrom.pveducation.org/MANUFACT/Images/FZ.jpg.
10. Saw wire, http://jfs.name/eng2006/products.html.

11. Single-crystal sapphire products, http://americas.kyocera.com/kicc/industrial/crystal.html.
12. ENF – introduction to report, http://www.enf.cn/reports/analysis2.html.
13. Ribbon-growth-on-substrate silicon wafer casting technology, http://www.ecn.nl/nl/units/zon/rd-programma/silicium-pv-technologie/ribbon-growth-on-substrate-silicium-wafer-technologie/.
14. A. Eyer, A. Raeuber and A. Goetzberger, Silicon sheet materials for solar cells, *Optoelectron. Tokyo*, 1999, **5**, 239–257.
15. B. Ai, H. Shen, Q. Ban, X. Wang, Z. Liang and X. Liao, Preparation and characterization of Si sheets by renewed SSP technique, *J. Crystal Growth*, 2004, **270**, 446–454.
16. A. Goetzberger, C. Hebling and H. W. Schock, Photovoltaic materials, history, status and outlook, *Mater. Sci. Eng. Rep.*, 2003, **40**, 1–46.
17. E. A. Schiff, Hole mobilities and the physics of amorphous silicon solar cells, *J. Non-Cryst. Solids*, 2006, **352**, 1087–1092.
18. L. L. Kazmerski, F. R. White and G. K. Morgan, Thin-film $CuInSe_2$/CdS heterojunction solar cells, *Appl. Phys. Lett.*, 1976, **29**, 268–270.
19. T. L. Chu, S. S. Chu, K. D. Han and M. Mantravadi, Mercury telluride as an ohmic contact to efficient thin film cadmium telluride solar cells, *Conference Record of the Twentieth IEEE*, 1988, pp. 1422–1425.
20. J. Bohland and K. Smigielski, First Solar's CdTe module manufacturing experience; environmental, health and safety results, *Proceedings of the 28th IEEE Photovoltaic Specialists Conference*, Anchorage, September 2000.
21. G. Yu, J. Gao, J. C. Hummelen, F. Wudl and A. J. Heeger, Polymer photovoltaic cells: enhanced efficiencies via a network of internal donor-acceptor heterojunctions, *Science*, 1995, **270**, 1789–1791.

Further Reading

Handbook of Photovoltaic Science and Engineering, ed. A. Luque and S. Hegedus, Wiley, New York, 2003.

T. Markvart and L. Castañer, *Practical Handbook of Photovoltaics: Fundamentals and Applications*, Elsevier, Amsterdam, 2003.

A. Goetzberger and V. U. Hoffmann, *Photovoltaic Solar Energy Generation*, Springer, Berlin, 2005.

B. Agrawal and G. N. Tiwari, *Developments in Environmental Durability for Photovoltaics*, PIRA International, Leatherhead, UK, 2008.

CHAPTER 3

Thermal Comfort

3.1 Introduction

Thermal comfort is defined in ASHRAE Standard 55 and International Standard ISO 7730 as "the state of mind that expresses satisfaction with the surrounding environment". Therefore the term "thermal comfort" describes a person's psychological state of mind and is usually referred to in terms of whether someone is feeling too hot or too cold.

Thermal comfort that will make people feel comfortable is influenced by (a) personal factors that include the activity level and the clothing thermal insulation of the subject and (b) physical factors that include a group of environmental parameters such as air temperature, mean radiant temperature, air velocity and air humidity. These factors make up what is known as the "human thermal environment". The best that you can realistically hope to achieve is a thermal environment that satisfies the majority of people in the workplace, or put more simply "reasonable comfort". Usually, 80% of occupants are considered as a reasonable limit for the minimum number of people who should be thermally comfortable in an environment.

The passive solar concepts need to be included in every step of a building's design. Although the conventional or active solar heating systems can be somewhat independent of the conceptual organization of a building, it is difficult to add a passive system to a building once it has been designed. Thus, the information regarding various passive concepts/systems must lead to a necessary degree of accuracy at each stage of building design. A judicious application of the passive solar concepts provides thermal comfort in an economical way. The thermal comfort chart, shown in Figure 3.1, gives the various factors influencing thermal comfort. The factors can be divided into three conditions, *viz*. (a) physical, (b) physiological and (c) intermediate.

The factors can further be classified as primary and dominant; additional; and secondary and imaginary factors. For example, in physical conditions, the

RSC Energy Series No. 4
Building Integrated Photovoltaic Thermal Systems: For Sustainable Developments
By Basant Agrawal and G. N. Tiwari
© B. Agrawal and G. N. Tiwari 2011
Published by the Royal Society of Chemistry, www.rsc.org

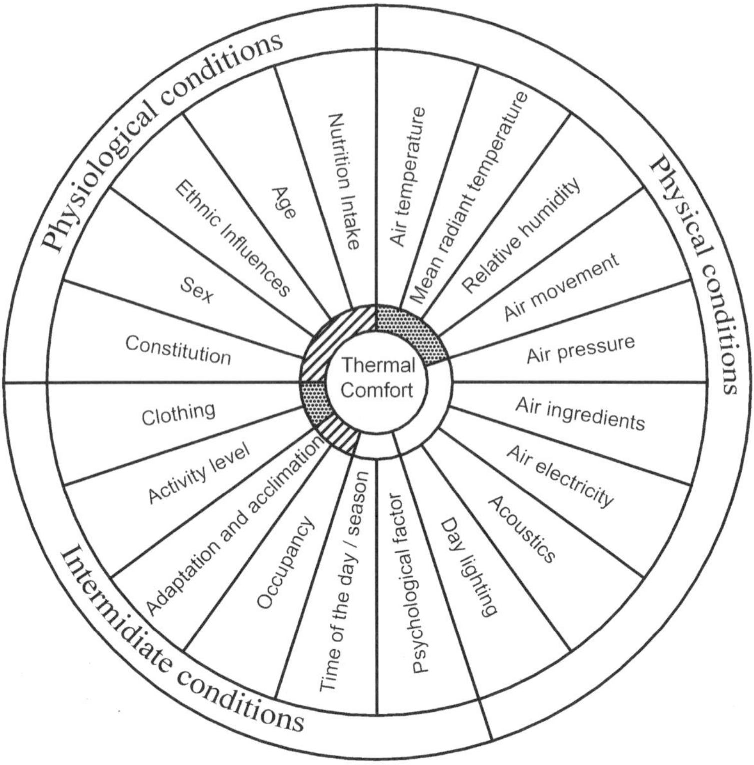

Figure 3.1 Figure showing thermal comfort chart.

primary factors are the air temperature, mean radiant temperature, the relative humidity and the air movement, whereas in intermediate conditions, clothing and activity level are the dominant factors. Similarly, in physiological conditions, all the factors are the additional factors, and in intermediate conditions, occupancy and adaptation and acclimatization are the additional factors.

3.2 Physical Parameters

3.2.1 Air Temperature

There are basically two temperatures generally used for thermal comfort. These are dry-bulb and wet-bulb temperatures. Dry-bulb temperature (DBT) is a temperature of the air measured by a mercury-in-glass thermometer with the bulb exposed. It is the indication of the sensible heat content of the air, *i.e.* it is capable of producing a sentient response in a person subjected to its effect. Wet-bulb temperature (WBT) is the temperature of the air measured by a mercury-in-glass thermometer having the bulb enclosed in a wetted cotton sleeve. The wet-bulb reading is reached when the rate of heat transfer and the rate of evaporation are stabilized. WBT is an indication of the enthalpy of moist air, *i.e.* the enthalpy is approximately the same for a constant wet-bulb temperature at constant atmospheric pressure.

The WBT is lower than the DBT due to evaporation of moisture from the wetted sleeve, the latent heat for evaporation being absorbed from the mercury reservoir. The recommended DBT for thermal comfort for summer and winter are 25±1 °C and 20±1 °C, respectively.

3.2.2 Relative Humidity

Humidity is related to moist air, a mixer of dry air and water vapour, and is defined as:

$$\text{Moist air} = \text{Dry air (fixed part)} + \text{Water vapour (variable part)} \qquad (3.1)$$

If the absorption of water reaches its maximum value in a given volume of dry air, then it is known as saturated air.

The *humidity ratio* or *specific humidity* (ω) is ratio of the mass of water vapour (m_v) to the mass of dry air (m_a) in a given volume of moist air (a mixture of dry and water vapour). It is expressed as:

$$\omega = \frac{m_v}{m_a} \qquad (3.2)$$

It is clear from the above equation that the numerical value of the specific humidity is always greater than one (1).

The *degree of saturation* is the ratio of actual specific humidity to the specific humidity of saturated air at a given temperature.

The *relative humidity* (γ) is the ratio of the mass of water vapour (m_v) in a given volume of moist air at a given temperature to the mass of water vapour (m_s) in the same volume of saturated air at the same temperature. It is expressed as:

$$\gamma - \frac{m_v}{m_s} \qquad (3.3)$$

The recommended value of the relative humidity for thermal comfort for both summer and winter condition is 50±10%. Its numerical value is always less than 100%.[1]

The variation of the humidity ratio (specific humidity) with dry-bulb temperature for different relative humidities at atmospheric pressure is shown in Figure 3.2. It is referred to as a psychometric chart.

3.2.3 Air Movement

The temperature of the body remains constant between 36.9 and 37.2 °C. A temperature above 40 °C may be fatal. As mentioned earlier, the thermal

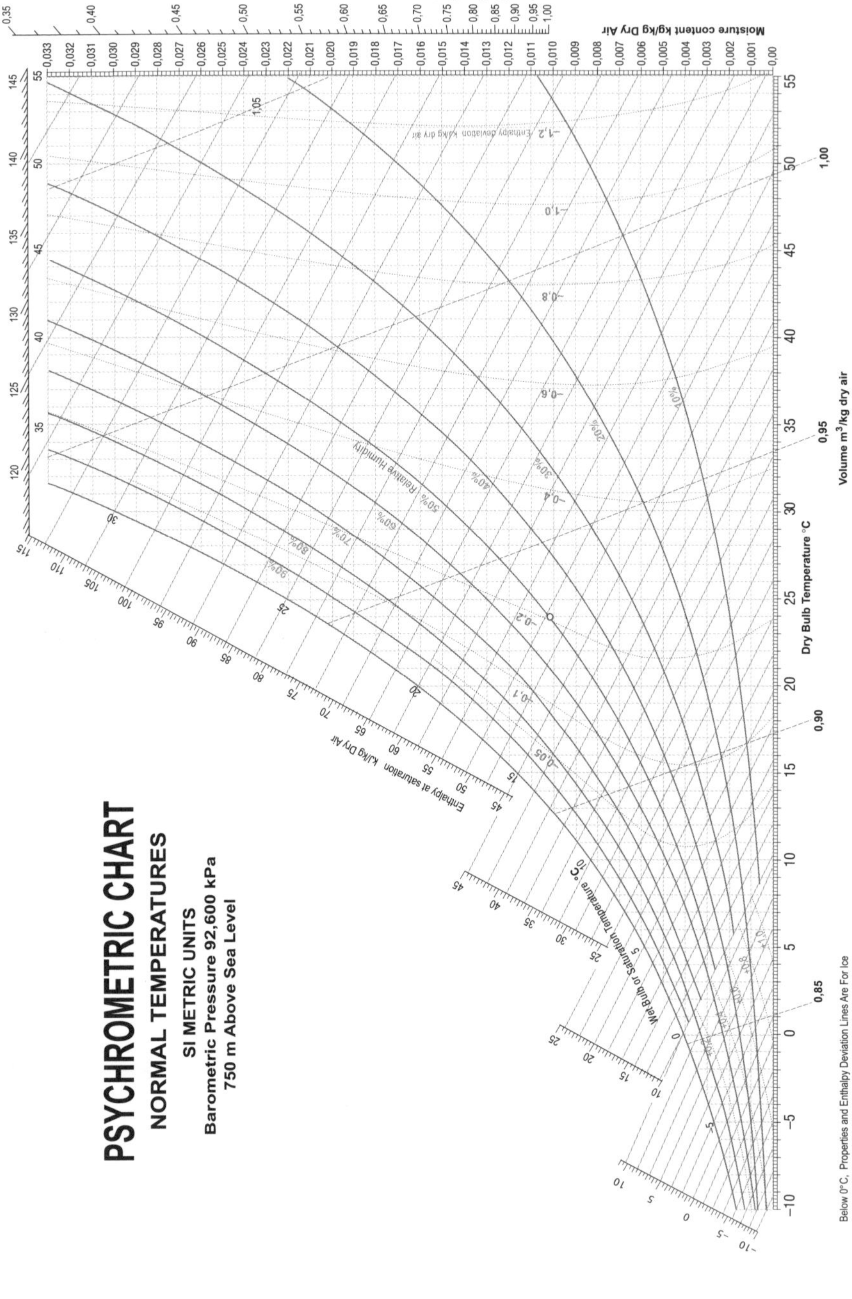

Figure 3.2 A psychometric chart.

comfort air temperature for humans is around 20–25 °C for both winter and summer.

If the air surrounding the body of a human is circulated with a certain movement, then there will be heat transfer (respiration) from the body to the environment. This reduces the thickness of air film near the body. As the air movement near the body increases, the level of discomfort reduces. This is true only when the air temperature of the surroundings is lower than the body temperature, otherwise the level of discomfort increases.

An increase of air movement helps in heat transfer (respiration). It will create discomfort for a higher air temperature and relative humidity. The recommended value of air movement is about 0.2 m s^{-1} and 0.4 m s^{-1} in winter and summer, respectively, inside a room. This condition is achieved in the summer by operating a ceiling or table fan.

3.2.4 Mean Radiant Temperature

The mean radiant temperature (MRT) is a concept arising from the fact that the net exchange of radiant energy between two objects is approximately proportional to their temperature difference multiplied by their ability to emit and absorb heat (emissivity). Technically, MRT is defined as the uniform temperature of a surrounding surface giving off blackbody radiation (emissivity $\varepsilon = 1$) that results in the same radiation energy gain on a human body as the prevailing radiation fluxes, which are usually very varied under open-space conditions. It is the most important parameter governing human energy balance, especially on hot sunny days. It also has the strongest influence on thermophysiological comfort indexes such as predicted mean vote (PMV), which is derived from heat exchange models. Physiologists have discovered that living human skin has an extraordinarily high absorptivity and emissivity (0.97), greater than almost any other known substance. Consequently, we are highly responsive to changes in mean radiant temperature.

Radiant temperature is usually measured with what is known as a globe thermometer. This is simply a normal dry-bulb thermometer encased in a 150 mm diameter matt-black copper sphere whose absorptivity approaches that of the skin. The MRT is calculated from the globe temperature (t_g) and the air temperature (t_a) using the following expression:

$$\bar{t}_r = \sqrt[4]{(t_g + 273)^4 + \frac{h_{cg}}{h_r}(t_g - t_a)} - 273 \qquad (3.4)$$

where h_r is the radiant heat transfer coefficient. For a globe:

$$h_{cg} = \text{max of} \begin{bmatrix} 18(v_{ar})^{0.55} & \text{Forced convection} \\ 3(|t_g - t_a|)^{0.25} & \text{Natural convection} \end{bmatrix} \qquad (3.5)$$

For an operative temperature transducer:

$$h_{cg} = \text{max of} \begin{bmatrix} 6.3\dfrac{(v_{ar})^{0.6}}{D^{0.4}} & \text{Forced convection} \\ 1.4\left(\dfrac{|t_g - t_a|}{D}\right)^{0.25} & \text{Natural convection} \end{bmatrix} \tag{3.6}$$

where v_{ar} is the relative velocity between the body and air in m s^{-1} and D is the diameter of a matt-black copper sphere.

3.2.5 Air Pressure

A person feels more comfortable at atmospheric pressure. At higher altitude the pressure is reduced and a human feels more discomfort due to high pressure within the body.

3.2.6 Air Ingredients

Each person requires 0.65 m^3 h^{-1} of oxygen (O_2) under normal conditions and produces 0.2 m^3 h^{-1} of carbon dioxide (CO_2). The human body requires at least this rate of oxygen for survival and to sustain its action inside as well outside of a building. In open conditions this requirement is met by plants, which release oxygen to maintain the level of oxygen in the environment. The carbon dioxide released by humans is consumed by the plants to balance the ecosystem. The optimum levels of oxygen and carbon dioxide in the atmosphere are 20.94×10^4 ppm and 320 ppm, respectively. However, the optimum level of oxygen and carbon dioxide should also be maintained inside a building either by passive or active modes. Conventionally, to maintain this level, a roof vent, roshandan, ventilator or exhaust fan was provided in the building to have direct contact between the environment and the living space.

3.2.7 Air Electricity

A person feels shocks at the skin when a thin layer of air passes over the skin. This can be observed while rubbing the skin in the presence of a thin layer of air. This is due to friction between the thin air layer and the skin.

3.2.8 Acoustics

A person feels more comfortable in a living space for effective working conditions up to a maximum sound level of 120 db (0.01 W m^{-2}). An average person can hear frequencies from 20 to 16 000 Hz. Unwanted sound creates a sense of irritation in a living space and is termed as noise pollution.

3.2.9 Day Lighting

In the absence of natural lighting in a building, artificial lighting is created by using electrical appliances, *e.g.* an electric bulb, lamp or tube. By using an electric bulb, the thermal heat gain in a building is significant because in a lamp only a fraction of the electrical energy is converted into light energy. The heating effect of day lighting is 1 watt per lumen.

3.3 Physiological Parameters

3.3.1 Nutritional Intake

Human comfort is influenced by physiological factors determined by the rate of heat generated inside body and the rate of heat dissipated to the environment. The rate at which the body produces heat is known as the metabolic rate. A healthy person produces heat at the rate of 60 W ($\sim$35 W m^{-2}) during sleeping, which can go up to 600 W ($\sim$350 W m^{-2}) during hard work.

It is well known that a person requires an optimum nutritional intake, which depends upon the composition of the body. It varies from person to person and from season to season. For example, the nutritional intake (healthy food) for a person increases during winter and *vice versa* during summer. Because of this, a person feels more thermal comfort after taking more caloric value food during the winter period in comparison to the summer period.

3.3.2 Age

With increasing age a person feels colder in winter and warmer in summer. This is due to the followings reasons: (1) a decrease of metabolic rate (the rate of generating heat within the body), (2) a change in food habit, (3) a decrease in nutritional intake and (4) a change in activity level, *etc.*

3.3.3 Ethnic Influences

This deals with the way that people live in a particular region. This includes food habits and clothing, *etc.*

3.3.4 Gender Differences

Although thermal comfort preferences between male and female subjects generally seem to be small, there are some differences especially in office environments.[1] Females are less satisfied with room temperatures than males, and feel both uncomfortably cold and uncomfortably hot more often. Moreover, females prefer significantly higher room temperatures than males. If the females are satisfied, it is most likely that males will also be satisfied. Thus, female subjects should primarily be used in field studies on thermal comfort.

3.3.5 Constitution

This refers to the body constitution of individuals. It is well known that a weak person feels more cold in winter than a healthy one and *vice versa* in summer.

3.4 Intermediate Parameters

3.4.1 Clothing

Clothing interferes with air movement across the skin and as a result it decreases the potential of heat transfer (conduction, convection and evaporation/respiration) from the body to the room air. The colour of the clothing is also important. For example, (1) in a desert, thin, loose fitting and light coloured clothing is suitable. This gives a lower absorption of solar radiation and heat transfer from the outside to the skin and provides air movement across the skin due to the loose fitting. (2) In a hot and humid climate, loose, light coloured and porous clothing is advised. This provides fast heat transfer from the skin to the environment air due to the porosity of the cloth. (3) In a cold climate, dark coloured, thick and tight fitting clothing is recommended for higher absorption of solar radiation and minimum heat transfer from the skin to the environment air.

Clothing insulation can be described in terms of its clo value. The clo value is a numerical representation of a clothing ensemble's thermal resistance; 1 clo = 0.155 $m^2\,°C\,W^{-1}$. A heavy two-piece business suit and accessories have an insulation value of about 1 clo, while a pair of shorts is about 0.05 clo. Clo values for common articles of clothing are listed in Table 3.1 The total insulation value of a clothing ensemble can be estimated as the sum of the individual garment's clo values.

3.4.2 Metabolism

This is the body heat production rate resulting from the oxidation of food. Its value for every person depends upon their diet and level of activity. It is estimated by:

$$M = 2.06 \times 10^4 \dot{V}(F_{oi} - F_{oe})\ \mathrm{W} \tag{3.7}$$

where $\dot{V}$ = air breathing rate per second and F_{oi} and F_{oe} are the fraction of oxygen in the inhaled and exhaled air, respectively. The value of F_{oi} is normally 0.209 and F_{oe} varies with the composition of the food used in the metabolism; for a fat diet, $F_{oe} \approx 0.159$; for carbohydrates, $F_{oe} \approx 0.163$.

The rate of metabolism per unit area of the body surface can be estimated by:

$$A_D = 0.202 m^{0.425} H^{0.725}\,\mathrm{m}^2 \tag{3.8}$$

Table 3.1 Clo values for individual items of clothing.[2,3]

Men *Clothing*		clo^a	*Women* *Clothing*		clo^a
Underwear	sleeveless	0.06	Underwear	girdle	0.04
	T-shirt	0.09		bra and panties	0.05
	briefs	0.05		half slip	0.13
	long underwear, upper	0.10		full slip	0.19
	long underwear, lower	0.10		long underwear, upper	0.10
				long underwear, lower	0.10
Shirt	light, short sleeve	0.14	Blouse	light, long sleeve	0.20
	light, long sleeve	0.22		heavy, long sleeve	0.29
	heavy, short sleeve	0.25	Dress	light	0.22
	heavy, long sleeve	0.29		heavy	0.70
(plus 5% for tie or turtleneck)					
Vest	light	0.15	Skirt	light	0.10
	heavy	0.29		heavy	0.22
Trousers	light	0.26	Slacks	light	0.10
	medium	0.32		heavy	0.44
	heavy	0.44			
Sweater	light	0.20	Sweater	light, sleeveless	0.17
	heavy	0.37		heavy, long sleeve	0.37
Jacket	light	0.22	Jacket	light	0.17
	heavy	0.49		heavy	0.37
Socks	ankle length, thin	0.03	Stockings	any length	0.01
	thick	0.04		panty hose	0.01
	knee high	0.10			
Shoes	sandals	0.02	Shoes	sandals	0.02
	Oxfords	0.04		Pumps	0.04
	boots	0.08		boots	0.08
Hat and overcoat		2.00	Hat and overcoat		2.00
Mask		0.03	Mask		0.03

^{a}1.0 clo $= 0.155\,m^2\,°C\,W^{-1} = 0.133\,m^2\,h\,°C\,kcal^{-1}$.

Table 3.2 Metabolic rates of different typical activities in cabins.

Activity		*Metabolic rate* *met*[a]	*W m*$^{-2}$
Rest	sleeping	0.7	40
	reclining	0.8	45
	seated, relaxed	1	60
	standing, relaxed	1.2	70
Walking	slow	2	115
	moderate	2.6	150
	fast	3.8	220
Affairs	typing, computer	1.1	65
	sedentary activity	1.2	70
	standing activity	1.4	80
	packing/transport	2.1	120
Equipment handling	light machine work	2.0–2.4	115–140
	heavy machine work	4	235
Other	cooking	1.6–2.0	95–115
	cleaning	2.0–2.4	115–200
	exercise[b]	4	235

[a]1.0 met = 58.2 W m^{-2} = 50 kcal m^{-2} h^{-1}.
[b]Exercise: running, table tennis, volleyball, *etc.*

where m = body mass in kg and H = body height in m. For an average man of mass 70 kg and height 1.73 m, $A_D = 1.83\,m^2$.

Metabolism is often given the unit "met", which corresponds to the metabolism of a relaxed, seated person, *i.e.* 1 met = 58.2 W m^{-2}. Usually, metabolism varies from 35 W m^{-2} to 350 W m^{-2}. Typical values of metabolism rate are given in Table 3.2. Hence, the human body can be considered as a heat engine converting thermal energy into mechanical energy with an efficiency of 20%. In a space of 6 m^3 the room air temperature may rise to 48 °C h^{-1} if a man's body dissipates energy at rate of 335 kJ h^{-1}. In order to provide thermal comfort, the heat generated should be removed by providing ventilation to the room.

3.4.3 Adaption and Acclimatization

Adaption and acclimatization mean that a person should first become immune with an artificial similar climate before he/she starts living in an actual harsh climatic condition.

3.4.4 Time of the Day or Season

The recommended thermal comfort air temperature is 20 °C. It is known that ambient air temperature changes with (1) time of the day due to changes in the level of solar intensity (insolation) and the month of the year from season to season. For example, there is a variation of ambient air temperature from 5 to 15 °C in the winter months and from 30 to 45 °C in the summer months for northern climatic conditions. Therefore, the requirement of a heating/cooling load depends on the time of the day or season.

3.4.5 Occupancy

As we have seen, a person can produce heat due to energy generated within the body. If the number of persons is increased for a given volume of enclosed space, then the total amount of heat produced will also be increased, depending upon numbers. This leads to an increase in room air temperature. This affects the living as well as the working conditions of people.

3.4.6 Psychological Factors

Sometimes by expressing the level of peak winter and summer conditions in a group, people feel comfortable by sharing their thoughts with others. This gives the pleasure of an individual talking in a group. This is due to psychological factors.

India has been divided into a number of climatic zones for the simplification of analysis in solar energy systems. The criteria for classification of these climatic zones in India are given in Table 3.3. The heating/cooling requirements of a building can be determined if the sol-air temperature, defined below, is known.

3.5 The Comfort Equation

The first condition to maintain thermal comfort for the human body is the existence of a heat balance. The human body is a thermodynamic system. It produces mechanical work and heat from food metabolism and oxygen consumption. This system requires maintaining a constant internal temperature around $37 \pm 0.5\,^{\circ}C$ in order to ensure health and good functioning of internal organs. It can then be assumed that, for long exposures to a constant (moderate) thermal environment with a constant metabolic rate, a heat balance will exist for the human body, so the heat production will equal the heat dissipation. This heat is dissipated to the environment by convection and radiation ("dry" heat loss) from the skin. The skin is thus the organ that is exposed to the environmental changes and reacts accordingly with these changes by varying its temperature, sweat coverage and hair position. If the dry heat loss is not enough to balance the metabolic rate, sweat is produced by the skin glands and the evaporation of that sweat provides the additional required cooling. The cooling efficiency of sweat evaporation decreases if humidity values are high. In this case the body produces more sweat than in an environment with lower humidity in order to obtain the required physiological cooling. Afterwards a tiny layer of sweat is formed on the skin, obstructing the evaporation process and therefore the cooling effect.

Table 3.3 Criteria for the classification of climates.

Climate	*Mean monthly temperature (°C)*	*Relative humidity (%)*	*Precipitation (mm)*	*Number of clear days*	*Example*
Hot and dry (HD)	>30	<55	<5	>20	Jodhpur
Warm and humid (WH)	>30	>55	>5	<20	Bombay
Moderate (MO)	25–30	<75	<5	<20	Bangalore
Cold and cloudy (CC)	<25	>55	>5	<20	Srinagar
Cold and sunny (CS)	<25	<55	<5	>20	Leh
Composite (CO)	This applies when six months or more do not fall within any of the above categories				New Delhi

The second condition is the fulfilment of the body's energy balance, *i.e.* the heat produced by the metabolism should be equal to the amount of heat lost from the body by conduction, convection, radiation, evaporation and respiration. These two conditions are combined in a single equation called the *comfort equation*. When the person is in thermal neutrality the body's energy balance is:

$$M - W = H + E_c + C_{res} + E_{res} \tag{3.9}$$

where M and W are the metabolic rate of subject and the effective mechanical work it does, respectively, H is the body heat loss by convection, radiation and conduction, E_c is the evaporative heat exchange at the skin when the subject experiences a sensation of thermal neutrality, C_{res} is the respiratory convective heat exchange and E_{res} is the respiratory evaporative heat exchange.

If the person is not in thermal neutrality, the heat will be either stored or released from the body. The rate of heat exchange by the human body with the surrounding temperature is shown in Figure 3.3. The expression for heat stored in the body (S) is:

$$S = (M - W) - (H + E_c + C_{res} + E_{res}) \tag{3.10}$$

In summer conditions the body temperature rises owing to the energy stored in the body (S is positive) and the blood flow rate increases through the extremities, which is known as vasodilation. In winter conditions the temperature

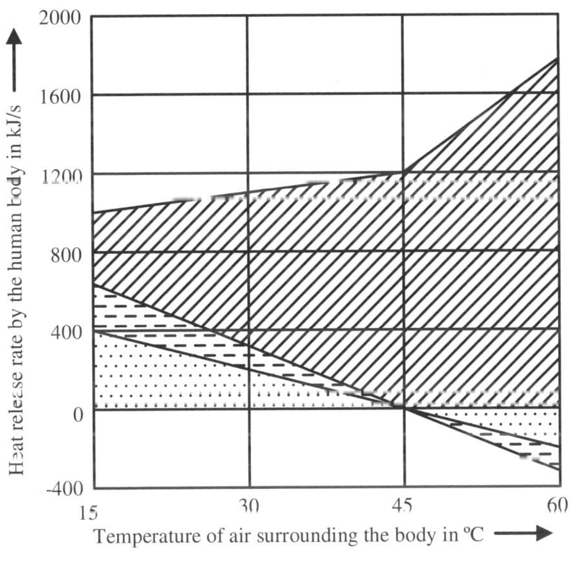

Figure 3.3 Effect of surrounding temperature on the rate of heat released by the body.

of the body tends to fall owing to loss of energy from the body (S being negative) and the blood flow rate decreases, which is known as shivering.

3.5.1 Radiation

The heat exchange by radiation that occurs between the surface of the body (clothing and skin) and the surrounding surfaces is calculated using the Stefan–Boltzmann equation:

$$R = f_{\text{eff}} f_{\text{cl}} \varepsilon \sigma \left\{ (t_{\text{cl}} + 273)^4 - (\bar{t}_{\text{r}} + 273)^4 \right\} \text{ W m}^2 \quad (3.11)$$

where f_{eff} is the factor of the effective radiation area, *i.e.* the ratio of the effective radiation area to the total surface area of the clothed body, f_{cl} is the factor of clothing area, *i.e.* the ratio of the surface area of the clothed body to the surface area of the nude body, ε is the emissivity of the clothed body, σ is the Stefan–Boltzmann constant (5.67×10^{-8} W m^{-2} K^{-4}), t_{cl} is the surface temperature of the clothing in °C and $\bar{t}_r$ is the mean radiant temperature, *i.e.* the effective temperature of the room surface in °C.

Because some parts of the human body act as a shield to other parts, the factor f_{eff} has a value of 0.696 for a seated person and 0.725 for a standing person; a mean value of 0.71 is usually used. The combined emissivity of skin and clothing, ε, is approximately 0.97. Thus, the equation can be reduced to:

$$R = 3.96 \times 10^{-8} f_{\text{cl}} \left\{ (t_{\text{cl}} + 273)^4 - (\bar{t}_{\text{r}} + 273)^4 \right\} \quad (3.12)$$

The range of temperatures in the indoor environment is usually small (typically 10–30 °C) and can be adequately replaced by a linear equation as:

$$R = f_{\text{eff}} h_{\text{r}} \left\{ (t_{\text{cl}} + 273)^4 - (\bar{t}_{\text{r}} + 273)^4 \right\} \quad (3.13)$$

where the radiant heat transfer coefficient, h_{r}, can be approximated by:

$$h_{\text{r}} = 4.6\{1 + 0.01 \times (\bar{t}_{\text{r}} + 273)\} \approx 5.7 \text{ W m}^{-2} \text{ K}^{-1} \quad (3.14)$$

3.5.2 Convection

The heat transfer between a body and the surrounding air is primarily by convection, which can be either free (natural), caused by buoyancy, or forced (mechanical), caused by relative movement between the body and air. The general heat convection equation is:

$$C = f_{\text{cl}} h_{\text{c}} (t_{\text{cl}} - t_{\text{a}}) \text{ W m}^{-2} \quad (3.15)$$

where t_a is the air temperature in °C and h_c is the convective heat transfer coefficient in $W\,m^{-2}\,K^{-1}$.

This convective heat transfer from skin or clothing results from an airstream perturbing the insulating boundary layer of air clinging to the surface of the body. Generally, the faster the flow of air around the body, the thinner the boundary layer of air on the body's surface, and hence the lower the thermal insulation afforded the subject. The process of convection from a heated surface such as human skin or clothing can be further classified into three distinct modes: natural convection, where the air movement is driven purely by thermally induced buoyancy and generally confined to low ambient air speeds; forced convection at speeds generally higher than $1.5\,m\,s^{-1}$; and a region of mixed-mode convection prevailing at air speeds between these two limits. Table 3.4 shows the human body part heat transfer coefficient values, for both natural and forced convection conditions, defined by a set of laboratory measurements by de Dear *et al.*[4]

The expression for the convective heat transfer coefficient is

$$h_c = \begin{cases} 2.38(t_{cl} - t_a)^{0.25} & \text{for } 2.38(t_{cl} - t_a)^{0.25} > 12.1\sqrt{v_{ar}} \quad \text{(natural convection)} \\ 12.1\sqrt{v_{ar}} & \text{for } 2.38(t_{cl} - t_a)^{0.25} < 12.1\sqrt{v_{ar}} \quad \text{(forced convection)} \end{cases} \tag{3.16}$$

where v_{ar} is the relative velocity between the body and air in $m\,s^{-1}$.

3.5.3 Conduction through Clothing

The heat conduction through clothing is given by:

$$K = h_{cl}(t_s - t_{cl})\ W\,m^{-2} \tag{3.17}$$

where h_{cl} = heat conductive coefficient of clothing in $W\,m^{-2}\,K^{-1}$ and t_s = average skin temperature in °C. The conductive coefficient is often replaced by the reciprocal of the thermal resistance of the clothing, I_{cl}, usually given the unit "clo" (1 clo = $0.155\,m^2\,K\,W^{-1}$). Hence:

$$h_{cl} = \frac{1}{0.1555 \times I_{cl}} = \frac{6.45}{I_{cl}} \tag{3.18}$$

Typical values of I_{cl} for different clothing ensembles are given in Table 3.1.
The clothing area factor is given by

$$f_{cl} = \begin{cases} 1.00 + 1.290 I_{cl} & \text{for } I_{cl} < 0.078\,m^2\,°C\,W^{-1} \\ 1.05 + 0.645 I_{cl} & \text{for } I_{cl} > 0.078\,m^2\,°C\,W^{-1} \end{cases} \tag{3.19}$$

The mean skin temperature, t_s (for a metabolic rate M between 1 and 4), may be estimated from:

$$t_s = 35.57 - 0.0275(M - W)\,°C \tag{3.20}$$

Table 3.4 Natural and forced convection heat transfer coefficients for a nude thermal man in standing and sitting conditions.

	Natural convection heat transfer coefficient ($W m^{-2} K^{-1}$)		*Forced convection heat transfer coefficient* ($W m^{-2} K^{-1}$)*			
Body part	*Standing*	*Seated*	*Standing* *B*	*n*	*Seated* *B*	*n*
Head	3.6	3.7	3.2	0.97	4.9	0.73
Chest	3.0	3.0	7.5	0.66	9.1	0.59
Back	2.9	2.6	7.7	0.63	8.9	0.63
Pelvis	3.4	2.8	8.8	0.59	8.2	0.65
Upper arm	2.9	3.4	10.0	0.62	11.4	0.64
Forearm	3.7	3.8	12.6	0.54	11.8	0.62
Hand	4.1	4.5	14.4	0.56	13.4	0.60
Thigh	4.1	3.7	10.1	0.52	8.9	0.60
Lower leg	4.1	4.0	12.9	0.50	13.2	0.57
Foot	5.1	4.2	12.0	0.50	12.9	0.54
Whole body	3.4	3.3	10.4	0.56	10.1	0.61

*Forced convection heat transfer coefficient $= B v^n$ where B is coefficient [$W m^{-2} K^{-1}$], v is ambient air speed [$m s^{-1}$], and n is a dimensionless exponent.

The surface temperature of clothing is given by:

$$t_{cl} = 35.7 - 0.028(M - W) - I_{cl} \times \left[3.96 \times 10^{-8} f_{cl} \times \left\{(t_{cl} + 273)^4 - (\bar{t}_r + 273)^4\right\} + f_{cl} \times h_c \times (t_{cl} - t_a)\right] \quad (3.21)$$

3.5.4 Evaporative Heat Loss

Heat loss by evaporation is partly due to the diffusion of water vapour through the skin tissues (E_d) and partly due to evaporation of sweat from the skin surface (E_{sw}). In both cases, heat is absorbed from the skin and this process controls the rise in body temperature. The water diffusion is a continuous process that occurs even in a cool environment, but the sweat evaporation only occurs in a hot environment and when the body activity is higher than normal.

When evaporation takes place, the water uses heat from the skin. The amount of water diffusion through the skin and the corresponding evaporative heat loss (E_d) is a function of the difference between the saturated water vapour pressure at skin temperature (p_s) and the water vapour pressure in the ambient air (p_a). Therefore:

$$E_d = 3.05 \times 10^{-3}[p_s - p_a]\,\mathrm{W\,m^{-2}} \quad (3.22)$$

where p_s and p_a are in Pa (pascal).

The saturated water vapour pressure at the skin surface temperature (t_s) in the range of 27–37 °C may be approximated as a linear function, given by:

$$p_s = 256t_s - 3373\ \mathrm{Pa} \quad (3.23)$$

Substituting the value of p_s from eqn (3.23) into eqn (3.22), we have:

$$E_d = 3.05 \times 10^{-3}[256t_s - 3373 - p_a]\,\mathrm{W\,m^{-2}} \quad (3.24)$$

Then, substituting the value of t_s from eqn (3.20) into eqn (3.24):

$$E_d = 3.05 \times 10^{-3}[5733 - 6.99(M - W) - p_a] \quad (3.25)$$

Water diffusion through the skin will normally result in a heat loss equal to approximately 10 W m^{-2}. A typical case is a skin temperature $t_s = 33$ °C, a water vapour pressure $p_a = 1400$ Pa at 23 °C ambient temperature and 50% relative humidity. This will result in a heat loss equal to 11.2 W m^{-2}.

The heat loss by water diffusion through the skin takes place all the time and is not controlled by the thermoregulatory system. Evaporation of sweat from the skin surface (E_{sw}) is one of the most effective ways by which the body can keep the internal temperature from increasing, even during hard work. The amount of this evaporation may change from 0 at rest to a maximum of 400 W m^{-2} with very hard work in a hot, dry environment. It is limited how much a person is able to sweat and there are great individual differences. Persons who are used to living and working in hot environments or performing hard work can improve the function of the sweat glands and obtain a better control of their body temperature. An acclimatized person is normally not able to sweat more than $1\,\mathrm{L\,h^{-1}}$, with a total amount of approximately 3.5 L. If all this sweat is evaporated, it is equal to a heat loss of $375\,\mathrm{W\,m^{-2}}$ and a total amount of energy of 8505 kJ. During hard work in hot environments it is important to drink water (plus salt) to be able to sweat enough. The estimation of the heat loss due to the evaporation of sweat is rather complicated and not fully understood yet. By excessive sweating, some of the produced sweat will drip and does not remove any heat from the body by evaporation. It is only the sweat which evaporates at the skin surface that removes heat from the body. Fanger[5] has produced a formula based on experimental measurements correlating E_{sw} to metabolism as:

$$E_c = 0.42(M - W - 58.15)\,\mathrm{W\,m^{-2}} \tag{3.26}$$

On combining eqns (3.25) and (3.26), the evaporative heat exchange at the skin is then given by:

$$E_c = 3.05 \times 10^{-3}[5733 - 6.99(M - W) - p_a] + 0.42(M - W - 58.15) \tag{3.27}$$

3.5.5 Respiration Heat Loss

Inspired air is both warmed and humidified by its passage through the respiratory system. The sensible and latent heat losses are proportional to the volume flow rate of air to the lungs that in turn is proportional to the metabolic rate.

The respiratory convective heat exchange is given by:

$$C_{res} = 0.0014M(34 - t_a)\,\mathrm{W\,m^{-2}} \tag{3.28}$$

where t_a is the ambient air temperature in °C.

The respiratory evaporative heat exchange is given by:

$$E_{res} = 1.72 \times 10^{-5}M(5867 - p_a)\,\mathrm{W\,m^{-2}} \tag{3.29}$$

where p_a is the ambient water vapour pressure in Pa. Respiration heat loss is only significant at high activity and under normal sedentary activity. It can be neglected in cases where the value is less than $6\,\mathrm{W\,m^{-2}}$.

3.6 Predicting the Thermal Comfort

Apart from establishing a conceptual frame for the thermal comfort issue, efforts have been made to find both the best way to measure and predict thermal comfort and also techniques to achieve it. According to ASHRAE 55 and ISO 7730, the simple way of predicting thermal comfort is to use figures and tables from manuals; another way is using numerical and more rigorous predictions by applying the PMV/PPD and two-node models. PMV/PPD indexes (see below) show the quantitative values of the degree of discomfort and the effectiveness of not only environmental factors but also human factors.

3.6.1 Predicted Mean Vote Index

Concerning the empirical–numerical models, the most widely used thermal model was proposed by Fanger in 1970 and is called the predicted mean vote (PMV).[5] PMV predicts the mean value of the votes of a large group of persons on a thermal sensation scale that has seven points. There are also the ASHRAE thermal scale and Bedford's comfort scale, which use the same number of points but with different semantics, as shown in Table 3.5. The PMV index has been accepted as an international standard since the 1980s (ISO 7730) and in ASHRAE 55-1992, and consequently a large number of researchers have taken this index as the reference for their studies.

For the calculation of the PMV index, both the activity (metabolic rate) and the clothing (thermal resistance) have to be estimated and the following environmental parameters have to be measured: air temperature, mean radiant temperature, relative air velocity and partial water vapour pressure. The PMV index was developed for steady-state conditions, but according to several researchers it can be applied with good approximation during minor fluctuations of one or more of the variables, provided that time-weighted averages of the variables are applied. The general expression is:

$$\mathrm{PMV} = \left(0.303\mathrm{e}^{-0.036M} + 0.028\right)[(M - W) - (H + E_\mathrm{c} + C_\mathrm{res} + E_\mathrm{res})] \quad (3.30)$$

Substituting the values from eqns (3.12), (3.15), (3.26), (3.28) and (3.29) into eqn (3.30):

Table 3.5 Relationship between PMV and a thermal sensation.

Scale	*ASHRAE thermal scale*	*Bedford's comfort scale*
−3	cold	much too cool
−2	cool	too cool
−1	slightly cool	comfortably cool
0	neutral	comfortable
1	slightly warm	comfortably warm
2	warm	too warm
3	hot	much too warm

Table 3.6 Limitations to the range of conditions over which PMV applies.

Variable	*Symbol*	*Lower limit*	*Upper limit*
Metabolic rate	M	0.8 met (46 $W\,m^{-2}$)	4.0 met (232 $W\,m^{-2}$)
Clothing insulation	I_{cl}	0 clo (0 $m^2\,K\,W^{-1}$)	2 clo (0.310 $m^2\,K\,W^{-1}$)
Air temperature	t_a	10 °C	30 °C
Radiant temperature	$\bar{t}_{ar}$	10 °C	40 °C
Relative air velocity	v_{ar}	0 $m\,s^{-1}$	1 $m\,s^{-1}$
Water vapour pressure[a]	p_a	0 Pa	2700 Pa
Predicted mean vote	PMV	–2	+2

[a]Inside the range for p_a it is furthermore recommended that the relative humidity be kept between 30% and 70%.

$$\mathrm{PMV} = \begin{pmatrix} 0.303 \times e^{-0.036M} \\ +0.028 \end{pmatrix} \left\{ \begin{array}{l} (M - W) - 3.05 \times 10^{-3} \\ \times \{5733 - 6.99(M - W) - p_a\} - 0.42 \\ \times \{(M - W) - 5815\} - 1.7 \\ \times 10^{-5} M(5867 - p_a) - 0.0014 \\ \times M(34 - t_a) - 3.96 \times 10^{-8} \times f_{cl} \\ \times \left\{(t_{cl} + 273)^4 - (\bar{t}_r + 273)^4\right\} - f_{cl} \\ \times h_c \times (t_{cl} - t_a) \end{array} \right\} \tag{3.31}$$

where f_{cl} is the ratio of a man's surface area while clothed to a man's surface area while nude, t_a is the air temperature in °C, $\bar{t}_r$ is the mean radiant temperature in °C, v_{ar} is the relative air velocity (relative to the human body) in $m\,s^{-1}$, p_a is the partial water vapour pressure in Pa, h_c is the convective heat transfer coefficient in $W\,m^{-2}\,K^{-1}$ and t_{cl} is the surface temperature of the clothing in °C.

From eqn (3.13) the PMV can be calculated for different combinations of metabolic rate, clothing, air temperature, mean radiant temperature, air velocity and air humidity. The equations for t_{cl} and h_c may be solved by iteration. The limitations to the range of conditions over which PMV applies are given in Table 3.6.[3]

The PMV index is derived for steady-state conditions but can be applied with good approximation during minor fluctuations of one or more of the variables, provided that time-weighted averages of the variables during the previous one-hour period are applied.

3.6.2 Predicted Percentage Dissatisfied Index

The PMV gives the mean value of the "thermal" votes of a large group of people exposed to the same environment. However, individual votes are scattered around this mean value and it is useful to predict the number of people likely to feel uncomfortably cold or warm (*i.e.* dissatisfied). This is exactly the

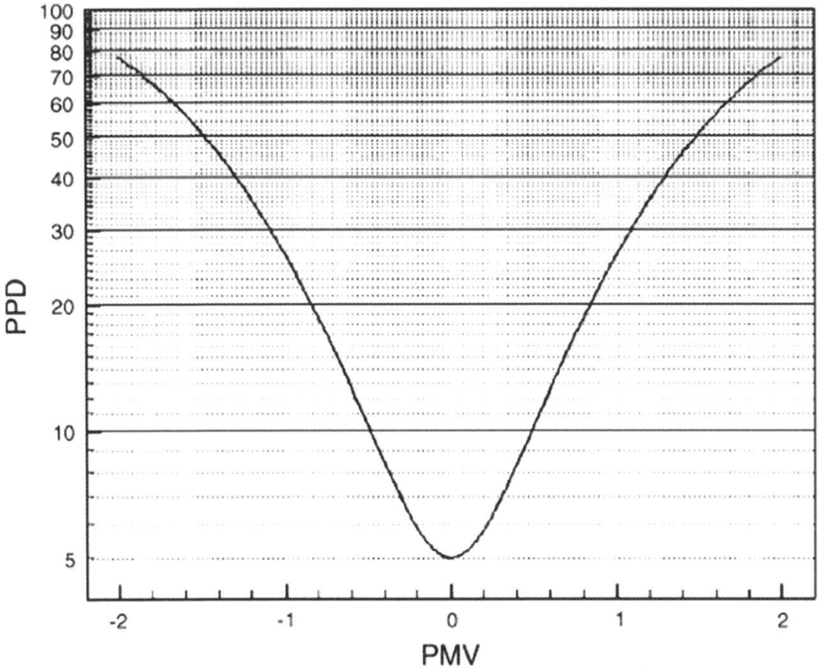

Figure 3.4 Predicted percentage dissatisfied (PPD) index as a function of predicted mean vote (PMV).

aim of the predicted percentage dissatisfied (PPD) index. It is found as a function of the PMV from eqn (3.31) or Figure 3.4, as follows:

$$\mathrm{PPD} = 100 - 95 \times \exp[-(0.03353 \times \mathrm{PMV}^4 + 0.2179 \times \mathrm{PMV}^2)] \quad (3.32)$$

It is characteristic of the PPD index that its value does not in practice fall below 5% for any value of the PMV. The reason for this fact is the difference in thermal sensation between individuals; the thermal neutrality for different people is achieved at environmental parameters, which are not identical. Thus, the PPD predicts the percentage of thermally dissatisfied people. PPD may be more reliable than PMV because individual votes show scatter due to human factors.

3.7 Recent Research and Conclusions

PMV and PPD form the basis of ISO 7730 for human comfort. It provides guidelines to determine the control settings of an existing air-conditioning system. However, there have been several comments on PMV from researchers. Some reported that PMV could not yield good a prediction for human thermal comfort. In 2004, Feriadi and Wong[6] found that PMV predicted warmer thermal perception compared to what people in Indonesia actually felt in their naturally ventilated houses. In the study they stated that people would perform adaptive actions to compensate for the less comfortable thermal conditions. These actions included drinking, changing clothes, taking a bath, walking around, *etc.* Instead of using air-conditioning systems, Indonesian people preferred to operate fans and windows to achieve a higher wind speed for comfort. In the same year, Borong *et al.*[7] reported that it was inadequate to apply PMV in Chinese vernacular dwellings. The buildings were also naturally ventilated at the time of conducting the study. A similar failure of PMV in this kind of building was observed by Ealiwa *et al.*[8] He also reported that PMV performed improperly in buildings with air-conditioning systems in North Africa. He thus called for modifications of the PMV equation for comfort assessment in naturally ventilated buildings. In addition, there was another situation causing the failure of the PMV. Chun *et al.*[9] stated that the PMV should not be used for human comfort prediction in transitional spaces. In their study, these spaces were characterized by dynamic, unstable and fluctuating thermal environments, in which people performed a variety of activities. They accounted for all these variations in thermal environments and human activities by PMV failure. Furthermore, two researchers have severely criticized the performance of the PMV. Nicol[10] compiled the results of previous field studies in thermal comfort. He summarized that people would find ways to make themselves more comfortable. One popular way was to turn on a fan to increase air velocity. This matched with one of the findings from Feriadi and Wong.[6] Nicol also emphasized that the PMV performed poorly in tropical countries, where hot climates are common. Humphreys and Nicol[11] stated that the PMV differed markedly and systematically from the actual mean vote for naturally ventilated buildings and even for air-conditioned spaces. Several origins of biases were discussed in this study. One of them was ambiguities in the assessment of metabolic rate and clothing insulation.

On the other hand, many researchers have supported the PMV. In 2002, Parsons[12] observed that there were very small differences in the thermal comfort responses of male and female subjects for neutral and slightly warm conditions. Changes in thermal comfort responses in neutral and slightly warm environments, due to acclimatization to heat, were not significant. This obviously contradicted the findings of Nicol[11] and Feriadi and Wong,[6] who suggested that people would try to adapt to less comfortable environments by performing some actions. Parson's work also confirmed that there were few group differences between thermal comfort requirements of people with and without physical disabilities. In 2003, Chamra *et al.*[13] reported that the PMV performed well in many of the cases in their study. The cases for proper PMV prediction fell within office environments having air-conditioning systems for comfort control. With activities up to 2.3 met, humidity at mid to high ranges and air velocities up to $0.25\,\mathrm{m\,s^{-1}}$, the prediction of PMV was confirmed to be valid. These situations were common in air-conditioned environments. More recently, there have been many researchers still employing PMV in their research for human comfort assessment. Baskin and Vineyard[14] employed PMV and PPD in thermal comfort assessment of conventional and high-velocity distribution systems for cooling seasons. These systems delivered high air speed to occupants to enhance cooling. The situations were similar to naturally ventilated buildings. Pan *et al.*[15] proposed an energy-saving scheme by using a personalized partition-type fan-coil unit which increased air velocity up to 0.4 m s^{-1}. It was a similar situation to that found by Baskin and Vineyard.[14] Once again, the PMV was applied in this study to evaluate the performance of this scheme. In 2004, Kulkarni and Hong[16] conducted a comfort survey in the situation of a transient pull-down, which was similar to a transitional space (Chun *et al.*[9]). In this study, a specific technique was developed to accurately measure PMV values to reflect the thermal environment. Even in air-conditioning control, PMV has been treated as a favourite candidate to be employed in advanced control algorithms. More than 10 years ago, Simmonds[17] replaced the dry-bulb temperature with PMV in his rule-based control strategy. Every actuator was commanded so as to push the PMV to a zero value that corresponded to the neutral comfort environment. Lower operating costs resulted. In 2004, Calvino *et al.*[18] introduced a fuzzy adaptive controller for indoor comfort control. Inside the complicated algorithms, the PMV was employed as a control index. A set of optimal control actions was derived based on minimization of the PMV. They reported that an effective and fast control of the indoor microclimate conditions was achieved. Lastly, Fanger and Toftum[19] revised the PMV equation. In new work, they proposed an extension of the PMV model to account for any discrepancies found in naturally ventilated buildings. Expectancy factors were introduced and improvements in such buildings were observed.

Thus, both positive and negative comments on the performance of the PMV on human comfort exist. Humans, being the end users of air-conditioning systems in air-conditioned spaces, solely determine the performance of this system through their thermal sensations.

3.8 Related Standards

The standards related directly or indirectly to thermal comfort are as follows.

1. Thermal comfort and related thermal environment:
    - ASHRAE 55: thermal environmental conditions for human occupancy.
    - ISO 7730: moderate thermal environments; determination of the PMV and PPD indices and specification of the conditions for thermal comfort (EN ISO 7730).
    - ISO 7993: hot environments; analytical determination and interpretation of thermal stress using calculation of required sweat rate.
2. Design of the indoor environment:
    - ASHRAE 62: ventilation for acceptable indoor air quality.
    - CR 1752: ventilation for buildings, design criteria for the indoor environment.

3. Measurement of the indoor thermal environment parameters:
   - ASHRAE 55: thermal environmental conditions for human occupancy.
   - ASHRAE 113: methods of testing for room air diffusion.
   - ISO 7726: ergonomics of the thermal environment; instruments for measuring physical quantities.
4. Determination of personal factors:
   - ISO 8996: ergonomics; determination of metabolic heat production.
   - ISO 9920: estimation of the thermal insulation and evaporative resistance of a clothing ensemble.

References

1. S. Karjalainen, Gender differences in thermal comfort and use of thermostats in everyday thermal environments, *Build. Environ.*, 2007, **42**, 1594–1603.
2. J. A. Miller, Wearing the right clothes, *Sci. News*, 1980, **117**, 396–397.
3. M. S. Jang, C. D. Koh and I. S. Moon, Review of thermal comfort design based on PMV/PPD in cabins of Korean maritime patrol vessels, *Build. Environ.*, 2007, **42**, 55–61.
4. R. J. de Dear, E. Arens, Z. Hui and M. Oguro, Convective and radiative heat transfer coefficients for individual human body segments, *Int. J. Biometeorol.*, 1997, **40**, 141–156.
5. P. O. Fanger, *Thermal Comfort*, McGraw-Hill, New York, 1972.
6. H. Feriadi and N. H. Wong, Thermal comfort for naturally ventilated houses in Indonesia, *Energy Build.*, 2004, **36**, 614–626.
7. L. Borong, T. Gang, W. Peng, S. Ling, Z. Yingxin and Z. Guangkui, Study on the thermal performance of the Chinese traditional vernacular dwellings in summer, *Energy Build.*, 2004, **36**, 73–79.
8. M. A. Ealiwa, A. H. Taki, A. T. Howarth and M. R. Seden, An investigation into thermal comfort in the summer season of Ghadames, Libya, *Build. Environ.*, 2001, **36**, 231–237.
9. C. Chun, A. Kwok and A. Tamura, Thermal comfort in transitional spaces – basic concepts: literature review and trial measurement, *Build. Environ.*, 2004, **39**, 1187–1192.
10. F. Nicol, Adaptive thermal comfort standards in the hot-humid tropics, *Energy Build.*, 2004, **36**, 628–637.
11. M. A. Humphreys and J. F. Nicol, The validity of ISO-PMV for predicting comfort votes in every-day thermal environments, *Energy Build.*, 2002, **34**, 667–684.
12. K. C. Parsons, The effects of gender, acclimation state, the opportunity to adjust clothing and physical disability on requirements for thermal comfort, *Energy Build.*, 2002, **34**, 593–599.
13. L. M. Chamra, W. G. Steele and K. Huynh, The uncertainty associated with thermal comfort, *ASHRAE Trans.*, 2003, **109**, 356–365.

14. E. Baskin and E. A. Vineyard, Thermal comfort assessment of conventional and high-velocity distribution systems for cooling season, *ASHRAE Trans.*, 2003, **109**, 513–519.
15. C. S. Pan, H. C. Chiang, M. C. Yen and C. C. Wang, Thermal comfort and energy saving of a personalized PFCU air-conditioning system, *Energy Build.*, 2004, **37**, 443–449.
16. M. R. Kulkarni and F. Hong, An experimental technique for thermal comfort comparison in a transient pull-down, *Build. Environ.*, 2004, **39**, 189–193.
17. P. Simmonds, Thermal comfort and optimal energy use, *ASHRAE Trans.*, 1993, **99**, 1037–1048.
18. F. Calvino, M. L. Gennusa, G. Rizzo and G. Scaccianoce, The control of indoor thermal comfort conditions: introducing a fuzzy adaptive controller, *Energy Build.*, 2004, **36**, 97–102.
19. P. O. Fanger and J. Toftum, Extension of the PMV model to non-air-conditioned buildings in warm climates, *Energy Build.*, 2002, **34**, 533–536.

Further Reading

G. N. Tiwari, *Solar Energy Fundamentals, Design, Modelling and Applications*, Narosa, New Delhi, 2002.

CHAPTER 4

Solar Heating and Cooling Concepts for Buildings

4.1 Introduction

A passive solar house is a new concept in architecture and describes a way to design a building while keeping under consideration the site, climate, local building materials and the Sun. It implies a special relationship of building design and natural processes that offer the potential for an inexhaustible source of energy. This concept, however, is not entirely new, as can be seen from its description in vernacular architecture. The statement of Socrates (400 B.C.): "*Now in houses with south aspect, the sun's rays penetrate into porticos in winter, but in summer the path of the sun is right over our heads and above the roof. If then, this is the best arrangement one should build the south side loftier to keep out the cold winds*", reflects this. Ancient Iranian architecture used the concepts of clustering (decreasing surface to volume ratio, to reduce thermal load), of thick walls (large thermal storage capacity to smoothen out temperature fluctuations) and of plantations for shade and living in basements (during extreme cold and heat). Iranians also introduced the concept of wind towers which, along with cooling by earth and water evaporation, made the buildings comfortable in summer. The concept of clustering and mass walls was also used in medieval India. American Indians used passive solar techniques as long ago as 1100 A.D. However, scientific application of solar energy for passive heating started in about 1881 when Professor E. L. Morse was granted a patent on glazed south-facing dark walls for keeping the house warm. This idea was not used for a long time until 1972, when Trombe repatented the concept and successfully applied it.

In recent years, however, people have chosen to abandon these long-standing considerations, becoming increasingly dependent on mechanical control of the indoor environment, rather than exploiting climatic and other natural processes to satisfy their requirements. As a result, even a minor power or

RSC Energy Series No. 4
Building Integrated Photovoltaic Thermal Systems: For Sustainable Developments
By Basant Agrawal and G. N. Tiwari
© B. Agrawal and G. N. Tiwari 2011
Published by the Royal Society of Chemistry, www.rsc.org

equipment failure can make these buildings uninhabitable. Furthermore, with the advent of the energy crisis, the effective utilization of energy resources has become the basis of building planning. Now there is a renewed interest in passive solar heating and cooling systems, because these systems operate on energy available in the immediate environment, are simple in concept and require little or no maintenance. Also, these systems do not generate thermal pollution, as they require no external energy input and produce no physical byproducts or waste. As solar energy is extensively distributed over the world, expensive transportation and distribution networks are eliminated. The chapter highlights the popular methods analyzed and suggested by researchers and designers for heating buildings by sunlight. However, the best methods need to consider the location and climatic conditions of the site.

4.2 Sol-air Temperature

The sol-air temperature is defined as the outside air temperature which, in the absence of solar radiation, would give the same temperature distribution and rate of heat transfer through a wall (or roof) as exists due to the combined effects of the actual outdoor temperature distribution plus the incident solar radiation. The concept facilitates predictions of transient and periodic heat flows through building structures. Thus, many organizations and institutions recommend the use of sol-air temperatures whose values have been appropriately listed for various conditions, *e.g.* physical location and time of year (and hence incidence angle for solar radiation). Unfortunately, many building services engineers have grown so accustomed to using the sol-air temperature concept that they now regard it as an intrinsic climatic property.

The steady-state rate of heat transfer per unit area through, say, a wall, may according to present custom be predicted from:

$$\dot{q} = \frac{Q}{A} = U(T_{sa} - T_b) \tag{4.1}$$

where U (in W m^{-2} K^{-1}) is the overall thermal transmittance through the wall, and T_{sa} and T_b (in kelvin) are respectively the average sol-air temperature and the constant room environmental temperature.

This relationship is similar in form to those by which transmission gains or losses can be estimated, the only difference being that by substituting the sol-air temperature for the outside environmental air temperature, the equation purports to make an appropriate allowance for solar radiation gain.

The sol-air temperature is defined as:

$$T_{sa} = T_a + R[\alpha I(t) - \varepsilon I(l)] \tag{4.2}$$

where R (in m^2 K^{-1} W^{-1}) is the thermal resistance of the interface between the external surface and the air, α is the absorptivity for solar radiation, ε is the emissivity for infrared radiation of the considered wall, T_a (in K) is the outdoor air temperature, $I(t)$ (in W m^{-2}) is the intensity of direct plus diffuse solar radiation on the outer surface of the wall, and $I(l)$ (in W m^{-2}) is the intensity of long-wave radiation from a thermally black body.

Hence the sol-air temperature combines the effect of solar radiation, ambient air temperature and long-wavelength radiant heat exchange with the environment. Physically, the sol-air temperature can be interpreted as the temperature of the surroundings that will produce the same heating effect as the incident radiation in conjunction with the actual external air temperature. It is a variable used to calculate the cooling load of a building and determine the total heat gain through exterior surfaces.

4.2.1 Sol-air Temperature for Bare Surfaces

Let us consider a wall or a roof, as shown in Figure 4.1, exposed to solar radiation, resulting in heat flow through conduction and loss through convection and radiation.

The energy balance at the bare surface $x = 0$, is given by:

$$\begin{bmatrix}\text{Rate of heat}\\ \text{absorbed by surface}\end{bmatrix} = \begin{bmatrix}\text{Rate of heat flow}\\ \text{through conduction}\end{bmatrix} + \begin{bmatrix}\text{Rate of heat loss}\\ \text{through convection}\end{bmatrix} + \begin{bmatrix}\text{Rate of heat loss}\\ \text{through radiation}\end{bmatrix}$$

or

$$\alpha I(t) = \left[-K\frac{\partial T}{\partial x}\bigg|_{x=0}\right] + \left[h_{ca}\left(T|_{x=0} - T_a\right)\right] + \left[\varepsilon\sigma\left\{\left(T|_{x=0}\right)^4 - \left(T_{sky}\right)^4\right\}\right] \quad (4.3)$$

where $\alpha = \alpha_b\ (1 - \rho)$ is the fraction of solar radiation absorbed of the bare surface, ρ is the reflectivity and α_b is the absorbtivity of the black surface; $I(t)$

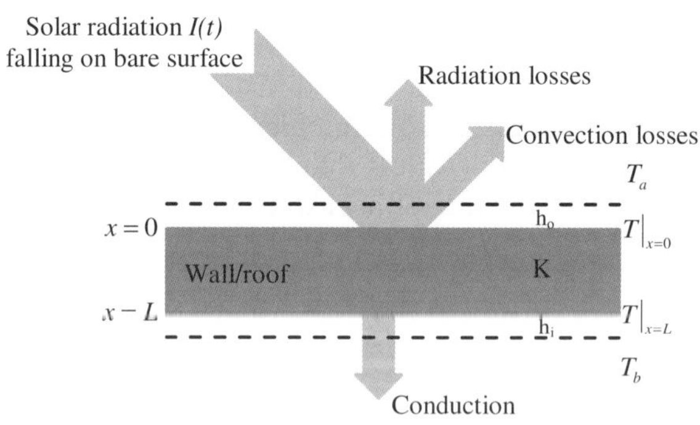

Figure 4.1 A bare surface exposed to solar radiation.

(in $\mathrm{W\,m^{-2}}$) is the intensity of the solar radiation, K (in $\mathrm{W\,m^{-1}\,K^{-1}}$) is the thermal conductivity, $\frac{\partial T}{\partial x}\big|_{x=0}$ is the temperature gradient at $x=0$, $T|_{x=0}$ (in K) is the temperature of the bare surface at $x=0$, ε is the emissivity of the surface, σ is the Stefan–Boltzmann constant, T_a (in K) is the temperature of the ambient air, T_{sky} (in K) is the temperature of the sky (usually 6–12 °C less than T_a) and h_{ca} (in $\mathrm{W\,m^{-2}\,K^{-1}}$) is the convective heat transfer coefficient of the interface between the external surface and the ambient air and is considered to be equal to $2.8+3.0V$ for wind flow velocity (V) less than $5\,\mathrm{m\,s^{-1}}$.

Equation (4.3) can be rewritten as:

$$\alpha I(t) = \left[-K\frac{\partial T}{\partial x}\bigg|_{x=0}\right] + \left[h_{ca}\left(T|_{x=0}-T_a\right)\right] + \left[\varepsilon\sigma\left\{\left(T|_{x=0}\right)^4 - T_a^4\right\}\right] + \left[\varepsilon\sigma\left\{T_a^4 - \left(T_{sky}\right)^4\right\}\right]$$

or

$$\alpha I(t) = \left[-K\frac{\partial T}{\partial x}\bigg|_{x=0}\right] + \left[h_{ca}\left(T|_{x=0}-T_a\right)\right] + \left[h_{ra}\left(T|_{x=0}-T_a\right) + \varepsilon\Delta R\right] \tag{4.4}$$

where $\Delta R = \sigma\,(T_a^4 - T_{sky}^4)$ is the intensity of the long-wave radiation emitted from a bare surface and $h_{ra} = \left[\varepsilon\sigma\left\{\frac{\left(T|_{x=0}\right)^4 - T_a^4}{T|_{x=0}-T_a}\right\}\right]$ (in $\mathrm{W\,m^{-2}\,K^{-1}}$) is the radiative heat transfer coefficient of the interface between the external surface and the ambient air. For a complete cycle of the day, h_{ra} can be evaluated for the known average surface temperature ($\bar{T}|_{x=0}$) and the average ambient temperature ($\bar{T}_a$).

Equation (4.4) can be rewritten as:

$$\alpha I(t) = \left[-K\frac{\partial T}{\partial x}\bigg|_{x=0}\right] + \left[\left(h_{ca}+h_{ra}\right)\left(T|_{x=0}-T_a\right)\right] + \varepsilon\Delta R$$

or

$$\alpha I(t) = \left[-K\frac{\partial T}{\partial x}\bigg|_{x=0}\right] + \left[h_o\left(T|_{x=0}-T_a\right)\right] + \varepsilon\Delta R \tag{4.5}$$

where $h_o = h_{ca} + h_{ra}$ is the overall heat transfer coefficient of the interface between the external surface and the ambient air. It is considered equal to $5.7+3.8V$ for a wind velocity (V) less than $5\,\mathrm{m\,s^{-1}}$.

Rearranging eqn (4.5), the rate of heat transfer through conduction at $x=0$ is given by:

$$\begin{aligned}\dot{q} = -K\frac{\partial T}{\partial x}\bigg|_{x=0} &= \alpha I(t) - h_o\left(T|_{x=0}-T_a\right) - \varepsilon\Delta R \\ &= h_o\left[\frac{\alpha I(t)}{h_o} + T_a - \frac{\varepsilon\Delta R}{h_o} - T|_{x=0}\right]\end{aligned}$$

or

$$\dot{q} = h_o\left(T_{sa} - T|_{x=0}\right) \tag{4.6}$$

where $T_{sa} = \frac{\alpha}{h_o} I(t) + T_a - \frac{\varepsilon \Delta R}{h_o}$ is the sol-air temperature for the bare surface, which is equivalent to eqn (4.2).

It can be seen that (a) for $T_{sa} > T_a$, $\{\frac{\alpha}{h_o} I(t) - \frac{\varepsilon \Delta R}{h_o}\} > 0$ or $I(t) > \frac{\varepsilon \Delta R}{\alpha}$. The sol-air temperature is greater than the ambient temperature if the intensity of the solar radiation is greater than $\frac{\varepsilon \Delta R}{\alpha}$. It may be noted that the factor $\frac{\varepsilon \Delta R}{\alpha}$ is the characteristic of material to absorb and emit long-wavelength radiation. (b) Similarly, for $T_{sa} < T_a$, $\left\{\frac{\alpha}{h_o} I(t) - \frac{\varepsilon \Delta R}{h_o}\right\} < 0$ or $I(t) < \frac{\varepsilon \Delta R}{\alpha}$. The sol-air temperature is then less than ambient temperature if the intensity of the solar radiation is less than $\frac{\varepsilon \Delta R}{\alpha}$, which occurs mostly when there is dew during sunshine hours.

4.2.1.1 Rate of Heat Transfer from a Bare Surface

Consider a wall/roof exposed to solar radiation as shown in Figure 4.1. In the steady-state condition the rate of heat transfer at the outer surface ($x = 0$) is given by eqn (4.6) as:

$$\dot{q} = h_o\left(T_{sa} - T|_{x=0}\right) \tag{4.7}$$

In the steady-state condition, the rate of heat conducted from $x = 0$ to $x = L$ is given by:

$$\dot{q} = \frac{K}{A}\left(T|_{x=0} - T|_{x=L}\right) \tag{4.8}$$

where $T|_{x=L}$ is the temperature of the bare surface at $x = L$.

Further, the rate of heat transfer from the inner surface (surface exposed to room) of the wall/roof to the room space through convection and radiation is given by:

$$\dot{q} = h_{si}\left(T|_{x=L} - T_b\right) \tag{4.9}$$

where T_b is the temperature of the room space air, h_{si} (in W m^{-2} K^{-1}) is the convective heat transfer coefficient of the interface between the inner surface and the room space air, which is usually considered equal to $2.8 + 3V$ owing to negligible heat transfer through radiation.

Equations (4.7) to (4.9) can be rearranged as:

$$\frac{\dot{q}}{h_o} = T_{sa} - T|_{x=0} \tag{4.10a}$$

$$\frac{\dot{q}}{K/L} = T|_{x=0} - T|_{x=L} \tag{4.10b}$$

$$\frac{\dot{q}}{h_{si}} = T|_{x=L} - T_b \tag{4.10c}$$

On adding the above three eqns (4.10a) to (4.10c), we have:

$$\dot{q}\left[\frac{1}{h_o} + \frac{1}{K/L} + \frac{1}{h_{si}}\right] = \left(T_{sa} - T|_{x=0}\right) + \left(T|_{x=0} - T|_{x=L}\right) + \left(T|_{x=L} - T_b\right)$$

or

$$\dot{q} = U(T_{sa} - T_b) \tag{4.11}$$

where $U = \left[\frac{1}{h_o} + \frac{L}{K} + \frac{1}{h_{si}}\right]^{-1}$ is the overall heat transfer coefficient for the wall/roof. If the external surface of the walls/roof are covered or insulated with a thermal conductivity of K_i for a thickness of L_i, mostly applied to reduce heat loss of the room space during off-sunshine hours, then the overall heat transfer coefficient $U = \left[\frac{1}{h_o} + \frac{L_i}{K_i} + \frac{L}{K} + \frac{1}{h_{si}}\right]^{-1}$.

The heat transfer per unit area for time interval $0 - t_T$ is given by

$$\begin{aligned}(Q_u)_i &= \int_0^{t_T} \dot{q}\,dt \\ &= \int_0^{t_T} U\left[\frac{\alpha}{h_o}\{I(t)\}_i + \{T_a\}_i - \frac{\varepsilon\Delta R}{h_o} - T_b\right]dt \\ &= U\left[\frac{\alpha\{\overline{I(t)}\}_i}{h_o} + \{\overline{T_a}\}_i - \frac{\varepsilon\Delta R}{h_o} - T_b\right]t_T\end{aligned} \tag{4.12}$$

where $\overline{I(t)} = \frac{1}{t_T}\int_0^{t_T} I(t)\,dt$ and $\overline{T_a} = \frac{1}{t_T}\int_0^{t_T} T_a dt$. Generally, t_T is considered as one hour.

The total heat transfer per unit area during sunshine hours (N), is given by:

$$\begin{aligned}Q_{ud} &= \sum_{i=1}^{N} (Q_u)_i \\ &= \sum_{i=1}^{N} U\left[\frac{\alpha\{\overline{I(t)}\}_i}{h_o} + \{\overline{T_a}\}_i - \frac{\varepsilon\Delta R}{h_o} - T_b\right]t_T\end{aligned} \tag{4.13}$$

The total heat transfer per unit area during off-sunshine hours (24 – *N*) is given by:

$$\begin{aligned} Q_{\mathrm{un}} &= \sum_{j=1}^{(24-N)} (Q_{\mathrm{u}})_j \\ &= \sum_{j=1}^{(24-N)} U\left[\{\overline{T_{\mathrm{a}}}\}_j - T_{\mathrm{b}}\right] t_{\mathrm{T}} \end{aligned} \tag{4.14}$$

Therefore, the total heat transfer per unit area during the complete cycle of 24 hours is given by adding eqns (4.13) and (4.14), *i.e.*:

$$\begin{aligned} Q_{\mathrm{T}} &= Q_{\mathrm{ud}} + Q_{\mathrm{un}} \\ &= \sum_{i=1}^{N} U\left[\frac{\alpha\{\overline{I(t)}\}_i}{h_{\mathrm{o}}} + \{\overline{T_{\mathrm{a}}}\}_i - \frac{\varepsilon\Delta R}{h_{\mathrm{o}}} - T_{\mathrm{b}}\right] t_{\mathrm{T}} + \sum_{j=1}^{(24-N)} U\left[\{\overline{T_{\mathrm{a}}}\}_j - T_{\mathrm{b}}\right] t_{\mathrm{T}} \end{aligned} \tag{4.15}$$

Example 4.1

Calculate the sol-air temperature of a wall when exposed to solar radiation of intensity 350 W m^{-2}. The absorptivity of the wall is 0.4, the emissivity is 0.9, the wind flows at a velocity of 3 m s^{-1} and the ambient temperature is 12 °C.

Solution

The overall heat transfer coefficient of the interface between the external surface and the ambient air is:

$$\begin{aligned} h_{\mathrm{o}} &= h_{\mathrm{ca}} + h_{\mathrm{ra}} \\ &= 5.7 + 3.8\,V \\ &= 5.7 + (3.8 \times 3) = 17.1\ \mathrm{W\,m^{-2}\,K^{-1}} \end{aligned}$$

Now $\Delta R = \sigma\ (T_{\mathrm{a}}^4 - T_{\mathrm{sky}}^4)$. If $T_{\mathrm{sky}} = T_{\mathrm{a}} - 6\,°\mathrm{C}$, then $\Delta R = 5.67 \times 10^{-8} \times \{(12+273)^4 - (6+273)^4\} = 30.52\ \mathrm{W\,m^{-2}}$

Thus the sol-air temperature for the bare surface is:

$$\begin{aligned} T_{\mathrm{sa}} &= \frac{\alpha}{h_{\mathrm{o}}} I(t) + T_{\mathrm{a}} - \frac{\varepsilon\Delta R}{h_{\mathrm{o}}} \\ &= \frac{0.4}{17.1} \times 350 + 12 - \frac{0.9 \times 30.52}{17.1} = 18.58\,°\mathrm{C} \end{aligned}$$

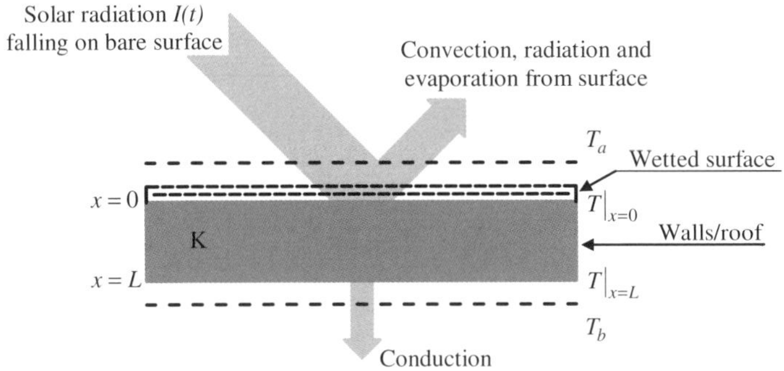

Figure 4.2 A wetted surface exposed to solar radiation.

4.2.2 Sol-air Temperature for Wetted Surfaces

Now let us consider that the external surface of the wall/roof is wetted, as shown in Figure 4.2. In this case, in addition to convection and radiation there will be evaporation.

Hence, the energy balance at the wetted surface at $x = 0$ is given by:

$$\alpha I(t) = \left[-K\frac{\partial T}{\partial x}\bigg|_{x=0}\right] + \left[h_{ca}\left(T|_{x=0} - T_a\right)\right] + \left[\varepsilon\sigma\left\{\left(T|_{x=0}\right)^4 - \left(T_{sky}\right)^4\right\}\right] + \left[h_{ea}\left(T|_{x=0} - T_a\right)\right] \tag{4.16}$$

where $h_{ea} = 16.276 \times 10^{-3} \frac{h_{cw}\left(\overline{P_w} - \gamma\overline{P_a}\right)}{\overline{T_w} - \overline{T_a}}$ is the evaporative heat transfer coefficient, $\overline{T_w}$ and $\overline{T_a}$ are the average water temperature and the outside ambient temperature, respectively, $\overline{P_w} = \exp(25.317 - \frac{5144}{\overline{T_w}})$ and $\overline{P_a} = \exp(25.317 - \frac{5144}{\overline{T_a}})$ are the average partial pressures at the water temperature and the ambient temperature, respectively, γ is the relative humidity of the ambient air, h_{cw} is the convective heat transfer coefficient for the wetted surface and is usually considered as $2.7 + 3V$ W m^{-2} K^{-1} (V is the wind velocity in m s^{-1}).

Equation (4.16) can be rewritten as:

$$\alpha I(t) = \left[-K\frac{\partial T}{\partial x}\bigg|_{x=0}\right] + \left[h_{ca}\left(T|_{x=0} - T_a\right)\right] + \left[h_{ra}\left(T|_{x=0} - T_a\right) + \varepsilon\Delta R\right] + \left[h_{ea}\left(T|_{x=0} - T_a\right)\right]$$

or

$$\alpha I(t) = \left[-K\frac{\partial T}{\partial x}\bigg|_{x=0}\right] + \left[(h_{ca} + h_{ra} + h_{ea})\left(T|_{x=0} - T_a\right)\right] + \varepsilon\Delta R$$

or

$$\alpha I(t) = \left[-K\frac{\partial T}{\partial x}\bigg|_{x=0}\right] + \left[h_1\left(T|_{x=0} - T_a\right)\right] + \varepsilon\Delta R \tag{4.17}$$

where $h_1 = h_{ca} + h_{ra} + h_{ea}$ is the overall heat transfer coefficient of the interface between the wetted external surface and the ambient air.

Rearranging eqn (4.17), the rate of heat transfer through conduction at $x = 0$ is given by:

$$\dot{q} = -K\frac{\partial T}{\partial x}\bigg|_{x=0} = \alpha I(t) - h_1\left(T|_{x=0} - T_a\right) - \varepsilon\Delta R$$
$$= h_1\left[\frac{\alpha I(t)}{h_1} + T_a - \frac{\varepsilon\Delta R}{h_1} - T|_{x=0}\right]$$

or

$$\dot{q} = h_1\left(T_{sa} - T|_{x=0}\right) \tag{4.18}$$

where $T_{sa} = \frac{\alpha}{h_1} I(t) + T_a - \frac{\varepsilon\Delta R}{h_1}$ is the sol-air temperature for the wetted surface, equivalent to eqn (4.2).

4.2.2.1 *Rate of Heat Transfer from a Wetted Surface*

Considering Figure 4.2, in the steady-state condition the rate of heat transfer through conduction from the wetted external surface is given by eqn (4.17) as:

$$\dot{q} = h_1\left(T_{sa} - T|_{x=0}\right) \tag{4.19a}$$

In the steady-state condition the rate of heat transfer from $x = 0$ to $x = L$ is given by eqn (4.8) as:

$$\dot{q} = \frac{K}{A}\left(T|_{x=0} - T|_{x=L}\right) \tag{4.19b}$$

Further, in the steady-state condition the rate of heat transfer from the inner surface (surface exposed to the room) to the room space is through convection and radiation and is given by eqn (4.9) as:

$$\dot{q} = h_{si}\left(T|_{x=L} - T_b\right) \tag{4.19c}$$

On rearranging eqns (4.19a) to (4.19c) and adding we have:

$$\dot{q} = U'(T_{sa} - T_b) \tag{4.20}$$

where $U' = \left[\frac{1}{h_1} + \frac{L}{K} + \frac{1}{h_{si}}\right]^{-1}$ is the overall heat transfer coefficient.

The heat transfer per unit area for time interval $0 - t_T$ is given by:

$$(Q_u)_i = \int_0^{t_T} \dot{q}\,dt = U'\left[\frac{\alpha\{\overline{I(t)}\}_i}{h_1} + \{\overline{T_a}\}_i - \frac{\varepsilon\Delta R}{h_1} - T_b\right]t_T \tag{4.21}$$

Therefore the total heat transfer per unit area during the complete cycle of 24 hours is given by:

$$Q_T = Q_{ud} + Q_{un} = \sum_{i=1}^{N} U'\left[\frac{\alpha\{\overline{I(t)}\}_i}{h_1} + \{\overline{T_a}\}_i - \frac{\varepsilon\Delta R}{h_1} - T_b\right]t_T + \sum_{j=1}^{(24-N)} U'\left[\{\overline{T_a}\}_j - T_b\right]t_T \tag{4.22}$$

Example 4.2

Calculate the sol-air temperature of a wall whose external surface is wetted and exposed to solar radiation. The intensity of the solar radiation is $350\ \mathrm{W\,m^{-2}}$, the surface temperature of the wall is 20 °C, the emissivity is 0.9, the absorptivity is 0.4, the ambient temperature is 12 °C, the relative humidity is 0.6 and the wind flows with a velocity of $3\ \mathrm{m\,s^{-1}}$. Also, find the rate of heat loss due to convection, radiation and evaporation.

Solution

The convective heat transfer coefficient for air and water is:

$$h_{ca} = h_{cw} = 2.8 + 3V = 2.8 + (3.0 \times 3) = 11.8\ \mathrm{W\,m^{-2}\,K^{-1}}$$

The radiative heat transfer coefficient is:

$$h_{ra} = \left[\varepsilon\sigma\left\{\frac{(T|_{x=0})^4 - T_a^4}{T|_{x=0} - T_a}\right\}\right] = 0.9 \times 5.67 \times 10^{-3}\left[\frac{(20+273)^4 - (12+273)^4}{20-12}\right] = 4.93\ \mathrm{W\,m^{-2}\,K^{-1}}$$

The evaporative heat transfer coefficient is:

$$h_{\text{ea}} = 16.276 \times 10^{-3} \frac{h_{\text{cw}}(\overline{P_{\text{w}}} - \gamma \overline{P_{\text{a}}})}{\overline{T_{\text{w}}} - \overline{T_{\text{a}}}}$$

where

$$\overline{P_{\text{w}}} = \exp\left(25.317 - \frac{5144}{T_{\text{w}}}\right) = \exp\left(25.317 - \frac{5144}{20 + 273}\right) = 2446.5\,\text{N m}^{-2}$$

N m^{-2} and

$$\overline{P_{\text{a}}} = \exp\left(25.317 - \frac{5144}{T_{\text{a}}}\right) = \exp\left(25.317 - \frac{5144}{12 + 273}\right) = 1433.5\,\text{N m}^{-2}$$

Therefore

$$\begin{aligned} h_{\text{ea}} &= 16.276 \times 10^{-3} \frac{h_{\text{cw}}(\overline{P_{\text{w}}} - \gamma \overline{P_{\text{a}}})}{\overline{T_{\text{w}}} - \overline{T_{\text{a}}}} \\ &= 16.276 \times 10^{-3} \left\{ \frac{11.8(2446.5 - 0.6 \times 1433.5)}{20 - 12} \right\} = 38.08\,\text{W m}^{-2}\,\text{K}^{-1} \end{aligned}$$

Hence, the net heat transfer coefficient for the wetted surface is:

$$h_{\text{l}} = h_{\text{ca}} + h_{\text{ra}} + h_{\text{ea}} = 11.8 + 4.93 + 38.08 = 54.81\,\text{W m}^{-2}\,\text{K}^{-1}$$

Now $\Delta R = \sigma\ (T_{\text{a}}^4 - T_{\text{sky}}^4)$. If $T_{\text{sky}} = T_{\text{a}} - 6\,°\text{C}$, then $\Delta R = 5.67 \times 10^{-8} \times \{(12 + 273)^4 - (6 + 273)^4\} = 30.52\,\text{W m}^{-2}$

Thus the sol-air temperature for the wetted surface is:

$$\begin{aligned} T_{\text{sa}} &= \frac{\alpha}{h_{\text{l}}} I(t) + T_{\text{a}} - \frac{\varepsilon \Delta R}{h_{\text{l}}} \\ &= \frac{0.4}{54.81} \times 350 + 12 - \frac{0.9 \times 30.52}{54.81} = 14\,°\text{C} \end{aligned}$$

Now the rate of heat loss due to convection is:

$$\dot{q}_{\text{ca}} = h_{\text{ca}}(T_{\text{w}} - T_{\text{a}}) = 11.80(20 - 12) = 94.4\,\text{W m}^{-2}$$

The rate of heat loss due to radiation is:

$$\dot{q}_{\text{ra}} = h_{\text{ra}}(T_{\text{w}} - T_{\text{a}}) = 4.93(20 - 12) = 39.4\,\text{W m}^{-2}$$

The rate of heat loss due to evaporation is:

$$\dot{q}_{\text{ea}} = h_{\text{ea}}(T_{\text{w}} - T_{\text{a}}) = 38.08(20 - 12) = 304.6\,\text{W m}^{-2}$$

4.2.3 Sol-air Temperature for Blackened and Glazed Surfaces

Now let us consider that the external surface of the wall or roof is blackened and glazed as shown in Figure 4.3. In this case, the energy balance at $x=0$ is given by:

$$\alpha\tau I(t) = \left[-K\frac{\partial T}{\partial x}\bigg|_{x=0}\right] + \left[U_t\left(T|_{x=0} - T_a\right)\right] - \varepsilon\Delta R \tag{4.23}$$

where $U_t = \left[\frac{1}{h_o} + \frac{L_g}{K_g} + \frac{1}{h_b}\right]^{-1}$ is the overall heat transfer coefficient from the blackened surface to an ambient one through a glass cover, L_g is the glass cover thickness and K_g is the thermal conductivity of the glass cover; h_b is the heat transfer coefficient of the interface between the blackened surface and the glass cover, which may be considered as $2.8\ \mathrm{W\,m^{-2}\,K^{-1}}$.

As the surface is glazed, $\varepsilon\Delta R = 0$. Substituting the value and rearranging eqn (4.23), the rate of heat transfer through conduction at $x=0$ is given by:

$$\dot{q} = -K\frac{\partial T}{\partial x}\bigg|_{x=0} = U_t\left[\frac{\alpha\tau}{U_t}I(t) + T_a - T|_{x=0}\right]$$

or

$$\dot{q} = U_t\left(T_{sa} - T|_{x=0}\right) \tag{4.24}$$

where $T_{sa} = \frac{\alpha\tau}{h_1}I(t) + T_a$ is the sol-air temperature for the blackened and glazed surface.

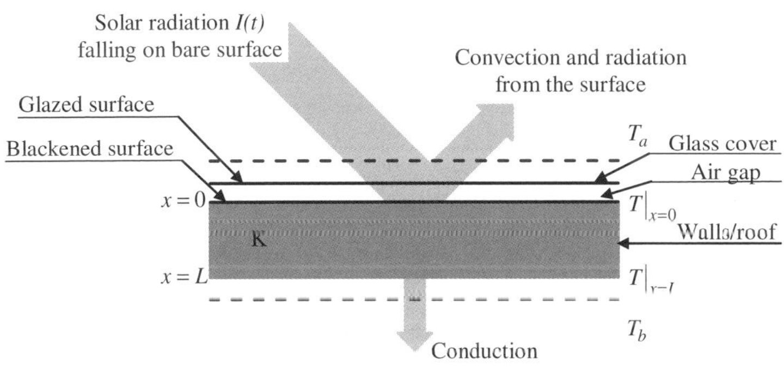

Figure 4.3 A glazed surface exposed to solar radiation.

4.2.3.1 Rate of Heat Transfer from a Blackened and Glazed Surface

Considering Figure 4.3, in the steady-state condition the rate of heat transfer through conduction from a blackened and glazed surface is given by eqn (4.24) as:

$$\dot{q} = U_{\mathrm{t}}\left(T_{\mathrm{sa}} - T|_{x=0}\right) \tag{4.25a}$$

In the steady-state condition the rate of heat transfer from $x=0$ to $x=L$ is given by eqn (4.8) as:

$$\dot{q} = \frac{K}{A}\left(T|_{x=0} - T|_{x=L}\right) \tag{4.25b}$$

Further, in the steady-state condition the rate of heat transfer from the inner surface (surface exposed to the room) to the room space through convection and radiation is given by eqn (4.9) as:

$$\dot{q} = h_{\mathrm{si}}\left(T|_{x=L} - T_{\mathrm{b}}\right) \tag{4.25c}$$

On rearranging eqns (4.25a) to (4.25c) and adding we have:

$$\dot{q} = U_L(T_{\mathrm{sa}} - T_{\mathrm{b}}) \tag{4.26}$$

where $U_L = \left[\frac{1}{U_{\mathrm{t}}} + \frac{L}{K} + \frac{1}{h_{\mathrm{si}}}\right]^{-1}$ is the overall heat transfer coefficient for the walls or roof having an external surface blackened and glazed.

The heat transfer per unit area for the time interval $0 - t_{\mathrm{T}}$ is given by:

$$\begin{aligned}(Q_{\mathrm{u}})_i &= \int_0^{t_{\mathrm{T}}} \dot{q}\,\mathrm{d}t \\ &= U_L\left[\frac{\alpha\tau\left\{\overline{I(t)}\right\}_i}{U_{\mathrm{t}}} + \left\{\overline{T_{\mathrm{a}}}\right\}_i - T_{\mathrm{b}}\right] t_{\mathrm{T}}\end{aligned} \tag{4.27}$$

Therefore the total heat transfer per unit area during the complete cycle of 24 hours is given by:

$$\begin{aligned}Q_{\mathrm{T}} &= Q_{\mathrm{ud}} + Q_{\mathrm{un}} \\ &= \sum_{i=1}^{N} U_L\left[\frac{\alpha\tau\left\{\overline{I(t)}\right\}_i}{U_{\mathrm{t}}} + \left\{\overline{T_{\mathrm{a}}}\right\}_i - T_{\mathrm{b}}\right] t_{\mathrm{T}} + \sum_{j=1}^{(24-N)} U_L\left[\left\{\overline{T_{\mathrm{a}}}\right\}_j - T_{\mathrm{b}}\right] t_{\mathrm{T}}\end{aligned} \tag{4.28}$$

4.3 Passive Solar Heating Systems

Passive solar heating systems are particularly well suited to applications which experience both low winter air temperatures and clear skies. Under these conditions, a passive solar heating system collects and stores the solar radiation during the sunshine hours for use at night time when heating is most required. Hence, all passive solar heating designs must have the equator facing the exposed glazing material (glass or plastic) to allow solar energy to enter and a material to absorb and store the heat for later use. With these two elements, basic approaches to design a passive solar heated system are direct thermal gain, indirect thermal gain and isolated thermal gain. In choosing a particular design approach, the site and climate conditions must be evaluated carefully so that the best approach or combination of approaches is incorporated. No one passive design approach is most advantageous in all climates or on all sites and situations.

Glazing transmits short-wavelength radiation (in the range of 0.38–2.5 μm) emitted by the Sun and blocks long-wavelength radiation (exceeding 2.5 μm) emitted by hot surfaces. This phenomenon is known as the greenhouse effect. Owing to the greenhouse effect, the temperature within a room space rises until the heat losses by conduction and convection equal the heat gain by radiation. In order to catch more solar radiation, all the passive heating systems should have a large glazed area on the side of the building facing the equator. To increase the glazed area, passive building designers usually raise the height of the glazed surface and incline the roof towards north, as shown in Figure 4.4.

As well as promoting the greenhouse effect within buildings, glazing also plays an important role in the self-regulation of solar heat gains. Glass transmits much more solar radiation when the angle of incidence is small compared to when it is large, as shown in Figure 4.5. When the angle of incidence is large, much of the incident solar radiation is reflected. In addition, because the Sun's angle of altitude is much higher in summer, vertical windows present a much smaller apparent area to solar radiation in summer compared with winter

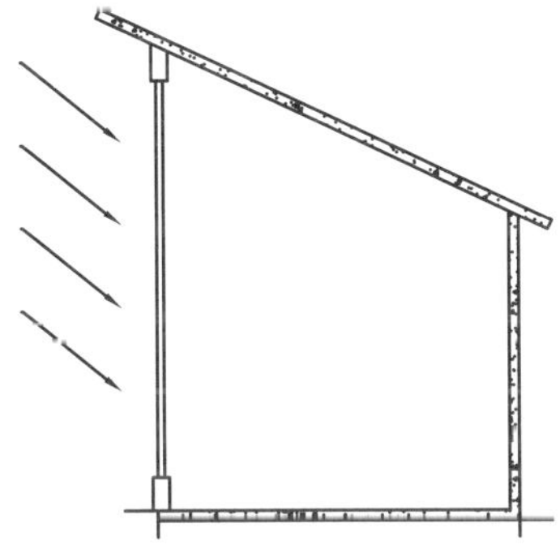

Figure 4.4 The roof is inclined towards the north to increase the glazed surface area on the south wall.

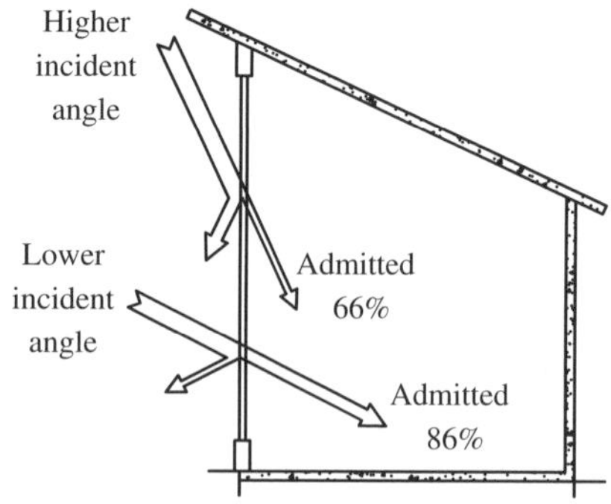

Figure 4.5 More transmission of solar radiation when the incidence angle is low.

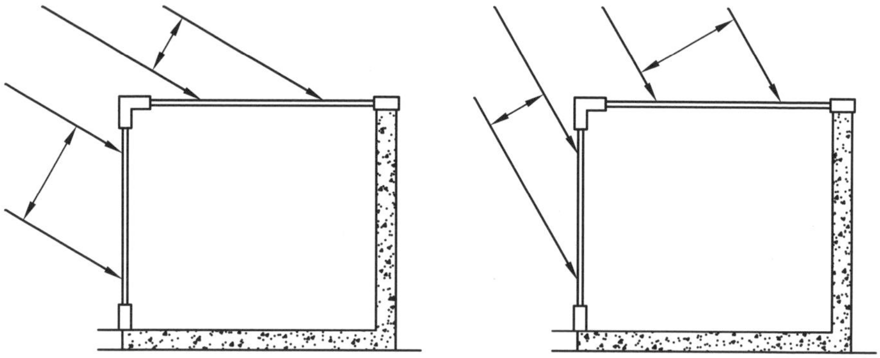

Figure 4.6 Solar radiation falling on walls and roofs of a building: *left*, during winter, more radiation falls on the wall than the roof; *right*, during summer, less radiation falls on the wall than the roof.

(Figure 4.6). As a result, the solar intensity on vertical glazing is often much lower in summer than in winter. This quality can be used to great effect by building designers.

A window helps to collect two to three times the solar energy during sunshine hours as it loses during off-sunshine hours. Anything that can be done to reduce heat losses or increase the solar gain will markedly improve the overall energy performance of the building. Useful strategies include increasing the amount of glazing, surface films, convection suppression, movable insulation and external reflectors. Each pane of normal glass loses about 14% of solar energy. Using low-iron glass the loss may be only 9% per pane, whereas with some plastic glazing materials the loss may be as low as 5% per pane. Inserting a plastic trap in the form of a honeycomb or simple horizontal layers of plastic between two layers of glazing may reduce both convective and radiative heat loses. The solar-control films can screen out as much as three-quarters of the incident solar radiation and are not recommended for any windows designed to collect solar energy in winter.

4.4 Direct Thermal Gain Systems

A direct thermal gain system is the simplest passive solar heating system, used generally in cold climatic conditions. In this system, the building is directly heated by sunlight admitted into the building through openings or glazed windows facing the equator, as shown in Figure 4.7 (*left*). The solar radiation entering through the glazing is stored in the form of thermal energy by the components of the building, such as floors, walls and/or furnishings, which then transmit the heat into the room space. At night, when the outside temperatures drop and the interior space cools, the heat flow into the storage masses is reversed, as shown in Figure 4.7 (*right*), and the heat is given up to the interior space in order to reach equilibrium. This re-radiation of collected daytime heat can maintain a comfortable temperature during cold nights and can extend through several cloudy days without recharging. Thus, the room space acts as a live-in solar collector, heat storage and distribution system.

The floor and walls must be made of a massive construction so that more heat is stored during sunshine hours. The room should have large areas of glazing on the side facing the equator. Translucent glazing, as shown in Figure 4.8 (*left*), may be used to scatter the solar radiation, thereby using all surfaces for thermal storage. Reflectors equivalent to the glazing size, shown in Figure 4.8 (*centre*),

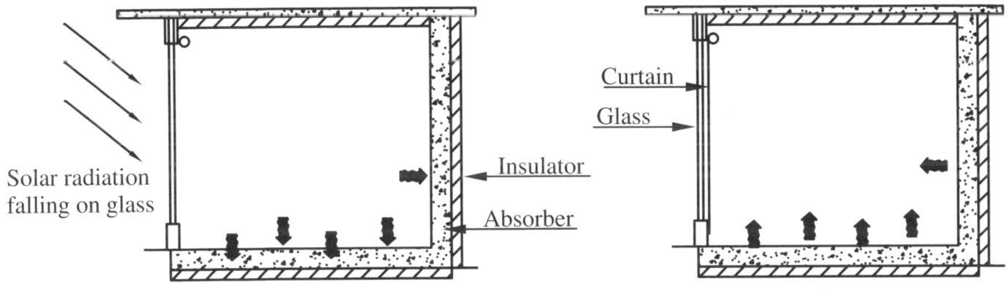

Figure 4.7 Direct thermal gain through glazing: *left*, solar radiation is stored by absorbers as heat during sunshine hours; *right*, heat is emitted by the absorbers during off sunshine hours.

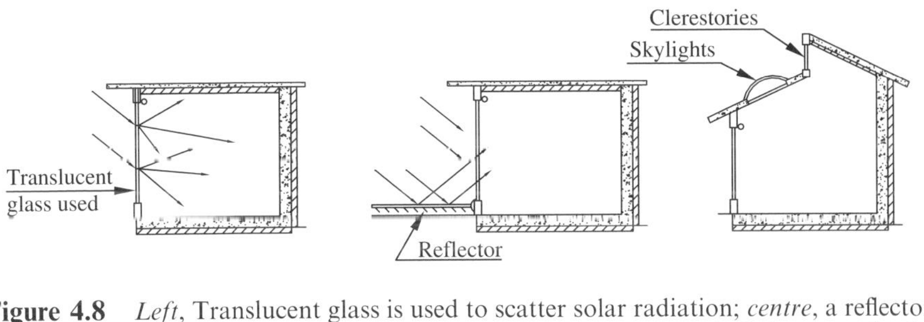

Figure 4.8 *Left*, Translucent glass is used to scatter solar radiation; *centre*, a reflector is used to increase direct gain; *right*, clerestories and skylights may also be used for direct gain.

may be provided outside the windows to increase the direct gain, but they may enhance a glare problem. Clerestories and skylights, shown in Figure 4.8 (*right*), are also adequate to increase the direct gain. Use of dark colours for internal surfaces helps in better absorption of solar radiation, whereas light colours help in better distribution of daylight. The designer must coordinate material selection with the tenant finish designer to minimize glare problems.

To reduce the loss of trapped thermal energy, especially at places that experience severe winters, double and sometimes triple glazing is preferred. Moreover, the windows are covered with insulation during the off-sunshine hours. If movable insulation is unacceptable for a particular application, infrared heat mirrors may be used. Overhangs, awnings and louvres may be provided to avoid overheating in the summer. Openable windows and vents can be employed for ventilation and for cooling when the temperature rises beyond the comfort level. Thus a direct gain system has the following components:

1. Glazing – to transmit and trap the incoming solar radiation.
2. Thermal mass – to store heat for night-time use.
3. Insulation – to reduce losses at night.
4. Ventilation – for summertime cooling.
5. Shading – to reduce overheating in summer.

Advantages:

1. They provide natural day lighting.
2. They permit outdoor views to the south.
3. They are simplest in design and inexpensive.

Disadvantages:

1. There is a possibility of a glare problem.
2. The colour of the furnishings or objects exposed to direct sunlight may fade.
3. The fibre materials may deteriorate in physical strength by extended exposure to ultraviolet radiation.

4.4.1 Sol-air Temperature and Heat Transfer for Single-glazed Windows

Consider a room has single glazing on the side facing the equator, as shown in Figure 4.7 (*left*). The rate of useful energy gain is given by:

$$\dot{q} = \tau I(t) - U_{\mathrm{t}}(T_{\mathrm{b}} - T_{\mathrm{a}}) \tag{4.29}$$

where τ is the transmissivity, $U_{\mathrm{t}} = \left[\frac{1}{h_{\mathrm{o}}} + \frac{L_{\mathrm{g}}}{K_{\mathrm{g}}} + \frac{1}{h_{\mathrm{si}}}\right]^{-1}$ is the overall heat transfer coefficient between the room and the atmosphere through the glazed window,

T_b is the temperature of the building/living space and T_a is the ambient air temperature.

From eqn (4.29) we have:

$$\dot{q} = U_t\left[\frac{\tau}{U_t}I(t) + T_a - T_b\right] = U_t[T_{sa} - T_b] \tag{4.30}$$

where $T_{sa} = \frac{\tau}{U_t}I(t) + T_a$, which is similar to eqn (4.2).

The total thermal gain per unit area during sunshine hours is given by:

$$Q_T = \sum_{i=1}^{N}\int_0^{t_T}\dot{q}\,dt = \sum_{i=1}^{N}U_t\left[\frac{\tau\{\overline{I(t)}\}_i}{U_t} + \{\overline{T_a}\}_i - T_b\right]t_T \tag{4.31}$$

where $\overline{I(t)} = \frac{1}{t_T}\int_0^{t_T} I(t)\,dt$ and $\overline{T_a} = \frac{1}{t_T}\int_0^{t_T} T_a\,dt$.

4.4.2 Sol-air Temperature and Heat Transfer for Double-glazed Windows

Consider a room has double glazing on the side facing the equator. The energy balance equation is given by:

$$\dot{q} = \tau^2 I(t) - U_t(T_b - T_a) \tag{4.32}$$

where τ is the transmissivity, $U_t = \left[\frac{1}{h_o} + \frac{L_g}{K_g} + \frac{L_a}{K_a} + \frac{L_g}{K_g} + \frac{1}{h_{si}}\right]^{-1}$ is the overall heat transfer coefficient between the room and the atmosphere through the glazed window, L_a is the distance between the glazed windows, K_a is the thermal conductivity of air, T_b is the temperature inside the building or living space and T_a is the ambient air temperature.

From eqn (4.32) we have:

$$\dot{q} = U_t\left[\frac{\tau^2}{U_t}I(t) + T_a - T_b\right] = U_t[T_{sa} - T_b] \tag{4.33}$$

where $T_{sa} = \frac{\tau^2}{U_t}I(t) + T_a$.

The total thermal gain per unit area during sunshine hours is given by:

$$Q_1 = \sum_{i=1}^{N}\int_0^{t_T}\dot{q}\,dt = \sum_{i=1}^{N}U_t\left[\frac{\tau^2\left\{\overline{I(t)}\right\}_i}{U_t} + \{\overline{T_a}\}_i - T_b\right]t_T \tag{4.34}$$

4.4.3 Sol-air Temperature and Heat Transfer for Single-glazed Windows with Reflectors

Consider a room reflector of size equal to the glazing, as shown in Figure 4.8 (*centre*). Not all the radiation reflected from the reflector will be received by the glazed area. If A_r is the ratio of the useful area of the reflector to the total glazed area, then the rate of useful energy gain is given by:

$$\dot{q} = \tau I(t) + A_r \rho_r \tau I(t) - U_t(T_b - T_a) \tag{4.35}$$

where τ is the transmissivity, ρ_r is the reflectivity of the reflector, $U_t = \left[\frac{1}{h_o} + \frac{L_g}{K_g} + \frac{1}{h_{si}}\right]^{-1}$ is the overall heat transfer coefficient between the room and the atmosphere through the glazed window, T_b is the reference temperature inside the building and T_a is the ambient air temperature.

From eqn (4.35) we have:

$$\dot{q} = U_t\left[\frac{\tau(1 + A_r\rho_r)}{U_t} I(t) + T_a - T_b\right] = U_t[T_{sa} - T_b] \tag{4.36}$$

where $T_{sa} = \frac{\tau(1+A_r\rho_r)}{U_t} I(t) + T_a$.

The total thermal gain per unit area during sunshine hours is given by:

$$Q_T = \sum_{i=1}^{N} \int_0^{t_T} \dot{q}\,dt = \sum_{i=1}^{N} U_t\left[\frac{\tau(1 + A_r\rho_r)}{U_t}\left\{\overline{I(t)}\right\}_i + \left\{\overline{T_a}\right\}_i - T_b\right] t_T \tag{4.37}$$

Example 4.3

Calculate the direct thermal energy gain (in kWh) through a glass window of surface area 3 m × 2 m irradiated with solar radiation of intensity 800 W m^{-2} for 4 h. The room space temperature is 20 °C, the ambient air temperature is 12 °C, the transmissivity of the glass is 0.9 and the overall heat transfer coefficient between the space and the ambient air is 1.864 W m^{-2} K^{-1}.

Solution

The total thermal gain during the sunshine hours is given by:

$$Q_T = \sum_{i=1}^{N} U_t\left[\frac{\tau\{\overline{I(t)}\}_i}{U_t} + \{\overline{T_a}\}_i - T_b\right] t_T \times \text{Area}$$

$$= 1.864 \times \left[\frac{0.9 \times 800}{1.864} + 12 - 20\right] \times 4 \times (3 \times 2)$$

$$= 16\,922\ \text{W or } 16.922\ \text{k Wh}$$

Example 4.4

Calculate the direct thermal energy gain through a double-glazed window of surface area 3 m × 2 m, irradiated with solar radiation of intensity 800 W m^{-2} for 4 h. The thickness of each glazing is 3 mm and they are 50 mm apart. The room space temperature is 20 °C, the ambient air temperature is 12 °C, the transmissivity of each glass is 0.9, the thermal conductivity of glass is 0.78 W m^{-1} K^{-1} and the thermal conductivity of air is 0.024 W m^{-1} K^{-1}.

Solution

Assuming the wind flow velocity as zero, the heat transfer coefficient on the external surface of the glazing $h_o = 5.7$ W m^{-2} K^{-1} and the heat transfer coefficient on the glazing surface inside the room $h_{si} = 2.8$ W m^{-2} K^{-1}. The overall heat transfer coefficient between the room and the atmosphere through the glazed window is given by:

$$U_t = \left[\frac{1}{h_o} + \frac{L_g}{K_g} + \frac{L_a}{K_a} + \frac{L_g}{K_g} + \frac{1}{h_{si}}\right]^{-1}$$

$$= \left[\frac{1}{5.7} + \frac{0.003}{0.78} + \frac{0.05}{0.024} + \frac{0.003}{0.78} + \frac{1}{2.8}\right]^{-1}$$

$$= 0.38 \text{ W m}^{-2}\text{ K}^{-1}$$

The total thermal gain during sunshine hours is then given by:

$$Q_T = \sum_{i=1}^{N} U_t \left[\frac{\tau^2\{\overline{I(t)}\}_i}{U_t} + \{\overline{T_a}\}_i - T_b\right] t_T$$

$$= 0.38 \times \left[\frac{0.9^2 \times 800}{0.38} + 12 - 20\right] \times 4 \times (3 \times 2)$$

$$= 15\,479 \text{ W or } 15.479 \text{ kWh}$$

4.5 Indirect Thermal Gain Systems

In an indirect thermal gain system, thermal storage materials are placed between the solar radiation and the living space so that there is no direct heating. Both the walls and the roof can be used as thermal storage materials, but a south-facing wall is preferred to the roof because of the abundance of solar radiation on the south wall in winter and the difficulty in supporting a heavy roof. A thermal storage wall may be either masonry or water. In this system, the solar radiation first strikes the outer surface of the thermal storage mass and becomes absorbed in the form of heat, which is then transferred to the living space by conduction.

The approaches used for indirect gain systems include the use of a Trombe wall, a water wall, a photovoltaic (PV) Trombe wall and solar roof ponds.

Advantages:

1. The system helps in maintaining a more uniform room-air temperature than the direct thermal gain system.
2. Owing to the time lag, the storage material emits heat to the living space at night when it is needed most.
3. The problems of glare and ultraviolet degradation of materials are eliminated.

Disadvantages:

1. Requires a larger collector area than a direct gain system.
2. The system is more expensive, owing to the cost of materials and the structural modifications required.
3. Thermal storage materials block the view and the daylight entering from the south.

4.5.1 Trombe Walls

A Trombe wall is a sun-facing wall patented in 1881 by its inventor, Edward Morse, and popularized in 1964 by the French engineer Félix Trombe and architect Jacques Michel. Figure 4.9 shows a Trombe wall, which is basically a thermal storage wall made of materials having high heat storage capacity such as concrete, adobe, stone or brick. To increase the absorption the external surface of the wall is blackened and glazed with an air gap in between. The solar radiation falling on the outer surface of the Trombe wall is absorbed as sensible heat and transferred to the interior of the storage mass by conduction or convection. During off-sunshine hours, heat stored in the thermal mass wall is then radiated and convected into the living space to be heated. Insulation may be provided between the glass and the Trombe wall to reduce potentially large

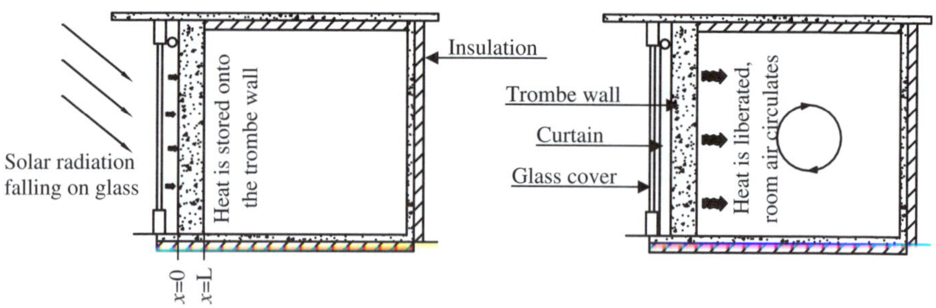

Figure 4.9 A Trombe wall placed behind glazing: *left*, stores heat during sunshine hours; *right*, emits heat during off-sunshine hours.

heat losses. If properly designed, the system can provide adequate heat to the living space throughout the night.

For each 10 cm of concrete,[1] there is a lag of about 2–2.5 hours between peak solar absorption and heat delivery inside. With a wall thickness below about 30 cm the temperature swing of the interior would be excessive. Increasing the thickness above about 40 cm would result in higher costs while having only a small affect on the indoor swing. Considering this aspect and also the time of peak heating, the optimal thickness for concrete in residential buildings is approximately 30–40 cm.

4.5.1.1 Sol-air Temperature for a Trombe Wall with Glazing at the Outer Surface

Consider a Trombe wall, as shown in Figure 4.9 (*left*), with a thickness L and a thermal conductivity K. Let the wall lie behind a glass cover of thickness L_g having a thermal conductivity K_g, absoptivity α and transmissivity τ. The wall is exposed to solar radiation of intensity $I(t)$ and the ambient air temperature is T_a. The useful thermal energy available for conduction on the outer surface (at $x=0$) is:

$$\dot{q} = \alpha\tau I(t) - U_t\left(T|_{x=0} - T_a\right) \tag{4.38}$$

where $U_t = \left[\frac{1}{h_o} + \frac{L_g}{K_g} + \frac{1}{h_b}\right]^{-1}$ and is the overall heat transfer coefficient from the blackened surface to the ambient air through a glass cover, h_o is the heat transfer coefficient at the interface of the ambient air and the glass cover, h_b is the heat transfer coefficient at the interface of the glass cover and outer surface (blackened) of the Trombe wall, which may be considered as 2.8 W m^{-2} K^{-1}, and $T|_{x=0}$ is the reference temperature at the outer surface ($x=0$) of the wall.

Rearranging eqn (4.38), we have:

$$\dot{q} = U_t\left[\frac{\alpha\tau}{U_t} I(t) + T_a - T|_{x=0}\right] = U_t\left[T_{sa} - T|_{x=0}\right] \tag{4.39a}$$

where $T_{sa} = \frac{\alpha\tau}{U_t} I(t) + T_a$ is the sol-air temperature.

The rate of heat transfer from $x=0$ to $x=L$ is given by

$$\dot{q} = \frac{K}{L}\left(T|_{x=0} - T|_{x=L}\right) \tag{4.39b}$$

If h_{si} is the inner surface heat transfer coefficient and T_b is the temperature maintained inside the building, then heat transfer from the inner surface to the room space is given by:

$$\dot{q} = h_{si}\left(T|_{x=L} - T_b\right) \tag{4.39c}$$

On rearranging and adding eqns (4.39a) to (4.39c), the net heat received through conduction is:

$$\dot{q} = U_L(T_{sa} - T_b) \tag{4.40}$$

where $U_L = [\frac{1}{U_t} + \frac{L}{K} + \frac{1}{h_{si}}]^{-1}$ is the overall heat transfer coefficient for the Trombe wall.

4.5.2 Water Walls

Water walls work on the same principle as that of a Trombe wall, except that they employ water as the thermal storage material. In this system a thermal storage wall is made up of drums of water stacked behind glazing, as shown in Figure 4.10. As the water is heated the convection process quickly distributes the heat throughout the mass and the interior space is warmed by heat radiated from the wall.

The outer surface of the wall is painted black to increase the absorption of solar radiation, whereas the inner surface of the wall can be painted with any colour or left unpainted.[2,3] Sometimes a thin concrete wall or insulating layer is also placed adjacent to the inner surface of the wall. Owing to a higher specific heat for water than concrete, a water wall stores more heat than a Trombe wall. As a water wall is a convective thermal storage unit, heat is transmitted to the living space rapidly compared with a Trombe wall. A translucent water wall may provide diffuse lighting.

4.5.2.1 Sol-air Temperature for a Water Wall with Glazing on its Outer Surface

Consider a water wall, as shown in Figure 4.10 (*left*), with thickness L and thermal conductivity K. Let the wall lie behind a glass cover of thickness L_g having thermal conductivity K_g, absoptivity α and transmissivity τ. The wall is exposed to solar radiation of intensity $I(t)$ and the ambient air temperature is T_a. The useful thermal energy available for conduction on the outer surface (at

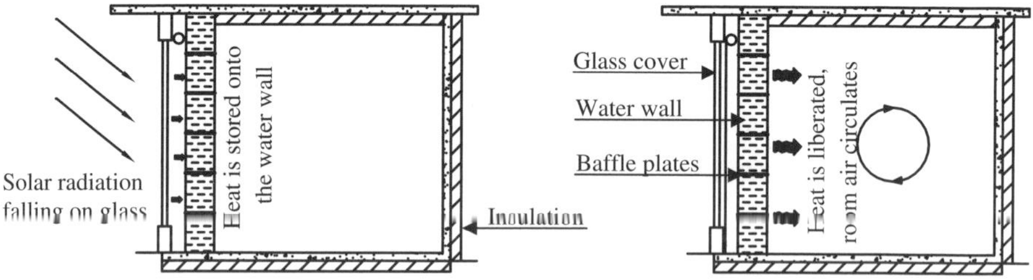

Figure 4.10 A water wall placed behind glazing: *left*, stores heat during sunshine hours; *right*, emits heat during off-sunshine hours.

$x=0$) is as follows. If $T|_{x=0}$ is the temperature of the metallic surface of the drums, the energy balance equations for the steady state can be written as:

$$\dot{q} = \alpha\tau I(t) - U_t(T|_{x=0} - T_a) \tag{4.41a}$$

$$\dot{q} = h'_1(T|_{x=0} - T_w) \tag{4.41b}$$

$$\dot{q} = h'_2(T_w - T_{si}) \tag{4.41c}$$

$$\dot{q} = h_{si}(T_{si} - T_b) \tag{4.41d}$$

where h'_1 and h'_2 are convective heat transfer coefficients from the outer surface of the water wall to water and from water to the inner surface of the water wall, respectively. On rearranging and adding eqns (4.41a) to (4.41d), the heat transfer through conduction is given by:

$$\dot{q} = U_L(T_{sa} - T_b) \tag{4.42}$$

where $U_L = \left[\frac{1}{U_t} + \frac{1}{h'_1} + \frac{1}{h'_2} + \frac{1}{h_{si}}\right]^{-1}$ is the overall heat transfer coefficient and $T_{sa} = \frac{\alpha\tau}{U_t} I(t) + T_a$ is the sol-air temperature.

4.5.3 Vented Trombe Walls

In the case of a Trombe wall, part of the heat generated in the air space between the glazing and the storage wall is lost back to the atmosphere through the glass. The hotter the air in the airspace, the greater is the heat loss. This heat loss can be reduced by venting the storage wall at the top and bottom, as shown in Figure 4.11 (*left*). The air in the space between the glazing and the wall becomes warm and enters the living room through the upper vents. Cool room air takes its place through the lower vents, thus establishing a natural convective loop (thermocirculation). At night the vents can be closed to keep cold air out and the interior space is then heated by the storage mass, which gives up

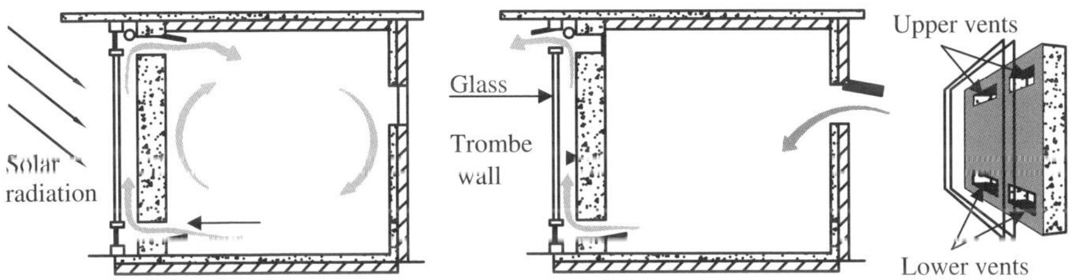

Figure 4.11 A vented Trombe wall: *left*, the heating space during sunshine hours in winter; *centre*, ventilation during summer; *right*, with double glazing (a three-dimensional view).

its heat by radiation. Such systems are suitable for buildings having daytime use, such as offices and shops.

The vented Trombe wall can provide induced ventilation for summer cooling of the space, as shown in Figure 4.11 (*centre*). Here, the heated air in the collector space flows out through exhaust vents at the top of the outer glazing, and air from outside enters the space through openings on the cooler side to replace the hot air. This continuous air movement cools the living space. Figure 4.11 (*right*) shows a perspective view of the arrangement of the system.

The overall efficiency of the vented Trombe wall is about 10% higher than the unvented Trombe wall.[4] If the vents are not closed effectively at night, the reverse air flow may lower the efficiency of vented walls below the level of unvented ones. Problems associated with vents are dust accumulation on the inner surface of the glazing and on the dark absorbing surface. Since it is impossible to remove the dust, it may eventually reduce the performance in addition to the aesthetic impact.

4.5.4 Phase-change Material Walls

Phase-change materials (PCMs) are those that possess thermal energy transfer by a change of state, from solid to liquid or liquid to solid. PCMs have melting temperatures between 20 and 32 °C and are useful for thermal storage in conjunction with both passive storage and active solar storage for heating and cooling in buildings.[5]

Traditionally, Trombe walls rely on sensible heat storage, but because of the potential for greater heat storage per unit mass, the PCM Trombe wall is an attractive concept still awaiting successful implementation. A wall filled with PCM is constructed on the south-side window of a house similar to a Trombe wall, vented or non-vented depending upon the passive heat required. The wall is heated during the day by incoming solar radiation, melting the PCM. At night the heat is withdrawn to warm the house. For a given amount of heat storage, the phase change units require less space than water walls or mass Trombe walls and are much lighter in weight. These are, therefore, more convenient to make use of in retrofit applications of buildings.

A large number of PCMs are known to melt with a heat of fusion in the required range. Salt hydrates and hydrocarbons are generally used as PCMs in Trombe walls. A calcium chloride hexahydrate (melting point 29 °C) wall with 8.1 cm thickness has thermal performance slightly better than a masonry wall with 40 cm thickness. Stearic acid has better transmittance than glass for the same thickness and can be used in windows or walls as a transparent insulating material. However, for their employment as latent heat storage materials they must exhibit certain desirable thermodynamic, kinetic and chemical properties. Moreover, economic considerations and easy availability of these materials have to be kept in mind.

Example 4.5

Calculate the rate of heat flow through a south-facing concrete wall of thickness 300 mm, thermal conductivity 0.72 W m^{-1} K^{-1} and absorptivity 0.6, when exposed with mean incident solar radiation of 250 W m^{-2}. The outside convective and radiative heat transfer coefficients are 8.7 W m^{-2} K^{-1} and 3.8 W m^{-2} K^{-1}, respectively, whereas the inside room heat transfer coefficient is 8 W m^{-2} K^{-1}. The ambient temperature is 13 °C and the mean room temperature is 20 °C. Also, calculate the rate of heat flux if the wall is covered during off-sunshine hours with 40 mm thick insulation of thermal conductivity 0.025 W m^{-1} K^{-1}.

Solution

Here the Trombe wall is without glazing. Given that $L = 0.3$ m, $K = 0.72$ W m^{-1} K^{-1}, $\alpha = 0.6$, $I(t) = 250$ W m^{-2}, $h_1 = h_c + h_r = 8.7 + 3.8 = 12.5$ W m^{-2} K^{-1}, $h_{si} = 8$ W m^{-2} K^{-1}, $T_a = 13$ °C, $T_b = 20$ °C, $L_i = 0.04$ m and $K_i = 0.025$ W m^{-2} K^{-1}, then during sunshine hours the overall heat transfer coefficient for the Trombe wall is given by:

$$U = \left[\frac{1}{h_1} + \frac{L}{K} + \frac{1}{h_{si}}\right]^{-1} = \left[\frac{1}{8.7 + 3.8} + \frac{0.30}{0.72} + \frac{1}{8}\right]^{-1} = 1.609 \text{ W m}^{-2} \text{ K}^{-1}$$

The sol-air temperature is:

$$T_{sa} = \frac{\alpha}{h_1} I(t) + T_a - \frac{\varepsilon \Delta R}{h_1}$$

As the Trombe wall is vertical, and assuming $\varepsilon \Delta R = 0$, then:

$$T_{sa} = \frac{0.6}{12.5} \times 250 + 13 = 25\,°\text{C}$$

The rate of heat flow is given by:

$$\dot{q} = U(T_{sa} - T_b) = 1.609 \times (25 - 20) = +8.045 \text{ W m}^{-2}$$

The positive sign indicates that the heat is gained by the room and its temperature will rise.

If the ambient air temperature is 5 °C, then the sol-air temperature during sunshine hours is:

$$T_{sa} = \frac{\alpha}{h_1} I(t) + T_a = \frac{0.6}{12.5} \times 250 + 5 = 17\,°\text{C}$$

Thus the rate of heat flow is:

$$\dot{q} = U(T_{sa} - T_b) = 1.609 \times (17 - 20) = -4.827 \text{ W m}^{-2}$$

The negative sign indicates that the heat is lost from the room to the atmosphere.

During off-sunshine hours when the wall is covered with insulation, the overall heat transfer coefficient is given by:

$$U = \left[\frac{1}{h_1} + \frac{L}{K} + \frac{L_i}{K_i} + \frac{1}{h_{si}}\right]^{-1}$$
$$= \left[\frac{1}{12.5} + \frac{0.04}{0.025} + \frac{0.3}{0.72} + \frac{1}{8}\right]^{-1} = 0.45\,\mathrm{W\,m^{-2}\,K^{-1}}$$

The rate of heat transfer is given by:

$$\dot{q} = U(T_a - T_b) = 0.45 \times (13 - 20) = -3.15\,\mathrm{W\,m^{-2}}$$

The negative sign indicates that there is a loss of heat from the room to the atmosphere.

Example 4.6

Consider a south-facing Trombe wall with thickness 305 mm, thermal conductivity 0.62 W m^{-1} K^{-1} and absorbtivity 0.8, placed behind a glass sheet with transmissivity 0.71. The heat transfer coefficients for the outer surface (through the glass) and the inner surface of the wall are 5 W m^{-2} K^{-1} and 8 W m^{-2} K^{-1}, respectively. The room is maintained at 18 °C. During sunshine hours the mean intensity of the solar radiation is 310 W m^{-2} and the average ambient temperature is 14 °C; during off-sunshine hours the average ambient temperature is 8 °C. Calculate the net heat flux for the Trombe wall when the Sun shines for (a) 8 hours (b) 4 hours.

Solution

Given that $L = 0.305$ m, $K = 0.62$ W m^{-1} K^{-1}, $\alpha = 0.8$, $\tau = 0.71$, $U_t = 5$ W m^{-2} K^{-1}, $h_{si} = 8$ W m^{-2} K^{-1}, $T_b = 18$ °C, $I(t) = 310$ W m^{-2}, $T_a = 14$ °C and $T_a' =$ 8 °C, then the overall heat transfer coefficient for the Trombe wall is given by:

$$U_L = \left[\frac{1}{U_t} + \frac{L}{K} + \frac{1}{h_{si}}\right]^{-1} = \left[\frac{1}{5} + \frac{0.305}{0.62} + \frac{1}{8}\right]^{-1} = 1.224\,\mathrm{W\,m^{-2}\,K^{-1}}$$

The mean heat flux for the Trombe wall during sunshine hours is:

$$\dot{q} = U_L(T_{sa} - T_b) = U_L\left[\frac{\alpha\tau}{U_t} I(t) + T_a - T_b\right]$$
$$= 1.224 \times \left[\frac{0.8 \times 0.71}{5} \times 310 + 14 - 18\right] = 38.21\,\mathrm{W\,m^{-2}}$$

The positive sign indicates that there is a heat gain during sunshine hours.

The mean heat flux for the Trombe wall during off-sunshine hours is:

$$\dot{q}' = U_L(T'_a - T_b) = 1.224 \times [8 - 18] = -12.24\,\mathrm{W\,m^{-2}}$$

The negative sign indicates that there is a heat loss during off-sunshine hours.

(a) The net heat flux when the Sun shines for 8 h is given by:

$$\dot{q}_T = \dot{q} \times \text{sunshine hours} + \dot{q} \times (24 - \text{sunshine hours})$$
$$= 38.21 \times 8 + (-12.24) \times (24 - 8) = 109.84\,\mathrm{Wh\,m^{-2}}$$

The positive sign indicates that solar energy is sufficient to maintain the room temperature for the whole day. No external means of heat source is required.

(b) The net heat flux when the Sun shines for 4 h is given by:

$$\dot{q}_T = \dot{q} \times \text{sunshine hours} + \dot{q} \times (24 - \text{sunshine hours})$$
$$= 38.21 \times 4 + (-12.24) \times (24 - 4) = -91.96\,\mathrm{Wh\,m^{-2}}$$

The negative sign indicates that solar energy is not sufficient to maintain the room temperature for the whole day. An external means of heat source is required for 91.96/12.24 = 7.5 h.

Example 4.7

Consider a south-facing water wall with the following parameters: absorbtivity 0.8, transmissivity 0.71, mean intensity of solar radiation 310 W m^{-2}, room temperature 18 °C, heat transfer coefficients for the outer surface (through glass) and the inner surface of the wall are 5 and 8 W m^{-2} K^{-1}, respectively, and the heat transfer coefficients from the outer surface of the water wall to water and from water to the inner surface of the water wall are 206 W m^{-2} K^{-1}; during sunshine hours the average ambient temperature is 14 °C, whereas during off-sunshine hours the average ambient temperature is 12 °C. Calculate: (a) the net heat flux into the room considering equal duration of day and night and (b) the rate of night losses if the glazing is covered with a 30 mm thick insulation of thermal conductivity 0.025 W m^{-1} K^{-1}.

Solution

Given that $\alpha = 0.8$, $\tau = 0.71$, $U_t = 5\,\mathrm{W\,m^{-2}\,K^{-1}}$, $h_{si} = 8\,\mathrm{W\,m^{-2}\,K^{-1}}$, $h'_1 = h'_2 = 206\,\mathrm{W\,m^{-2}\,K^{-1}}$, $T_b = 18\,°\mathrm{C}$, $I(t) = 310\,\mathrm{W\,m^{-2}}$, $T_a = 14\,°\mathrm{C}$, $T'_a = 12\,°\mathrm{C}$, $L_i = 0.03$ mm and $K_i = 0.025\,\mathrm{W\,m^{-1}\,K^{-1}}$, then:

(a) The overall heat transfer coefficient for the water wall is given by:

$$U_L = \left[\frac{1}{U_t} + \frac{1}{h'_1} + \frac{1}{h'_2} + \frac{1}{h_{si}}\right]^{-1} = \left[\frac{1}{5} + \frac{1}{206} + \frac{1}{206} + \frac{1}{8}\right]^{-1} = 2.988\,\mathrm{W\,m^{-2}\,K^{-1}}$$

The mean heat flux for the water wall during sunshine hours is:

$$\dot{q} = U_L(T_{sa} - T_b) = U_L\left[\frac{\alpha\tau}{U_t}I(t) + T_a - T_b\right]$$
$$= 2.988 \times \left[\frac{0.8 \times 0.71}{5} \times 310 + 14 - 18\right] = 93.26\,\mathrm{W\,m^{-2}}$$

The positive sign indicates that there is a heat gain during sunshine hours.

The mean heat flux for the water wall during off-sunshine hours is:

$$\dot{q}' = U_L(T'_a - T_b)$$
$$= 2.988 \times [12 - 18] = -17.926\,\mathrm{W\,m^{-2}}$$

The negative sign indicates that there is a heat loss during off-sunshine hours.

The net heat flux when the Sun shines for 12 h is given by:

$$\dot{q}_T = \dot{q} \times \text{sunshine hours} + \dot{q} \times (24 - \text{sunshine hours})$$
$$= 93.26 \times 12 + (-17.926) \times (24 - 12) = 904\,\mathrm{Wh\,m^{-2}}$$

The positive sign indicates that solar energy is sufficient to maintain the room temperature for the whole day. No external means of heat source is required.

(b) Considering insulation during off-sunshine hours, *i.e.* at night, the overall heat transfer coefficient for the water wall is given by:

$$U_L = \left[\frac{1}{U_t} + \frac{L_i}{K_i} + \frac{1}{h'_1} + \frac{1}{h'_2} + \frac{1}{h_{si}}\right]^{-1}$$
$$= \left[\frac{1}{5} + \frac{0.03}{0.025} + \frac{1}{206} + \frac{1}{206} + \frac{1}{8}\right]^{-1} = 0.652\,\mathrm{W\,m^{-2}\,K^{-1}}$$

The mean heat flux for the water wall during off-sunshine hours (at night) is:

$$\dot{q} = U_L(T_{sa} - T_b) = U_L(T_a - T_b) = 0.652 \times [12 - 18] = -3.9\,\mathrm{W\,m^{-2}}$$

4.6 Isolated Thermal Gain or Active Solar Collectors

The isolated thermal gain passive solar collector and storage are thermally isolated from the inner space of the building and function independently. This results in a greater flexibility in the design and operation. The concept is contrasted with the direct-gain passive solar concept where the collection and storage are integral with the living space, and with the indirect-gain concept where the collection and storage are separate from the living space but directly linked thermally. A few of them are as follows.

4.6.1 The Thermosyphon System

Thermosyphon (or thermosiphon) refers to a method of passive heat exchange based on natural convection that circulates a fluid without the necessity of a mechanical pump. This circulation can either be an open-loop or a closed-loop circuit. The intention is to simplify the pumping of fluid and/or heat transfer by avoiding the cost and complexity of a conventional liquid pump and/or air blower.

Figure 4.12 shows a schematic diagram of a thermosyphon passive heating system. It consists of a flat plate collector or photovoltaic array, comprising either an absorber mesh for air heating or tubes fitted on a plate for liquid heating. The collector is placed in a glass-fronted box painted matt black so that the solar absorption to the surface of the collector is maximized. To minimize the conduction losses from the back surface of the collector, it is well insulated from the back. During sunshine hours the fluid warms up in the collector and moves up to the basement to store thermal energy in the rock bed, water or PCMs. When the heating is required in the living space, especially during night time, the air from the living space is circulated through the basement.

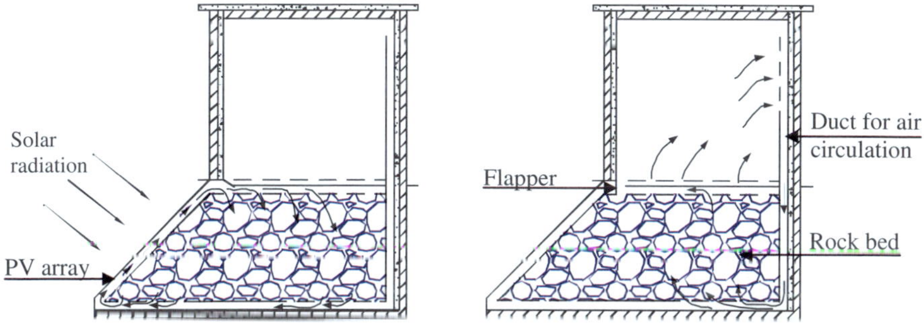

Figure 4.12 Thermosyphon passive heating system: *left*, heating of rock bed during sunshine hours; *right*, space heating from the energy stored in the rock bed during off-sunshine hours.

4.6.2 The Barra System

This system was developed by Horazio Barra in Italy.[6] It consists of an insulated glazed solar wall on the sunny side of the house and a concrete ceiling storage, as shown in Figure 4.13.

Looking at the outside and moving through a cross section, there is outside glazing. The robust polycarbonate glazing passes visible light but blocks infrared rays to reduce losses and blocks ultraviolet rays to protect building materials. Next to the glazing, there are one or more absorption meshes suspended vertically with an air gap on both the front and back sides. The mesh itself can hold very little heat and warms up rapidly in light. The heat is absorbed by the air passing around and through the mesh. Finally, a layer of insulation materials, such as polyisocyanurate foam, rock wool, foil or polystyrene, sits between the absorption space and the living space. With insulation the unwanted heat losses in winter (especially during extended cloudy periods) and heat gains in summer from the living space are minimized in comparison with a Trombe wall.

The hot air emerging from the absorption meshes flows horizontally within channels in the concrete ceiling. The ceiling acts as a thermal storage, where part of the heat of the air is absorbed while the still warm air exits from the distant channels of the building. Thus, the air warms the distant rooms before flowing back through the building space to the inlets at the lower part of the sun-facing collecting wall. This assures an even temperature distribution throughout the whole house, which is better than the temperature distribution achievable with other passive solar systems. An automatic back-draft damper acting as a one-way valve is fitted to both inlet and outlet vents, to prevent reverse convection at night. The system can be applied to multistory buildings and even to buildings where the main rooms do not face the sun.

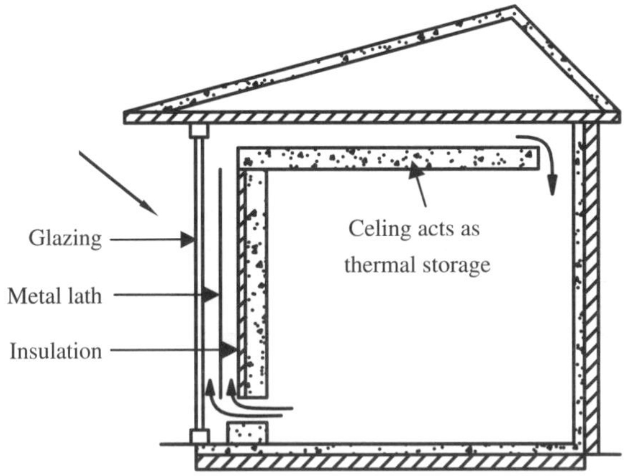

Figure 4.13 The Barra system.

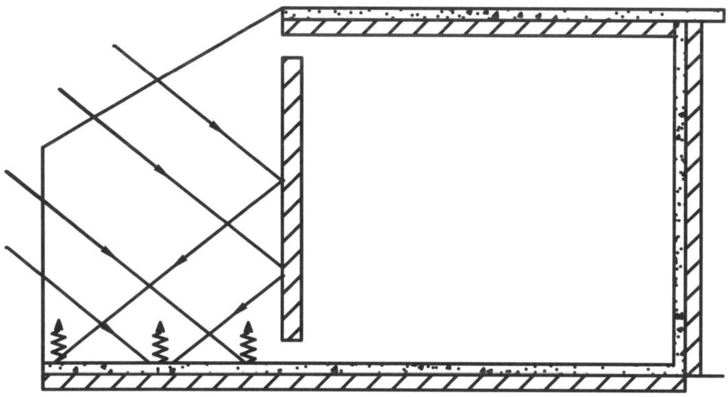

Figure 4.14 Sunspaces.

To store the thermal energy from the collector, the Barra system suspends a "spancrete" slab of concrete as a ceiling to store heat. This is fairly expensive and requires strong support. An alternative is to use water or PCMs.

4.6.3 Sunspaces

Figure 4.14 shows a sunspace which is basically a glazed room attached to a building but completely closed off from the main occupied areas. Sunspaces work in a similar manner to a Trombe wall, only the space between the glass and the absorbing surface is so large it is habitable.[7] They are also commonly called conservatories.

Solar radiation enters through the equator-facing windows and warms the air and the thermal mass floor. This warm sunspace air is then circulated to the building's interior for heating. Techniques can be employed to store the heat and prevent daytime overheating. Being separated from the main spaces of the building, a much greater temperature swing may be acceptable within sunspaces, more than can be tolerated in non-isolated direct gain spaces. Night-time heat loss is not as critical in the sunspace as with direct gain systems since it can be closed off from the rest of the building. However, night insulation or double-glazing is recommended if the sunspace serves as living area after sundown. The overall cost of a sunspace is higher, the energy collection efficiency per unit area of glazing is lower and the payback period of the investment in the sunspace construction is longer compared with direct gain systems.

4.7 Combined Thermal Gain Systems

Systems that use a combination of the direct, indirect and the isolated passive thermal gain systems can be classified as combined thermal gain systems. Some of the popular systems are as follows.

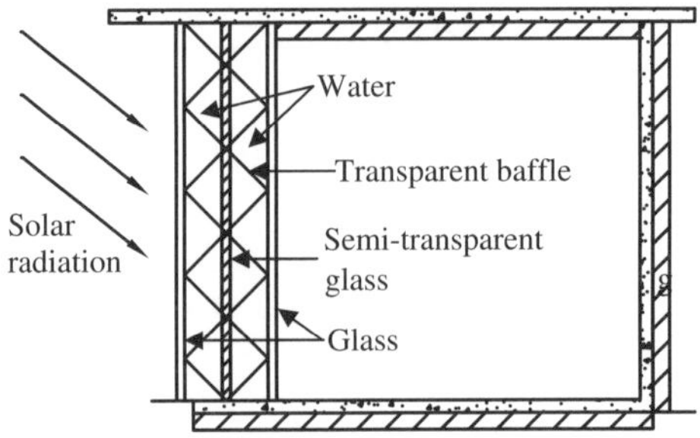

Figure 4.15 A transwall system.

4.7.1 A Transwall

Figure 4.15 shows a schematic diagram of a transwall integrated into the building. It is essentially a transparent modular water wall that allows vision to the exterior and, at the same time, provides thermal storage for the solar gain. It consists of water in a container made of parallel glass walls set in a metal frame and a semitransparent absorbing plate at the centre, parallel to the glass walls. The semitransparent plate absorbs 80% and transmits 20% of the incident solar energy. Thus, a transwall combines the attractive features of both direct as well as indirect gain systems and minimizes their disadvantages.[8–11] They absorb marginally less energy compared to the pure Trombe wall and find applications where the daytime heating load is significant.

Convective heat transfer in the transwall lowers its thermal performance efficiency but can be reduced by a combination of baffles. The viscosity of the water-based thermal storage medium can be increased by adding a gelling agent to the water. To prevent the growth of micro-organisms, an inhibiting agent may be added to the water.

The rate of useful energy gain for a transwall can is given by:

$$\dot{q} = \tau I(t) - U_{\mathrm{t}}(T_{\mathrm{b}} - T_{\mathrm{a}}) \tag{4.43}$$

where

$$U_t = \left[\frac{1}{h_{\mathrm{o}}} + \frac{L_{\mathrm{g}}}{K_{\mathrm{g}}} + \frac{1}{h_1} + \frac{1}{h_1'} + \frac{L_{\mathrm{t}}}{K_{\mathrm{t}}} + \frac{1}{h_2} + \frac{1}{h_2'} + \frac{L_{\mathrm{g}}}{K_{\mathrm{g}}} + \frac{1}{h_{\mathrm{si}}}\right]^{-1}$$

and

$$\tau = (1 - R_{\mathrm{g}})(1 - \alpha_{\mathrm{g}})(1 - \alpha_{\mathrm{w}})(1 - \alpha_{\mathrm{t}})(1 - \alpha_{\mathrm{w}})(1 - \alpha_{\mathrm{g}})$$

where α_{g}, α_{w} and α_{t} are the absorptivity of the glass cover, water and semitransparent absorbing plate, respectively; R_{g} is the reflectivity of the glass cover. The other symbols have their usual meanings.

Equation (4.43) can be rewritten as:

$$\dot{q} = U_t \left[\frac{\tau}{U_t} I(t) + T_a - T_b \right] \quad (4.44)$$

or

$$\dot{q} = U_t [T_{sa} - T_b] \quad (4.45)$$

4.7.2 A Solarium

A solarium (see Figure 4.16) is an integration of direct/indirect thermal gain concepts with a sunspace.[12–14] The thermal link between the collector space and the storage mass helps in heat retention and distribution and also enhances the efficiency of the system. The overheating of a sunspace during the summer can be avoided by the use of shading. The movable insulation or shutters minimize the heat losses.

4.7.2.1 Sol-air Temperature for Solarium (Combined Sunspace and Trombe Wall)

There is no direct heat transfer between the absorbing surface and the ambient air. The net heat flux gain for zone I, the partition wall and zone II can be expressed as:[15,16]

$$\dot{q} = \alpha\tau I(t) - h_{TS}\left(T|_{x=0} - T_{ss}\right) \quad (4.46)$$

$$\dot{q} = \frac{K}{L}\left(T|_{x=0} - T|_{x=L}\right) \quad (4.47)$$

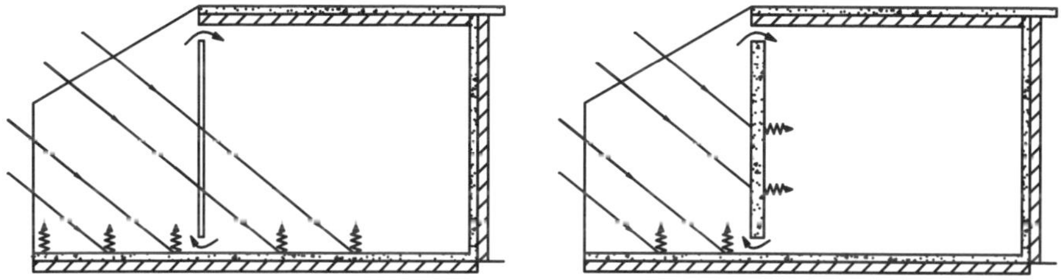

Figure 4.16 Schematic design of a solarium: *left*, combined isolated and direct gain of sunspace and window; *right*, combined isolated and indirect gain of sunspace and Trombe wall.

and

$$\dot{q} = h_{si}\left(T|_{x=L} - T_b\right) \tag{4.48}$$

The unknown air temperature of zone 1, T_{ss}, is related to the ambient air temperature through the following relation:

$$h_o(T_{ss} - T_a) = h_{TS}\left(T|_{x=0} - T_{ss}\right) \tag{4.49}$$

where h_o is the overall heat loss coefficient from the air temperature of zone 1 to the ambient air through the glazing and h_{TS} is a convective and radiative heat transfer coefficient from the wall's outside surface to the sunspace.

Simplification of the above equations gives:

$$T_{ss} = h_m\left[\frac{1}{h_o}T|_{x=0} + \frac{1}{h_{TS}}T_a\right] \tag{4.50}$$

where

$$h_m = \left[\frac{1}{h_o} + \frac{1}{h_{TS}}\right]^{-1}$$

After substituting the value of T_{ss} in eqn (4.46), we obtain:

$$\dot{q} = U\left[\frac{\alpha\tau I(t)}{h_m} + \frac{h_o}{h_m}\left(1 - \frac{h_m}{h_{TS}}\right)T_a - T_b\right] \tag{4.51}$$

where

$$U = \left[\frac{1}{h_m} + \frac{L}{K} + \frac{1}{h_{si}}\right]^{-1}$$

Substitution of the value of h_m from eqn (4.50) into eqn (4.51) and simplifying gives:

$$\dot{q} = U\left[\frac{\alpha\tau I(t)}{h_m} + T_a - T_b\right] \tag{4.52}$$

The above equation can be written as:

$$\dot{q} = U[T_{sa} - T_b] \tag{4.53}$$

where

$$T_{sa} = \frac{\alpha\tau I(t)}{h_m} + T_a.$$

4.8 Use of Photovoltaic Arrays as a Façade

For a self-sustainable building, especially in remote areas where there is no power grid connection, electrical power generation is of prime importance. Use of a PV array may fulfil the requirement to some extent. It is well know that only 20% of the solar energy irradiant on a PV cell's surface is converted into electrical energy and the remaining 80% is converted into thermal energy. This thermal energy is sufficient to fulfil the space heating requirement. With advancement in technology, PV arrays are now integrated into façades facing the equator to generate electrical power for household applications along with thermal energy for passive heating of buildings. Some of the popular systems are described below.

4.8.1 Semi-transparent Photovoltaic Arrays as a Façade

Figure 4.17 shows the schematic diagram of a semi-transparent (glass-to-glass type) PV array.[17–19] The radiation passing through the non-packing area of the cell helps in direct space heating, whereas the radiation falling on the solar cell is partially converted into electrical energy and the remainder helps in indirect space heating. During off-sunshine hours, a curtain may be used to prevent heat losses.

4.8.1.1 Sol-air Temperature and Heat Transfer for a PV Façade

Consider a room has a glass-to-glass PV array integrated into the building as a window facing the equator. The energy gain through the non-packing area of

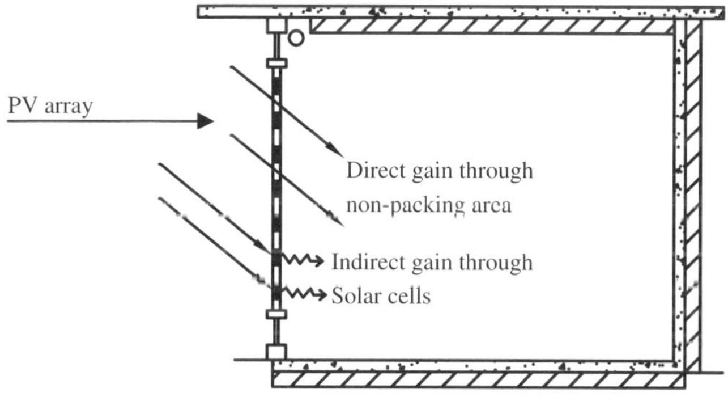

Figure 4.17 Semi-transparent photovoltaic façade.

the PV module is given by:

$$\dot{Q}_{\rm np} = \left[(1-\beta)\tau^2 I(t) - U_{\rm np}(T_{\rm b} - T_{\rm a})\right] A_{\rm m} \tag{4.54}$$

where $U_{\rm np} = \left[\frac{1}{h_{\rm o}} + \frac{L_{\rm g}}{K_{\rm g}} + \frac{L_{\rm EVA}}{K_{\rm EVA}} + \frac{L_{\rm g}}{K_{\rm g}} + \frac{1}{h_{\rm si}}\right]^{-1}$ is the overall heat transfer coefficient between the room and the atmosphere through the glazed window, $A_{\rm m}$ is the area of the module, β is the packing factor, $L_{\rm EVA}$ is the thickness of the poly(ethylene–vinyl acetate) (EVA) between the glasses, $K_{\rm EVA}$ is the thermal conductivity of EVA, $T_{\rm b}$ is the temperature inside the building/living space and $T_{\rm a}$ is the ambient air temperature.

The energy gain through the packing area of the PV array is given by:

$$\dot{Q}_{\rm p} = \left[\beta\tau I(t) - U_{\rm p}(T_{\rm b} - T_{\rm a})\right] A_{\rm m} \tag{4.55}$$

where $U_{\rm p} = \left[\frac{1}{h_{\rm o}} + \frac{L_{\rm g}}{K_{\rm g}} + \frac{L_{\rm c}}{K_{\rm c}} + \frac{L_{\rm g}}{K_{\rm g}} + \frac{1}{h_{\rm si}}\right]^{-1}$ is the overall heat transfer coefficient between the room and the atmosphere through the glazed window, $L_{\rm c}$ is the thickness of the solar cell, $K_{\rm c}$ is the thermal conductivity of the solar cell, $T_{\rm b}$ is the temperature inside the building/living space and $T_{\rm a}$ is the ambient air temperature.

Adding eqns (4.54) and (4.55), the net heat gain through the PV array is given by:

$$\begin{aligned}
\dot{Q} &= \dot{Q}_{\rm np} + \dot{Q}_{\rm p} \\
&= \left[(1-\beta)\tau^2 I(t) - U_{\rm np}(T_{\rm b} - T_{\rm a})\right] A_{\rm m} + \left[\beta\tau I(t) - U_{\rm p}(T_{\rm b} - T_{\rm a})\right] A_{\rm m} \\
&= \left[\{(1-\beta)\tau + \beta\}\tau I(t) - (U_{\rm np} + U_{\rm p})(T_{\rm b} - T_{\rm a})\right] A_{\rm m} \\
&= (U_{\rm np} + U_{\rm p})\left[\frac{(1-\beta)\tau + \beta}{U_{\rm np} + U_{\rm p}}\tau I(t) + T_{\rm a} - T_{\rm b}\right] A_{\rm m} \\
&= U_{\rm pv}[T_{\rm sa} - T_{\rm b}]
\end{aligned} \tag{4.56}$$

where

$$T_{\rm sa} = \frac{(1-\beta)\tau + \beta}{U_{\rm np} + U_{\rm p}}\tau I(t) + T_{\rm a}.$$

The total thermal gain through a module during sunshine hours is given by:

$$Q_{\rm T} = \sum_{i=1}^{N} \int_0^{t_{\rm T}} \dot{Q}\, {\rm d}t \tag{4.57}$$

4.8.2 Photovoltaic Trombe Walls

Figure 4.18 (*left*) shows a schematic diagram of a vented Trombe wall with a PV array. The air flow through the vented space lowers the cell temperature and thereby helps in providing higher electrical energy for household applications. If the natural convection is insufficient, then forced convection can be implemented by introducing a dc fan. The heat carried by the flowing air will enter the room, providing heating during daytime and partially stored in the massive wall for heating at night, giving higher aesthetic value. During a hot summer the circulation of air through a vent can be made by blowing the atmospheric air from the bottom vent of the PV array and discharging it from the upper vent to maintain the electrical efficiency of the array, as shown in Figure 4.18 (*right*). A curtain may also be used to avoid unnecessary heating of the space.

4.8.3 Photovoltaic Integrated Phase-change Material Walls

Since PCMs have greater potential for heat storage per unit mass and the use of a photovoltaic panel can produce sufficient electrical power necessary for household purposes, a photovoltaic integrated phase-change material (PV-PCM) system would be advantageous. Figure 4.19 shows a PV-PCM wall both

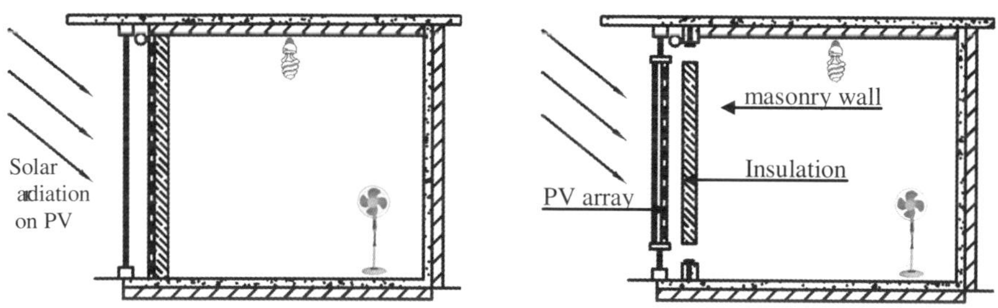

Figure 4.18 A photovoltaic vented Trombe wall: *left*, winter heating; *right*, summer cooling (the arrows show the air circulation path; the fan and bulb shown are powered by the PV array).

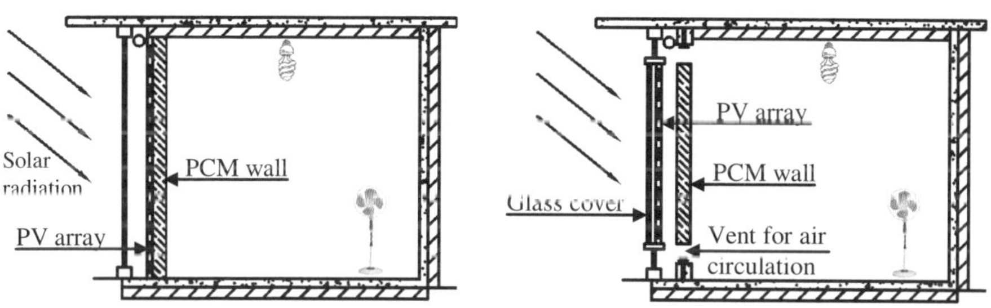

Figure 4.19 PV-PCM walls: *left*, unvented type; *right*, vented type.

without and with vents. The PV array may be semitransparent type (glass-to-glass) or may be opaque type (glass-to-Tedlar).

Example 4.8

Calculate the direct thermal energy gain through a semi-transparent PV module of area $0.6\,m^2$ and packing factor 0.8, irradiated with solar radiation of intensity $800\,W\,m^{-2}$ for 4 h. The transmissivity of both the glasses of the module is 0.9, the thermal conductivity of glass is $0.78\,W\,m^{-1}\,K^{-1}$, the thickness is 3 mm and they are 0.3 mm apart. The room space temperature is 20 °C, the ambient air temperature is 12 °C and the thermal conductivity of EVA is $0.04\,W\,m^{-1}\,K^{-1}$.

Solution

Assume the wind flow velocity as zero, the heat transfer coefficient on the external surface of the glazing as $h_o = 5.7\,W\,m^{-2}\,K^{-1}$ and the heat transfer coefficient on the glazed surface inside the room as $h_{si} = 2.8\,W\,m^{-2}\,K^{-1}$. The overall heat transfer coefficient between the room and the atmosphere through the glazed window is given by:

$$
\begin{aligned}
U_{np} &= \left[\frac{1}{h_o} + \frac{L_g}{K_g} + \frac{L_{EVA}}{K_{EVA}} + \frac{L_g}{K_g} + \frac{1}{h_{si}}\right]^{-1} \\
&= \left[\frac{1}{5.7} + \frac{0.003}{0.78} + \frac{0.0003}{0.04} + \frac{0.003}{0.78} + \frac{1}{2.8}\right]^{-1} \\
&= 1.825\,W\,m^{-2}\,K^{-1}
\end{aligned}
$$

The direct thermal gain during sunshine hours is given by:

$$
\dot{Q}_{np} = \left[(1-\beta)\tau^2 I(t) - U_{np}(T_b - T_a)\right] A_m
$$

$$
\begin{aligned}
Q_T &= \sum_{i=1}^{N} U_{np}\left[\frac{(1-\beta)\tau^2\{\overline{I(t)}\}_i}{U_{np}} + \{\overline{T_a}\}_i - T_b\right] A_m t_T \\
&= 1.825 \times \left[\frac{(1-0.8)0.9^2 \times 800}{1.825} + 12 - 20\right] \times 4 \times 0.6 = 276\,W
\end{aligned}
$$

As the thickness and thermal conductivity of the solar cells are almost equal to that of the EVA, then indirect gain through the PV module during sunshine hours is:

$$
\begin{aligned}
Q_T &= \sum_{i=1}^{N} U_p\left[\frac{\beta\tau\{\overline{I(t)}\}_i}{U_p} + \{\overline{T_a}\}_i - T_b\right] A_m t_T \\
&= 1.825 \times \left[\frac{0.8 \times 0.9 \times 800}{1.825} + 12 - 20\right] \times 4 \times 0.6 = 1347.4\,W
\end{aligned}
$$

The net thermal gain through the PV module is:

$$\dot{Q} = U_{pv}\left[\frac{\{(1-\beta)\tau+\beta\}\tau\{\overline{I(t)}\}_i}{U_{pv}} + \{\overline{T_a}\}_i - T_b\right] A_m t_T$$

$$= 1.825 \times \left[\frac{\{(1-0.8)\times 0.9 + 0.8\}\times 0.9 \times 800}{1.825} + 12 - 20\right] \times 4 \times 0.6$$

$$= 1658.4\,\mathrm{W}$$

4.9 Integration of Photovoltaic Arrays on the Roof

A photovoltaic array may also be integrated on the roof of a building. Two of the popular systems are described below.

4.9.1 Opaque Photovoltaic Arrays Integrated on the Roof

Figure 4.20 shows a schematic diagram of an opaque (glass-to-Tedlar type) PV array integrated on the roof of a building. A part of the solar radiation incident on the array is converted into electrical power and the remainder is converted into heat. The forced circulation of air between the array and the insulation helps to heat the living space. Thus the system provides both electrical and thermal gain.

Instead of using insulation concrete, PCM material may be used to store the thermal energy during the sunshine hours and release heat during the off-sunshine hours.

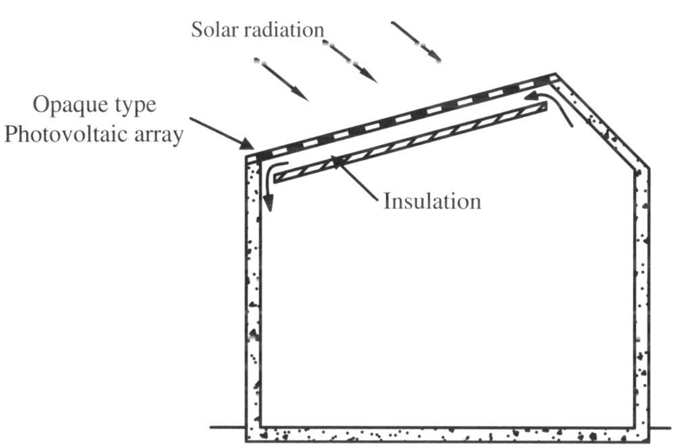

Figure 4.20 Opaque-type photovoltaic array integrated on the roof

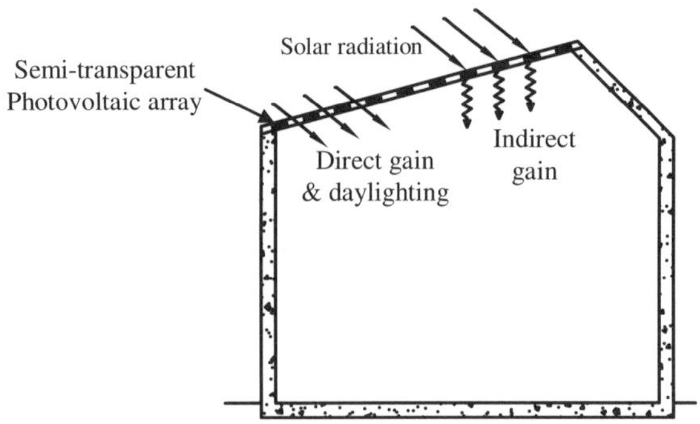

Figure 4.21 Semi-transparent photovoltaic array integrated on the roof.

4.9.2 Semi-transparent Photovoltaic Arrays Integrated on the Roof

Figure 4.21 shows a schematic diagram of a semi-transparent (glass-to-glass type) PV array integrated on the roof of a building. The solar radiation falling on the solar cells of the module/array provides both electrical and thermal energy, whereas radiation passing through the space between the solar cells provides thermal energy and the day lighting.

References

1. S. Ram and H. P. Garg, Heat flux through a Trombe wall/roof, *Appl. Energy*, 1985, **19**, 61–71.
2. J. K. Nayak, Thermal performance of a water wall, *Build. Environ.*, 1987, **22**, 83–90.
3. S. C. Kaushik and S. Kaul, Thermal comfort in buildings through a mixed water-mass thermal storage wall, *Build. Environ.*, 1989, **24**, 199–207.
4. P. Torcellini and S. Pless, *Trombe Walls in Low-Energy Buildings: Practical Experiences*, National Renewable Energy Laboratory, Golden, CO, USA, July 2004.
5. G. N. Tiwari, A. Madhuri and Y. P. Yadav, Thermal load levelling of heat flux in PCCM collection/storage walls/roof system, *Energy Convers. Manage.*, 1988, **28**, 75–79.
6. O. A. Barra, G. Artese, L. Franceschi, R. K. Joels and A. Nicoletti, *The Barra Thermosyphon Air System: Residential and Agricultural Applications in Italy, UK, and in the Sahara*, presented at the International Conference of Building Energy Management, Lausanne, 1987.
7. G. J. Schoenau, A. J. Lumbis and R. W. Besant, Thermal performance of four sunspaces in a cold climate, *Energy Build.*, 1990, **14**, 273–286.
8. R. Fuchs and J. F. McClelland, Passive solar heating of buildings using a transwall structure, *Sol. Energy*, 1979, **23**, 123–128.

9. M. Upadhya, G. N. Tiwari and S. N. Rai, Optimum distribution of water-wall thickness in a transwall, *Energy Build.*, 1991, **17**, 97–102.
10. M. S. Sodha, N. K. Bansal and S. Ram, Periodic analysis of a transwall: a passive heating concept, *Appl. Energy*, 1983, **14**, 33–48.
11. J. K. Nayak, Transwall versus Trombe wall: relative performance studies, *Energy Convers. Manage.*, 1987, **27**, 389–393.
12. G. N. Tiwari, Y. P. Yadav and S. A. Lawrence, Performance of a solarium: an analytical study, *Build. Environ.*, 1988, **23**, 145–151.
13. G. N. Tiwari and S. Kumar, Thermal evaluation of solarium-cum-passive solar house, *Energy Convers. Manage.*, 1991, **32**, 303–310.
14. M. S. Sodha, J. K. Nayak, N. K. Bansal and I. C. Goyal, Thermal performance of a solarium with removable insulation, *Build. Environ.*, 1982, **17**, 23–32.
15. J. K. Nayak, N. K. Bansal and M. S. Sodha, Analysis of passive heating concepts, *Sol. Energy*, 1983, **30**, 51–69.
16. M. S. Bhandari and N. K. Bansal, Solar heat gain factors and heat loss coefficients for passive heating concepts, *Sol. Energy*, 1994, **53**, 199–208.
17. J. Jie, Y. Hua, P. Gang and L. Jianping, Study of PV-Trombe wall installed in a fenestrated room with heat storage, *Appl. Thermal Eng.*, 2007, **27**, 1507–1515.
18. J. Jie, Y. Hua, H. Wei, P. Gang, L. Jianping and J. Bin, Modeling of a novel Trombe wall with PV cells, *Build. Environ.*, 2007, **42**, 1544–1552.
19. G. N. Tiwari, *Solar Energy Fundamentals, Design, Modelling and Applications*, Narosa, New Delhi, 2002.

CHAPTER 5

Implementation of Building Integrated Photovoltaic Thermal Systems and Case Studies

5.1 Introduction

The recent rapid expansion in installed photovoltaic (PV) capacity is largely due to the increase in grid-connected photovoltaic systems mounted on buildings. The term "building integrated" refers to PV systems that constitute part of a building envelope, but has also been used to describe systems that are simply mounted on the rooftops of buildings. For this reason, it is best to describe building integrated photovoltaic (BIPV) systems as those that are readily integrated with the physical building or with the building's grid connection. The integration of such systems usually requires the advice of professional civil engineers, architects and PV system designers during the design of the system and the building. In this case, a good evaluation of the installation site is required so as to maximize solar coverage and electricity output. BIPV systems are usually installed on facades, building window systems and as flexible rolls on roofs. Consequently, BIPV systems often have restricted views of the Sun, and their orientation must be optimized for the particular circumstances of their installation site.[1]

BIPV technologies can achieve significant cost reductions when they are used as part of the building envelope and thereby offset the cost of the building materials they replace. Many modern exterior claddings can have costs per square metre comparable to the price of PV modules. At the leading edge of BIPV systems are the three main thin-film photovoltaic technologies (a-Si, CdTe and CIGS) that are at present commercially available. The most important issues for the successful integration of thin-film BIPV technologies include gaining experience on the design and operation of such systems as well

RSC Energy Series No. 4
Building Integrated Photovoltaic Thermal Systems: For Sustainable Developments
By Basant Agrawal and G. N. Tiwari
© B. Agrawal and G. N. Tiwari 2011
Published by the Royal Society of Chemistry, www.rsc.org

as acquiring knowledge of their life-cycle costs. This chapter provides an overall view of BIPV technologies in countries where its implementation is at mature stage and summarizes ongoing research.

5.2 Implementation in Germany

The cumulative solar PV-installed capacity in Germany has increased from 195 MW in 2001 to 5337 MW in 2008, making Germany the largest market for PV installations worldwide with a global market share of 35% in the year 2008.[2] A further 2.5 GW was added in that year, taking the total to 7.8 GW at end of the year 2009. Solar power now meets about 1% of the country's electricity demand. The capacity factors are relatively low due to the low insolation for most of Germany, about 80–90 W m^{-2}.

Germany has positioned itself as an excellent location for solar energy investments due to strong governmental support, the availability of a highly qualified workforce, and a plentiful supply of scientific research centres and universities, such as the Fraunhofer and Max Planck Institutes. All these factors put together guarantee rapid and smooth implementation and development of PV projects in the country. The German support mechanism for PV systems is a feed-in tariff. There are four tariffs altogether and they vary, depending on the size. The highest tariff is for installation of a system with a size of less than 30 kW$_p$ capacity. In addition, there is an automatic increase or decrease of the degression rate if the installed capacity is above or below certain values in a year. The façade tariff bonus was withdrawn in 2009. The tariffs are granted on the day the contract starts and they are guaranteed for 20 years.

Waldpolenz Solar Park is the world's largest thin-film PV power system, installed by JUWI Group in Waldpolenz. It was fully operational by the end of 2008 and electricity producing capacity is 40 MW$_p$. 550 000 cadmium telluride (CdTe) modules made by First Solar are being used, which supply about 40 000 MWh of electricity per year. Among the BIPV installations, a 5 MW$_p$ PV system on the rooftop at Bürstadt and 3.9 MW$_p$ on a single roof at Muggensturm are among the largest in the world. Other solar roofs of the megawatt category can be seen on four factory stores of the private company Franz Fischer in Dingolfing (Bavaria; 3.7 MW$_p$) and in the Michelin Solar Park Homburg (Saarland; 3.5 MW$_p$). On the roof of the new Trade Fair in Munich an impressive 2.1 MW$_p$ have been installed, the roofs of the Solar Park Pfersdorf achieve 2 MW$_p$ and a logistics centre in Relzow (Mecklenburg, West Pomerania) has 1.5 MW$_p$.

5.2.1 BIPV Systems on the Rooftop and as a Façade of the Mont-Cenis Academy

The town of Herne possesses a 1 MW rated capacity BIPV system on the roof and as a façade of a spectacular building, making it one of the largest BIPV systems in the world. Figure 5.1 shows an aerial view of the BIPV system on the

Figure 5.1 The BIPV system at Mont-Cenis Academy: *top*, aerial view of the rooftop BIPV array; *bottom left*, inside view; *bottom right*, BIPV façade and free ventilation system in the glass walls.[3,4]

rooftop, an inside view and the façade, respectively, of Mont-Cenis Academy (51°32′29″N 7°15′14″E) at Herne Sodingen. The academy has average winter and summer ambient temperatures of 4.8 °C and 14.35 °C, respectively. The annual sunshine lasts for nearly 1454 hours, *i.e.* 3.98 hours per day.

The 2900 semi-transparent PV modules with an overall 925 kW_p rated output combine to make 9800 m^2 for the BIPV roof. The use of modules with different cell-area ratios and different levels of translucency creates a pattern resembling a cloud-patterned sky. It is possible to walk on the roof for maintenance and every point on the roof can be reached easily. Another 284 modules with overall 75 kW_p rated output are integrated in the southwest façade and cover about 800 m^2 area of the building. The solar cells (poly- and monocrystalline type) have efficiencies of 12.5% and 16%, respectively. The PV modules are up to 3.2 m^2 in size and tilted due south at an angle of 5° in order to let the rain wash the surface and leave the modules clean.

PV modules and glass panes of the overhead glazing rest on aluminium profiles and are held in place with aluminium pressure plates. The vertical PV and glass façade is carried out as a structural glazing façade. Their glass panes and PV modules are glued onto aluminium profiles. All the aluminium profiles are mounted on the load-bearing wooden substructure designed especially for this project. The interconnecting plugs and the required cabling are integrated in the aluminium profiles which hold the PV modules and glass panes into place. Like this they are invisible and protected against weather conditions and ultraviolet light.

The measured energy output is 600–650 MWh per year, which is nearly twice as much power as the building needs, and the CO_2 reduction is 525 tonnes per annum.[5] The surplus energy of the system is used in nearby homes. Owing to failure of the natural ventilation system, the air temperatures below the PV modules are much higher than expected. If site conditions had been taken into consideration while designing the system, such as optimized inclination and orientation, it could have resulted in much higher output. Although the PV modules are mounted only with a slope of 5°, the saw-tooth roof causes some shading on the PV modules in the early morning and evening hours as well as in winter, when the Sun is low at the horizon.

5.2.2 A BIPV System as a Rooftop in Burstadt

Figure 5.2 shows a 5 MP_p BIPV system built on a local logistics business's roof at Burstadt (49°38′50″N 8°27′50″E) in Hessen and has been in operation since 2005.[6] There are 29 182 monocrystalline modules (made by BP Solar) covering roughly 45 000 m^2 of roof area.[7] The plant produces about 4200 MWh of electricity per year at an efficiency of 10%. It accounts for 9% of the annual electricity consumption of Burstadt, which corresponds to the consumption of about 1200 households. The power is fed through three inverters into the public grid.

Some junction boxes prone to overheating were replaced in 2006 and 2007 as a precautionary measure. However, on 21st June 2009, fire exploded and destroyed nearly 200 m^2 of the BIPV system. As a precautionary measure and to enhance security the whole plant was shut down.[8] On the roof, 18 modules

Figure 5.2 A BIPV system as a rooftop at Burstadt, Hessen.

Figure 5.3 A BIPV system as a rooftop at Muggensturm, Rastatt.

are connected to a cable. Since each module can produce 44 V, the total could be close to 800 V and cause tension in each cable. Therefore it was decided to separate the modules and resolve the cable connections. This work is currently being done on the roof.

5.2.3 A BIPV System as a Rooftop in Muggensturm

On the flat roofs of the logistics company Hartmann in Muggensturm (48°40′50″N 08°58′30″E), the company Tauber-Solar from Tauberbischofsheim installed PV modules over 80 000 m^2 of area (Figure 5.3).[8] Almost 32 740 Conergy solar PV modules were installed on stands on a flat saddle roof (double-layer steel trapezium profile) with an integrated insulation layer. At a peak performance of 3.84 MW_p and at an efficiency of 11%, an annual output of 3.5×10^6 kWh is expected.[9] The annual solar power output is sufficient for the consumption of nearly 900 four-person households.

5.2.4 A PV Façade Integration at Lehrter Station

Figure 5.4 shows views of the PV façade integration from the outside and inside at Lehrter Station (52°31′29″N 13°22′8″E) in Berlin. The roof is 320 m long and consists of 780 glass–glass PV modules equipped with a total of 78 000 solar cells. The PV façade has a capacity of 189 kW_p electrical output.[10] The PV modules consist of two discs and an in-between layer of synthetic resin, into which the solar cells are embedded. Because of the shape and geometry of the glass hall, no module has the same dimensions as any other and each of the 780 modules is unique. Integrated into the mains net the facility is expected to generate approximately 150 MWh of electricity annually.

The solar modules are connected in series in strings of six modules each. This results in a system voltage of approximately 320 V dc.[10] The electric system of

Figure 5.4 Façade integration at Lehrter station in Berlin: *top*, aerial view; *bottom*, inside view.

connection, which is firmly laminated into the glass/glass network, is hidden completely by the post-crossbar structure underneath. Thus the complete dc cabling is invisible, an optical necessity for buildings with solar facilities integrated into transparent areas. The further routing of the cable to the 117 decentralized stringer inverters by the manufacturer SMA (type Sunny Boy), arranged at the southerly support points of the glass hall, is also hidden inside the post-crossbar structure. From the stringer inverters, which are arranged in groups of three, the three-phase current is routed to the public mains network in a concrete cable channel *via* a bus bar system encapsulated in metal.

5.3 Implementation in Spain

Spain is the fourth largest manufacturer in the world of solar power technology and exports 80% of this output to Germany. The year 2006 was somewhat of a transition in Spain, with the activity developed under a 2004 Royal Decree being superseded by the announcement in another Royal Decree of a new feed-in tariff scheme.[11] However, it has been demonstrated that the new scheme is working satisfactorily enough and the deployment of PV systems has gained strong momentum. In 2006, there were 63 MW of PV systems installed. A record 2 GW of solar power was added in 2008 and by the end of 2009 the total solar power in Spain was 3 GW. At present, solar energy covers 2.8% of the electricity demand. The distribution of PV installations in Spain is about 85% grid-connected systems and 15% off-grid systems.

The Spanish national programmes favour the installation of grid-connected PV systems and this has led to fewer installations of off-grid applications. Installation and system companies, and the great majority of manufacturers, focus on the installation of grid-connected systems. Within the grid-connected PV power applications, the distributed systems dominate. The main support framework in Spain is the feed-in tariff scheme, with the following main characteristics: long term, with a fixed tariff for the full duration of the contract, providing a current internal rate of return $>7\%$, revision every four years to evaluate the market progress, review of the tariffs and the internal rate of return for the coming years, a simple system, and without differences between locations with differences of irradiation.[11] Much of the growth that will be encouraged by this scheme will come from Huertas Solares projects—large PV farms with many individual PV systems ranging from 5 kW to 100 kW.

In addition, a new Royal Decree was approved that provides a technical building code (TBC), establishing obligatory requirements to be met by buildings. A section of the TBC regulates the incorporation of solar PV energy and enforces the installation of PV systems on new large buildings, such as offices, government buildings, hospitals, *etc*. In given buildings the PV electricity may be for own use or supplied to the grid. The minimum PV capacity required in each case will depend on the climatic zone, the building floor area and the building use. The peak capacity P to be installed is given by the formula $P\,(\mathrm{kW}) = C(AS + B)$, where C is the defined coefficient for each climatic zone, A and B are the coefficients defined for each kind of use and S is the floor area built in square metres. The minimum PV system to be installed will be 6.25 kW.

The Olmedilla Photovoltaic Park in Olmedilla de Alarcón is the world's largest PV plant. Built in 2008, the plant uses more than 160 000 solar photovoltaic panels to generate 60 MW_p. Puertollano Park Solar Power Plant and Arnedo Solar Plant are the other two largest plants, completed in 2008 with a capacity of 50 MW_p and 30 MW_p, respectively. Over 25 years during production, the energy produced here will replace the theoretical disposal of 2.1×10^6 and 3.75×10^5 tonnes of CO_2, respectively.

5.3.1 Integration of PV Systems as Parking Canopies, Pergolas and Façades at Jaén University

The Univer Project was developed at Jaén University (37.730°N, 3.780°W) under the Thermie Programme of the European Union. The location is characterized by a continental climate, relatively cool in winter and extremely hot in summer; in the latter season the ambient temperature rises to a maximum of about 45 °C. The average yearly peak solar hours are 4.9 h per day.

This project consists of a grid-connected PV system with a total power of 200 kW_p. PV modules are made up of monocrystalline silicon cells (made by Isofoton), an anodized aluminium outline frame, frontal tempered glass and white transparent Tedlar in the back side, making the structure semi-transparent. The system is designed to produce about 8% of the electricity demand of the university, which is estimated to be around 280 MWh per year. With respect to plant PV integration, different traditional architectural solutions, such as parking canopies, pergolas and façades, have been used in order to be able to compare their performance ratio. Figure 5.5 shows aerial views of the integration of PV systems on parking canopies, pergolas and façades on the university campus.[12,13]

According to analyzed data,[12] the average annual energy production registered was 168.12 MWh per year, which represents 6.40% of the total consumption of the university campus. In order to improve the performance, that part of the system which was responsible for the reduction of performance was replaced. At present, the whole PV systems are in their second phase of operation and it is under a new monitoring and supervision system.

Figure 5.5 Integrated PV systems at Jaén University: *top*, parking canopies with 140 kW_p output; *bottom*, pergolas and façades with 60 kW_p output.

Considering the 30 years lifetime, the cost of electricity generation is estimated to be about €0.22 per kWh.[13]

5.3.2 A BIPV System as a Rooftop and Façade at Telefonica Business Park Complex

Figure 5.6 shows the Telefonica Business Park Complex in Madrid (40°30′50″N 3°39′40″W). Although this building has both a glass façade and a canopy, neither holds the PV power. Fundamentally, the entire roof of this building is integrated with a PV system of 2.9 MW_p. There are about 16 600 PV modules occupying more than 21 000 m^2 of roof area and are distributed in such a way that they form the letter "C" that constitutes the name of the building ensemble. The installed power reaches an approximate 4 MW_p, generating over 3.6 GWh per year, 18% of the total consumed on the premises.[14] This represents a reduction of 2000 tonnes of CO_2 emitted into the atmosphere. The PV plant is divided into 24 independent 100 kW_p PV units, connected to the distribution grid through five transformation centres of 630 kW each. The roof's shadow, cast over the façades, cloisters and entrances, unifies, protects and refreshes.

5.3.3 A BIPV System as a Rooftop and Façade at Torre Garena

Figure 5.7 shows a solar BIPV system as the rooftop and façade of a building at Torre Garena (40°28′52″N 3°23′38″W), giving a total power of 85 kW_p.[15] The

Figure 5.6 Flat roof-mounted PV system at Telefonica Business Park Complex.

Figure 5.7 A BIPV system as the rooftop and façade of a building at Torre Garena.

building's façade and roof include two different solar PV power generation plants. The PV installation on the façade has been built using 948 photovoltaic modules manufactured by BP Solar. This installation has a total power of 75.8 kW_p, representing a saving of 85 tonnes of CO2 emissions per year. These modules have been installed at a 60° gradient to take full advantage of the available sunlight. The PV modules have been manufactured using state-of-the-art polycrystalline technology and consist of 36 cells with an antireflection coating and connected in series. Each panel has a nominal power of 80 W_p. The roof application incorporates an array of 93 glass-to-glass modules of polycrystalline solar cells, each with nominal power of 100 W_p. They are located on the south-facing roof with a gradient of 30°, being perfectly integrated into the building's design.

5.4 Implementation in Japan

A series of programmes were implemented in Japan to promote the residential installation of PV systems. The New Energy Foundation managed the first programme "Monitoring Programme for Residential PV Systems" from 1994 to 1996. The programme paid 50% of the installation costs and the number of households participating rose from 539 in 1994 to 1986 in 1996. The total annual cost of subsidies doubled from 2 to 4 billion yen, but unit costs declined as hoped. In 1997, a larger programme for the development of the infrastructure for the introduction of residential PV systems was launched with a larger budget (rising from 11.1 billion yen in 1997 to 23.5 billion yen in 2001). However, individual subsidies declined from 340 000 yen per kW_p to 120 000 yen per kW_p over the same period. With average annual electricity production levels of 950 kWh per kW_p, the average electricity savings amounted to 23 400 yen per kW_p.[16] These savings of approximately 70 000 yen per year for a 3 kW_p system were modest compared to investment costs, so lowering of the initial capital costs was a critical feature to successfully stimulate the market.

In Japan, subsidies were seen as a temporary measure to stimulate demand in the market so that new technologies could be introduced and production levels increased to a scale where unit costs were reduced. The resultant industry would be expected to be viable without subsidies. The planned reduction of national subsidies has been underway for over a decade (1994–2005), with 2006 marking the end of subsidies under the residential (RPVDP) programme. Although the subsidy has decreased substantially over the decade, the actual price has fallen by a similar amount, with the result that the net price paid by customers has remained largely unchanged, just under 2 million yen for a typical system.[16] In this context, the fear that the market might collapse with the end of the national subsidy programme is unlikely to be realized. Instead, capital incentives stimulated demand and increased the size of the market and resulted in reduced unit costs.

In 2006, 88.5%, or 254 MW_p, of the new installations were grid-connected residential systems, bringing the accumulated power of solar systems under the Japanese PV Residential Programme to 1617 MW_p (out of 1709 MW_p total installed PV capacity). However, in the year 2007 the Japanese market declined to 210 MW_p and only recovered slightly to 230 MW_p in 2008. At the end of 2008, the total cumulative installed capacity was 2.15 GW_p, less than half of the original 4.8 GW_p goal for 2010.

In order to boost the Japanese market and achieve a goal of 70% of new homes integrated with PV systems, the Japanese Ministry for Economy, Trade and Industry (METI) announced that they wanted to reinstate an investment subsidy for residential PV systems in the first quarter of 2009. In addition to the measures taken by the national government, over 300 local authorities have introduced measures to promote the installation of PV systems. One of the largest programmes was announced by the Tokyo Metropolitan Government, which plans to support the installation of 1 GW_p BIPV systems over 40 000 households in 2009–2010. The Federation of Electric Power Companies of Japan (FEPC) announced that they intend to install PV plants with a cumulative installed capacity of 10 GW_p by 2020.

Figure 5.8 The PV system on the rooftop and walls of the Sharp Corporation LCD Plant.

5.4.1 A BIPV System as the Rooftop and Façade of the Sharp Corporation LCD Plant at Kameyama

Figure 5.8 shows the PV system installed on the rooftop (polysilicon solar cells) and walls (semi-transparent thin film) of the Sharp Corporation LCD panel plant (34°52′17″N 136°24′21″E) in Kameyama. The PV array covers approximately 47 000 m^2 of surface area and has a capacity of 5.2 MW_p output.[17,18] The semi-transparent wall façades have a structure in which a crystalline thin-film silicon solar cell is stacked on top of an amorphous silicon solar cell on the same glass substrate. This design makes it possible to effectively use light from a wider band of wavelengths, improving the conversion efficiency to 11%. The electricity generated is used in the LCD TV assembly processes in addition to providing electric power for lighting inside the manufacturing facility. This PV system makes it possible to cut CO_2 emissions by about 3400 tonnes annually.

5.4.2 A PV Integrated System on the Louvres of a Roof Shelter in the Itoman City Government Building

Figure 5.9 shows the PV array mounted on southern louvres and flat roof of the Itoman city government building (26°7′20″N 127°39′55″E) on Okinawa. The PV cells used were EFG polycrystalline silicon and produced by RWE Schott Solar (now, Schott Solar). The total capacity of the solar cells is 195.6 kW_p and has been under operation since May 2002.[19] For effective operation, the system is designed to interact with the power utility so that excess power can be reversed. In the weekends and holidays, surplus power is sold to the Okinawa Power Company. An auxiliary 200 Ah battery is installed in the system to

Figure 5.9 A PV system integrated on the southern horizontal louvres and roof shelter of the Itoman City Government Building: *left*, aerial view; *right*, closer view of the southern louvres.

operate during power failure in case of disaster. As the Itoman City Hall is in the coastal area, consideration of corrosion protection against briny air had been taken into account. The solar cells are protected with two layers of tempered glass provided on both the front and back sides to have a self-washing effect by rainwater.

5.4.3 A PV Integrated Shelter for the Bus and Taxi Terminal at Kanazawa Station

Figure 5.10 shows the PV system of capacity 110 kW_p installed as a shelter for a bus and taxi terminal at Kanazawa station (36°34′41″N 136°38′51″E) in Ishikawa and has been in operation since 2005.[20] Amorphous silicon modules made by the Kaneka Corporation were used in the array. The electricity generated by the system is used for lighting of underground passages and for displays.

5.4.4 The Solar Ark Building at Gifu Prefecture

Figure 5.11 shows the Solar Ark building constructed by Sanyo at the Gifu Prefecture (35°19′50″N 136°40′17″E). Around 5046 PV panels were incorporated and have a cumulative capacity of about 630 kW_p to produce electricity.[21] The Ark is 315 metres long, weighs 3000 tonnes and can produce 530 MWh of clean energy per year. It includes 412 lighting units between the solar panels, each with 51 red, green and blue LEDs. The entire system is computer controlled and can create a variety of images and characters. The outstanding fact is that most of the monocrystalline modules used were factory rejects headed for the scrap pile. The Ark stands as one of the best examples of BIPV design.

Figure 5.10 A PV integrated system for the shelter of a bus and taxi terminal at Kanazawa station.

Figure 5.11 The Solar Ark building constructed by Sanyo at the Gifu Prefecture.

5.5 Implementation in the United States

In the United States, there is no single market for PV systems but a conglomeration of regional markets and special applications for which PV modules offer the most cost-effective solution. Since 2002, the grid-connected market has been growing much faster owing to a wide range of "buy-down" programmes, sponsored either by individual states or utilities. The main breakthrough was reached when the 2005 Energy Bill was passed by the Senate. In January 2006, the California Public Utilities Commission (CPUC) passed the "California Solar Initiative" to offer rebates on solar PV systems.

In 2008, the US was the third largest market, with 342 MW of PV installations, 292 MW of which were grid connected.[22] California, New Jersey and Colorado account for more than 75% of the US grid-connected PV market. In 2008 the cumulative installed capacity was around 1.15 GW (768 MW grid connected). Production grew by 53% to 414 MW, mainly driven by the production increase of thin-film manufacturers United Solar (a-Si) and First Solar

(CdTe). The US market share in the thin-film market is around 28% and much higher than the overall market share of 6%.

Despite the increase of grid-connected PV system installations during recent years, with growth rates of around 40%, much still needs to be done to reach the targets of the "One Million Roofs" initiative by the end of 2010. According to the Solar Program,[22] the residential and commercial price targets are based on current retail electricity prices and take into consideration the rather optimistic projection of the Energy Information Administration (EIA) that electricity prices will remain fairly constant (in real terms) through 2025. With these assumptions, the Program predicts that meeting the solar market cost goals will result in 5–10 GW of PV systems installed by 2015 and 70–100 GW by 2030 in the US.

5.5.1 The PV Mounted Roof at Toyota's NAPCC Manufacturing Plant

Figure 5.12 shows the PV mounted roof of Toyota's North American Parts Center California (NAPCC) manufacturing plant (32°02′41″N 113°33′14″W) in Ontario. It is the biggest single-roof solar installation in North America and has been in operation since September 2008. The array was built by SunPower using 10 417 PV modules and covers around 22 500 m^2 of surface area. With a total capacity of 2.3 MW_p, the installation is capable of generating 3.7×10^6 kWh of electricity annually. It is expected[23] that the energy generated would meet 60% of the demand of the manufacturing plant. This would avoid about 6.4 million pounds of carbon dioxide emissions annually, which is the

Figure 5.12 The PV mounted roof at Toyota's NAPCC manufacturing plant.

equivalent to the emissions from the energy use of approximately 255 homes in a year.

5.5.2 The PV Mounted Roof at Google Corporate Headquarters (Googleplex)

Figure 5.13 shows the Google Corporate headquarters in Mountain View (37°25′20″N 122°5′0″W), California, with a PV panel mounted over the roofs of the buildings and several car park shades. The array consists of 9212 PV panels produced by Sharp Electronics and has a capacity of 1.6 MW_p. The system stated operation officially on 18th June 2007. In one day the system generated 9.468 kWh of electricity. It is estimated that the electricity produced would be sufficient to meet about 30% of the energy demands for the site. That is enough energy to reduce CO_2 emissions by over 1650 tonnes annually. The panels have an average lifespan of 25 years, while the projected payback time is 7.5 years.[24] If the PV array functions even at half its capacity over the projected lifespan of the panels, Google will have saved over $5 million in that time.

5.5.3 The PV Mounted Roof of the California State University Buildings at Hayward Campus

Figure 5.14 shows the PV modules mounted over the roof of the four buildings of the California State University in East Bay at Hayward campus (37°39′24″N

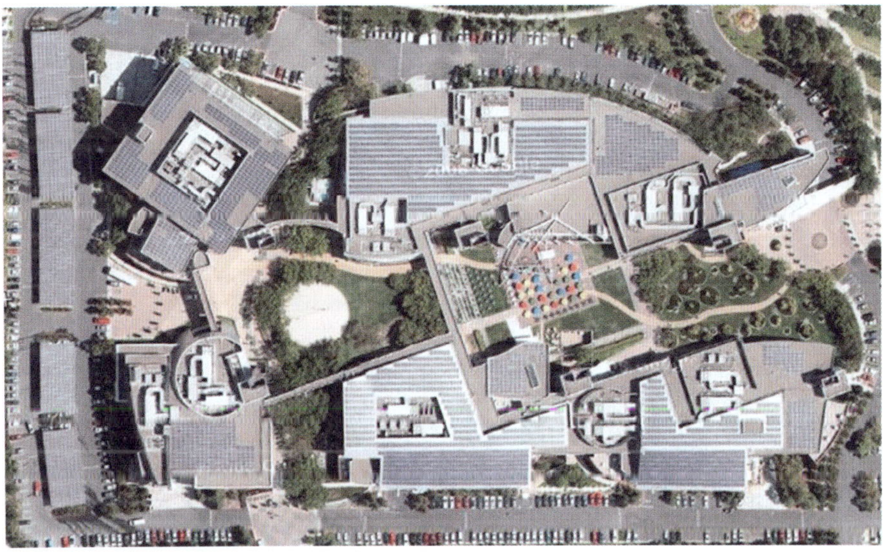

Figure 5.13 The PV system mounted on the roof of Google's headquarters in Mountain View.

Figure 5.14 PV modules mounted over four roofs of California State University at Hayward.

122°3′20″W). The array covers around 7000 m^2 surface area of the roof and has a capacity of producing 1.01 MW_p of electricity. The BIPV system can meet nearly 30% demand of the campus' peak electricity. At the time of its completion in March 2004, it was the largest solar electrical system at any university in the nation.

The BIPV system installed uses SunPower's PowerGuard Roof Tile technology, a lightweight BIPV roofing assembly that is installed over an existing roof membrane.[25] PowerGuard tiles fit together with interlocking tongue-and-groove side surfaces that enable the assembly to resist wind uplift without roof penetration. In addition to generating power, SunPower's solar roof tiles provide thermal insulation to decrease building energy consumption and reduce heating and air-conditioning costs. SunPower's roofing system also extends the roof's life by protecting the roof membrane from harsh UV rays and thermal exposure.

5.5.4 The Farmingdale Rooftop PV System on Long Island

Figure 5.15 shows a flat roof mounted with a PV system in Farmingdale (40°44′47″N 73°25′33″W). LIPA and the Fala Direct Mail group installed the 1.01 MW_p PV system containing 13 424 panels made by PowerGuard and covering 10 270 m^2 of surface area. The system generates around 1 GWh of power annually, reducing Fala's consumption by 20% and its peak load

Figure 5.15 A flat roof mounted with a PV system in Farmingdale.

electricity demand by an estimated 33%. In addition to generating electricity the PV system insulates the buildings, thereby reducing the costs of heating and air conditioning. Over the course of 25 years, the system will avoid emitting over 20 000 tonnes of CO_2, more than 30 tonnes of N_2O and more than 75 tonnes of SO_2 from burning fossil fuels.

The dc output from the PV modules is converted to ac electricity by inverters located at each building and then stepped up to 480 V, three-phase ac electricity by isolation transformers for connection to the buildings' service panels.[26] The modular tiles interconnect easily with no roof penetration, and come with a 20 year limited warranty on electricity output. The aerodynamic design is stable under wind speeds of up to 225 km s^{-1}. There are virtually no maintenance requirements and hence near-zero operating costs. The tiles also protect the roof for 30 years from thermal cycling and UV degradation.

5.6 Implementation in Korea

Before the 1980s, PV activities in Korea were in an embryonic stage. It was in December 1987 when the Korean Government announced the "Promotion Act for the New and Renewable Energy Development" that some sizable budget started to support PV R&D. In 1997, the Government formulated the "Ten-year National Plan for Energy Technology Development (1997–2006)", which included renewable energy, energy saving and clean energy technology. However, the economic crisis in late 1997 forced the funding to remain at an insufficient level. A renewed effort was initiated in 2001 by the "First Basic Plan for Renewable Energy Development and Dissemination". The Government magnified the effort in 2003 by enacting the "Second Basic Plan for Renewable Energy Development and Dissemination", which contained an ambitious plan for PV dissemination of 1.3 GW_p by 2012.[27] Korea sees PV systems as a promising industry which could create a large number of jobs by utilizing the infrastructure and the resources of semiconductor and display industries, where Korea has a world-class leadership.

The market expanded accordingly to about 6 MW in 2005, which was about 140% growth from 2.5 MW in 2004. In 2006, 2452 systems of total capacity 6.5 MW_p were installed on single-family houses. PV systems on multifamily apartments were a new application in 2006 and it is expected that this will be a popular market. Under the programme, beneficiaries pay 30% of the total system price. In 2008, Korea came fourth in the list of installed PV capacity according to EPIA statistics.[28] This position was achieved owing to a favourable feed-in tariff system with a cap of 500 MW_p in 2008. In that year the Korean government changed its successful feed-in tariff system. As a result of this change, the recently ignited Korean PV market is likely to shrink over the next three to four years compared with the success in 2008. The forecast market size in 2009 is around 98 MW_p, increasing to 132 MW_p by 2010. Further growth to more than 160 MW_p is foreseen for 2011. The government aims to reach 1.3 GW_p of cumulative installation by 2012. The emphasis is on the development of rooftop BIPV systems for residential use. This includes 100 000 rooftop systems each with a capacity of 3 kW_p for the roof of single-family house, 40 000 systems each with a capacity of 10 kW_p for public buildings and 30 000 systems each with a capacity of 20 kW_p for the commercial sector.

5.6.1 A Sun Room Integrated with Semi-transparent PV Modules

Figure 5.16 shows external and internal views of semi-transparent PV modules as glazing components in a sun room and covered balcony used by Park *et al.*[29] in Korea. The system consists of layers of glass, encapsulation material (EVA sheet), PV cells, EVA sheet, glass, an air gap with a spacer and glass. Figure 5.16 also shows the monitoring results of irradiance, outdoor temperature and temperature variation of semi-transparent PV modules during spring (a typical day during May).[29] It was found that the temperature of an air gap among layers of the PV module was the highest and was followed by that of the PV cell-supporting glass.

The PV module with bronze glass had a higher temperature compared to the module with clear glass. However, the clear glass used at the backside of the PV module provides more daylight inside the building as its light transmittance is 92%, compared to the bronze glass' rate of 43%. With the assumption that the surface temperature of the backside glass is equivalent to the PV module temperature, the temperature of the PV module rose by 55 °C on a clear day.

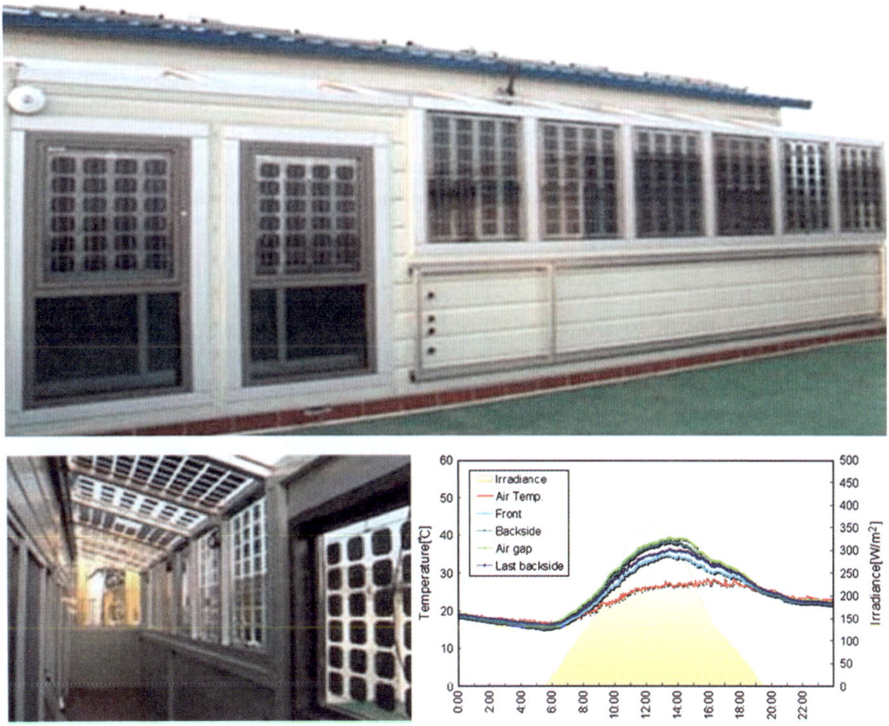

Figure 5.16 Semi-transparent PV integrated sunroom: *top*, external view; *bottom left*, inside view; *bottom right*, temperature variation of different layers of modules on a typical day in May.

The power generation of the PV module decreased by about 0.48% per °C increase in the indoor test (standard test conditions) and decreased approximately 0.52% per °C increase in the outdoor test (under 500 W m^{-2} of irradiance).

5.6.2 The BIPV System at the Samsung Institute of Engineering & Construction Technology

Figure 5.17 shows the PV system integrated on the roof (marked as A and B) and south façade (marked as C) of the Samsung Institute of Engineering & Construction Technology.[30] The PV cells are mounted on the south façade and on the roof of the Instiute in the Giheung area in the southern part of Seoul, Korea. Special care was taken in the building design to have the PV modules shade the building in the summer so as to reduce cooling loads, while at the same time allowing solar energy to enter the building during the heating season and provide daylight. As part of certain design compromises that took into account aesthetics, safety and cost considerations, non-optimal tilt angles and

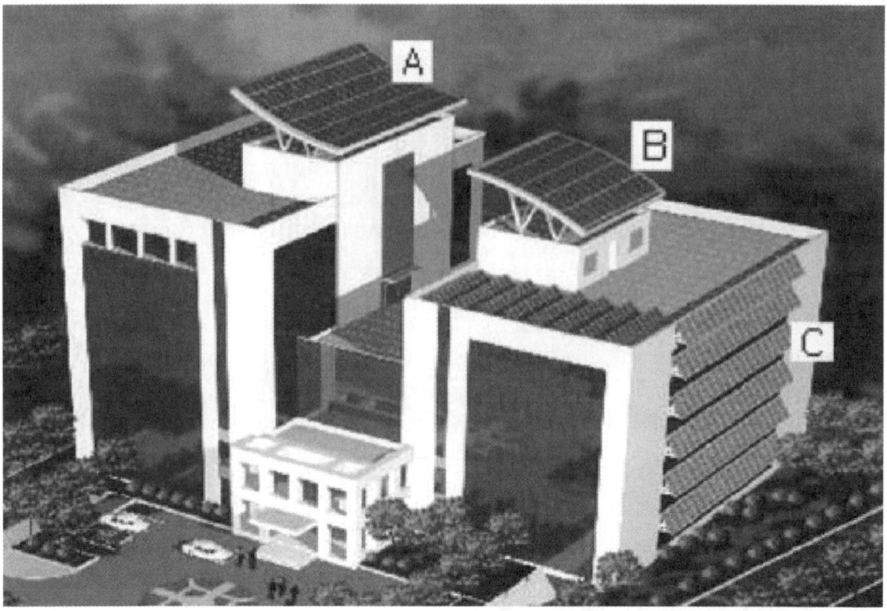

Figure 5.17 The BIPV system in the Samsung Institute in Giheung, Seoul.

occasional shading of the PV modules made the efficiency of the PV system lower than the peak rating of the cells. The yearly average efficiency of the sunshade solar panel is 9.2% (average over 28.6 °C surface temperature), with a minimum of 3.6% (average over 27.9 °C surface temperature) in June and a maximum of 20.2% (average over 19.5 °C surface temperature) in December.

5.6.3 The PV System on the Parking Lot of Hampyeong World Butterfly and Insect Expo

A 2 MW_p solar plant is located in the grounds of the 2008 Hampyeong World Butterfly and Insect Expo, held annually in Hampyeong County. The solar plant is raised on steel beam structures above the parking lot and occupies about 47 000 m^2 of the exhibition grounds and has an output of 2.4 GWh per year.[31] The plant also reduces CO_2 emissions by as much as 1500 tonnes per annum. The expo draws about 2 million visitors per year, making it a good location for raising the public's awareness of alternative energy.

5.7 Implementation in Italy

The Italian Ministry of Environment in cooperation with regional administrations launched in 2001 the National Tetti Fotovoltaici (Photovoltaic Rooftops) Programme to support the installation of grid-connected photovoltaic plants (1–20 kW_p) integrated in buildings (roofs, façades and other elements, *e.g.* urban infrastructures).[32] The first demonstrative part of the programme was restricted to PV plants realised in public buildings; afterwards the subsidies were also extended to private buildings through regional grants. The financial support was provided by a subsidy grant to the initial investment of up to 75% of the eligible cost of the plant, including components, support structures, integration, planning and labour. Based on the market price, an acceptable cost has been fixed in the different areas directly managed by the regions.

A feed-in law for renewable energies, introduced in July 2005, offers a very attractive support scheme, mixing net-metering and a well-segmented premium feed-in tariff paid on the electricity produced by PV technologies. This feed-in tariff (FiT) is paid by GSE (Gestore dei Servizi Elettrici) and changes according to the plant size and level of building integration. The incentives are granted for 20 years up to a limit of 1200 MW.[33] With the issue of the "Conto Energia" the PV market rose to a cumulative capacity of 430 MW at the end of 2008, of which 338 MW_p was installed in 2008 alone. In September 2009 the policy separated installations into two segments: ground mounted and rooftop.[34] Within these two segments, installations are separated into five different categories depending on system size. The proposal lowers FiTs between 5% for the smallest installations and 30% for the largest installations. The proposal states that the installations for the smallest rooftop and ground-mounted systems (ranging from 1 to 6 kW_p) would decline by 5% *vs.* 2010 FiTs. For projects ranging from 6 to 20 kW_p, tariffs would decline by 7% from 2010 levels; for 20–200 kW_p, tariffs would decline by 14%. Ground-mounted systems tariffs for projects from 200 kW_p to 1 MW would decline by 16% while tariffs for rooftop projects would decline by 22.5%. For projects greater than 1 MW, tariffs for ground-mounted projects would decline by 30%, while tariffs for rooftop projects would decline by 27% from 2010 FiTs. The strongest market segment in 2010 will be commercial customers realizing medium to large BIPV systems.

5.7.1 The Roof-mounted PV Plant at the Politecnico di Milano

Figure 5.18 (*left*) shows the PV plant at the Politecnico di Milano (45°17′N 9°10′E), installed on the roof of the Faculty of Architecture in the year 2001.[35] The 150 multicrystalline silicon modules of size 1220 mm × 580 mm (made by Shell Solar), each with 75 W_p of rated power, form a 11.25 kW_p PV array. Each PV module is a glass–Tedlar laminated wafer which incorporates 36 multicrystalline cells and offers an efficiency of about 10.6%. The PV modules are fixed to the sheds by means of special steel supports, L-shaped (edges) and U-shaped (middle), oriented towards the south with a tilt of 60°. The 10 cm air gap between the external shed surfaces and the PV modules allows convective air flow and thus the natural ventilation of both the surfaces, especially in the hot season. The modules are connected together in order to constitute eight strings altogether: six 20-module strings and two 15-module strings. Three 3 kW

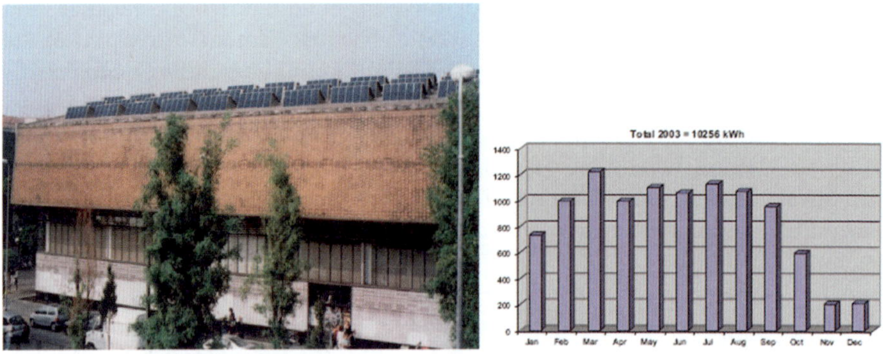

Figure 5.18 Roof-mounted PV system at the Politecnico di Milano: *left*, aerial view; *right*, monthly energy production.

inverters convert the power from the six 20-module strings and two 1.1 kW inverters convert the power from the two 15-module strings.

Figure 5.18 (*right*) represents the monthly energy production of the PV plant, measured during 2003.[35] The complete working-record database for the five years operative life of the PV plant allowed development of a simulation model, very useful to elaborate reliable forecasts of electrical production at similar installations. The total electrical production of the Politecnico PV installation over a period of 25 years is estimated at about 24.9 MWh. It should be mentioned that the useful life of the plant is considered to be 25 years; however, the average life of a PV installation could be considerable higher.

5.7.2 The Roof-mounted System with a Hybrid PVT Façade at Orbassano

In 2001, a hybrid PVT façade was commissioned over the roof of a building at Orbassano, Turin, by the Fiat Research Centre (CRF). Figure 5.19 (*top*) shows the solar roof inclined at 53° to the horizontal with a surface area of about 426 m^2, containing a BIPVT system over 160 m^2. The BIPVT system has about 130 PV modules (BP SX150L) containing multicrystalline silicon solar cells. The PV module has a total power capacity of 19.5 kW_p. All the modules are provided with a ventilation gap and thermal insulation at the base. The modules are configured in 22 strings connected in series; 21 are composed of six modules and one is of four modules. The inverters (two units) are installed inside the building to convert the current from dc to ac.

The PV elements produce electrical energy while the forced air circulation through the ventilation gap recovers the thermal heat and cools the cells as well, resulting in an increment in the PV conversion efficiency. The electricity generated by the PV modules is used to feed a heat pump and internal electrical requirements. The heat produced is used by the air-conditioning plant in the kitchen, in winter for pre-heating of the renewal air and in summer to feed the

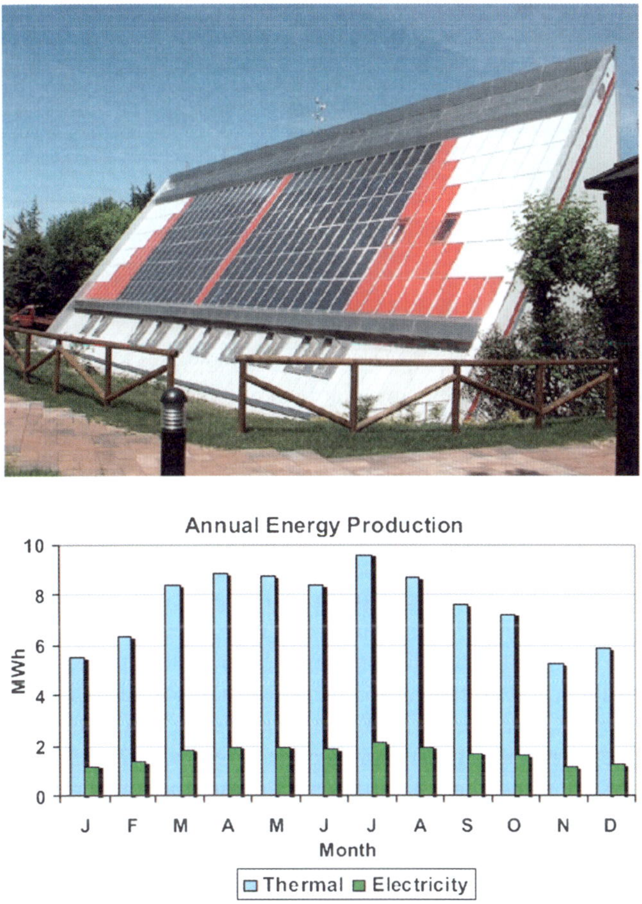

Figure 5.19 Hybrid PVT façade at the Fiat Research Centre: *top*, aerial view; *bottom*, annual thermal and electrical output.

dehumidification system. This integrated solar façade is monitored on-line by a highly sophisticated system, managed by the technicians of CRF.

With variable airflow of $9000\,m^3\,h^{-1}$, the air reaches temperatures up to $60\,^{\circ}C$, according to the specific climatic conditions. The annual thermal and electrical performance results are shown in Figure 5.19 (*bottom*). Butera *et al.*[36] estimate that the performance is around 10% higher with respect to a traditional PV installation. The thermal efficiency varies in average from 20% to 40% during the day (higher values are obtained in the early hours of the morning and late hours in the evening). However, the average electrical efficiency obtained is around 9–10% during the day. The annual production of electricity and heat from the hybrid façade contributes to total primary energy savings of about 185 MWh, corresponding to 36 tonnes of CO_2 emission reduction to the atmosphere annually.

Figure 5.20 PV modules installed on rooftops in the town of Serravalle Scrivia.

5.7.3 The PV Modules Installed on the Rooftop in the Town of Serravalle Scrivia

Figure 5.20 shows PV modules installed on the rooftop of buildings in the town of Serravalle Scrivia (44°45′15″N 8°50′40″E) in the Piedmont region of Italy. At present, this is the largest PV facility in Italy, covering approximately 38 000 m^2 of roof area. The PV modules were supplied by Trina Solar and the installation work was completed by ErgyCapital in February 2009. The plant has the capacity to generate 4.7 MW_p of electricity. The system is expected to generate approximately 5 GW of electricity per year and will benefit the environment by reducing up to 2500 tonnes of CO_2 emissions.[37]

5.7.4 The PV Modules on the Warehouse Roof of Coop's New CNNA-Prato Logistic Centre in Prato

Figure 5.21 shows a PV system mounted on the rooftop of a single building of Coop's new CNNA-Prato logistic centre in Prato. The system was specifically designed with the higher feed-in tariff rate for rooftop installations compared to those on the ground. The installation was completed by Mitsubishi Electric in November 2009.[38] There are 15 710 lead-free solder PV modules, having 2.906 MW_p capacity. Out of these, 15 650 modules have been installed on the warehouse roof, covering a surface of 21 000 m^2. The remaining 60 modules have been installed on the parking lot grounds. The system is expected to generate 3.2×10^6 kWh per annum, which will help to meet the energy needs of the logistic centre and an estimated 500 MWh of excess electricity to be distributed into the national grid network.

Figure 5.21 PV modules on the warehouse roof of Coop's new CNNA-Prato logistic centre.

5.8 Implementation in the People's Republic of China

China started solar cell research in 1958 and the first successful application of solar cells was in a satellite called the "No. 2 East Red Satellite" in 1971. The terrestrial application of solar cells started in 1973. Limited by high price, the initial development of the PV market was quite slow, with the main application being to provide power for satellites. During the period 1981–1990, PV applications were promoted in specialized industries and rural areas, for example in solar-powered microwave relay stations, military communication systems, cathode protection systems for sluice gates and oil pipelines, carrier wave telephone systems in rural areas, small-scale solar household systems and central power supply systems in villages.

Between 2003 and 2005 the European PV market, especially the German market, further encouraged the rapid development of production capacity in China. By the end of 2005, total production capacity of solar PV modules had reached 400 MWp, of which 140 MWp was manufactured in 2005 alone. Most of these modules are exported to the European market, with only 5 MW_p in 2005 and 10 MW_p in 2006 installed in China. In 2008, 40 MW_p of new PV installations were made, making the cumulative installed capacity 145 MW_p.

Although a series of laws and standards have been launched for motivation in BIPV projects, such as the Renewable Energy Law, Civil Building Energy Saving and Energy-Saving Design Standard of Various Regions, the high cost remained a large barrier.[39] Neither enterprises themselves nor government can afford such a high cost in the long run. Therefore, BIPV systems capture a considerably small market now in China and are mainly used in some demonstrative projects which enjoy government subsidies. However, this situation has changed since 2009. In March 2009, the government implemented the Solar Roof Program to promote the integrated application of solar photovoltaic buildings in cities and give fixed-amount subsidies to photovoltaic utilization in buildings in rural and remote regions. The following Golden Sun policy issued in July 2009 brought about a real demand of 642 MW_p through 294 projects.[40] Guided by government policies, several large-scale grid-connected power plant projects were approved in 2009. These include the 1 GW photovoltaic power plant project in Qaidam, a 200 MW photovoltaic power plant project in Golmud and a 500 MW photovoltaic power plant project in Wuzhong. BIPV systems are widely applied on exposed walls, sun-shading shelters, patios, tiles, roofs and sound-proof walls, as well as in fields of private apartments, schools, hospital buildings, airports, platforms of subway stations and large workshops. Thus, the BIPV market will see a bright prospect in the future. It is estimated that the accumulated installation capacity of the Chinese PV industry will reach 20 GW_p by 2020 and this prediction is just the same with the energy bureau.

In the last few years, a number of medium-scale BIPV systems were completed in Hong Kong, including government buildings, commercial buildings, schools and institutional buildings. Most of these BIPV systems are of the gird-connected type to eliminate the bulky and troublesome battery banks. Some of these projects are retrofitted on existing buildings, while others are parts of new buildings.

5.8.1 The BIPV System on the Shaw Amenities Building of Hong Kong Polytechnic University

Figure 5.22 (*top*) shows the grid-connected BIPV system installed on the roof and walls of the Shaw Amenities Building (22°18′16″N 114°10′45″E) at Hong Kong Polytechnic University, with the aim to investigate the cost-effective design under Hong Kong weather conditions and demonstrate the technology to the local industries and market.[41] Figure 5.22 (*bottom*) shows the circuit diagram of the installation. The system consists of 100 PV panels (made by BP) each of 80 W_p and containing monocrystalline silicon solar cells. PV systems integrated on the three walls and the roof of the building cover 55 m^2 of surface area. The number of PV panels integrated on the horizontal roof, vertical east facades, vertical west facades and vertical south facades are 40, 20, 18 and 22, respectively. In order to increase the dc voltage, 7 PV modules are connected in series. An air gap is maintained between the massive wall and the vertical façades to facilitate natural ventilation. This would keep the modules' temperature lower and provide relatively higher efficiency during hot summer conditions. Owing to the grid connection, battery banks were not required, saving approximately 20% of the initial investment compared with remote PV systems. The system is rated at 8 kW with an output dc voltage of 75–105 V and an output ac voltage of 220 V.

As the PV systems are integrated into the rooftop and façades at three different orientations, all of them have different power outputs. Table 5.1 gives the results of the monthly energy output from the BIPV systems at different orientations.[41] On sunny days the system produces sufficient electricity for the lighting circuit of a 250 m^2 floor area. The horizontal roof PV panels produce more power compared with the same surface area of PV panels on the walls,

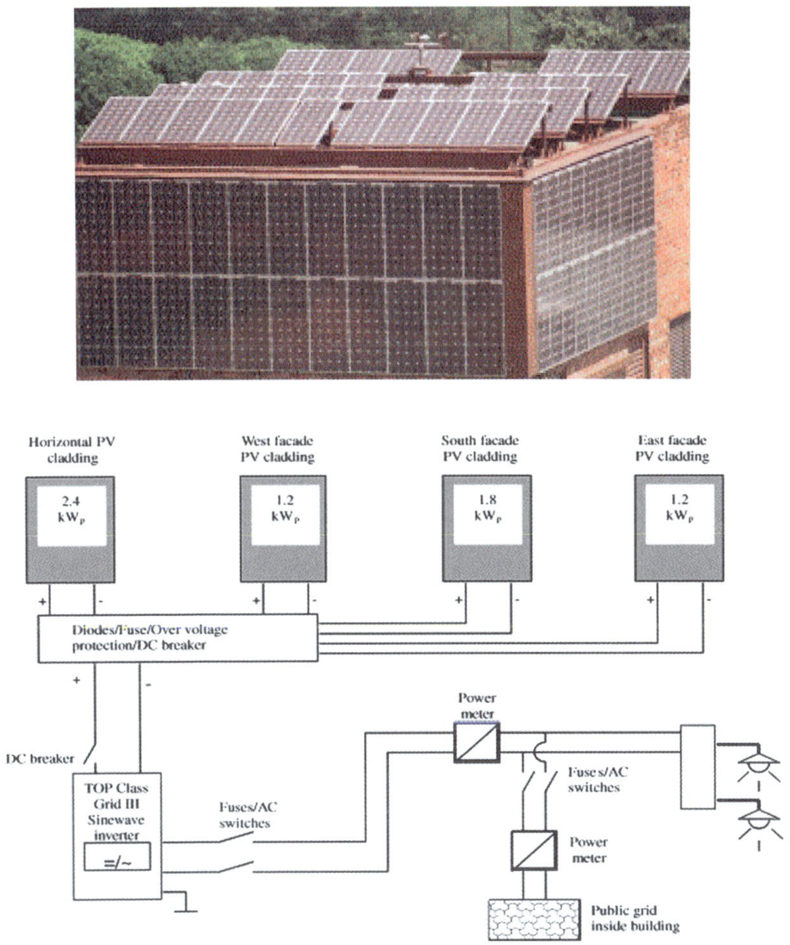

Figure 5.22 The BIPV system at Hong Kong Polytechnic University campus: *top*, aerial view; *bottom*, circuit diagram.

Table 5.1 Monthly power output from PV arrays for different orientations (in kWh)

Façade	*Jan*	*Feb*	*Mar*	*Apr*	*May*	*Jun*	*Jul*	*Aug*	*Sep*	*Oct*	*Nov*	*Dec*	*Total*
Roof	294	220	231	252	294	286	367	347	346	389	346	336	3708
South	156	90	72	48	36	31	36	54	90	150	163	180	1106
West	72	60	63	72	88	88	116	106	99	99	86	83	1032
West	72	60	63	72	88	88	116	106	99	99	86	83	1032
Total	592	430	429	444	506	493	635	613	634	737	681	682	6878

since the annual average solar incident angle is the smallest compared with the solar incident angles of the other three facades. Out of the total annual energy required for the lighting, 16 700 kWh, 41% of the energy is supplied by the installed BIPV system with an overall efficiency of 9%. The total harmonic from the BIPV system is less than 12% for most of the time, even when the incident solar irradiation is very weak.

The power price of the BIPV system is HK$ 1.5–2.0 per kWh, while the average price of electricity purchased from the two local power companies is about HK$ 0.90 per kWh. The cost of the BIPV systems with monocrystalline silicon PV modules is about HK$ 40 per W_p, including installation and other component costs (inverters, safety control and cables). When the cost reduction of the building façade outer skins due to PV integration is considered, the payback period for the roofs is about 20–30 years. Hence, the analysis shows that a grid-connected BIPV application is still not economical, but the technology should be promoted due to its huge potential in terms of environmental protection and future development, *e.g.* greenhouse emission costs, business losses due to pollution in urban areas and medical cost increases caused by pollution from conventional power generation plants.

5.8.2 The PV Mounted Roof of the Hong Kong EMSD Headquarters

Figure 5.23 shows the aerial view of the PV mounted roof of the headquarters of the Electrical and Mechanical Services Department (EMSD) (22°19′33″N 114°12′12″E) of the Hong Kong Special Administrative Region.[42] The figure also shows the view of the semi-transparent modules fitted on the glazing roof from the inside. The horizontal roof contains nearly 2300 PV modules tilted at 20° to the horizontal facing south and covers 7900 m^2 of surface area. Each PV module consists of 72 series-connected monocrystalline silicon PV cells. There are in total 20 sets of semi-transparent PV modules integrated into the roof of the gallery for day lighting in addition to electrical power, each containing 100 series-connected monocrystalline PV cells sandwiched between two sheets of glass. The total installed power capacity of the two PV systems is 350 kW_p. The dc output of the installation is converted into ac by a number of inverters. The ac power generated by the PV installation supplements the electricity supply from the power company to meet the electrical power requirement of the building. Figure 5.23 also shows the power output for a year from October 2005 to September 2006.[43] It is observed that the annual energy output is around 400 MWh. This would save 280 tonnes of CO_2 emission into the atmosphere annually.

5.8.3 The PV System Integrated with Permanent Buildings at Shanghai World Expo 2010

Figure 5.24 shows the Theme Pavilion, the China Pavilion and the Expo Center buildings, respectively, at the Shanghai World Expo 2010 (31°11′38″N

Figure 5.23 PV panels fitted over the horizontal roof of the Hong Kong EMSD Headquarters: *top left*, aerial view of the building; *top right*, aerial view of the roof fitted with PV modules; *bottom left*, inner view of the glazing roof of the gallery; *bottom right*, annual yield for 2005–2006.

Figure 5.24 Permanent buildings of the Shanghai World Expo 2010 integrated with PV systems: *top left*, the Theme Pavilion; *top right*, the China Pavilion; *bottom*, the Expo Center buildings.

121°29′10″E) that have adopted grid-connected BIPV applications.[44] The Theme Pavilion boasts a 2.8 MW integrated solar system comprising 96 custom-designed triangular sections and covers an area of 31 104 m^2. The China Pavilion contains 0.3 MW$_p$ PV modules combined with traditional Chinese architectural characteristics. Under the current plan, the entire exposition of solar PV power generation will reach a total installed capacity of 4.5 MW, an estimated annual saving of about 420×10^6 kWh and about 4100 tonnes of CO_2 emissions.

The 2010 Shanghai World Expo is not only a large-scale BIPV grid power plant display but also a demonstration of the applications of solar energy technology in a variety of landscapes and displays. There are also plans for other applications, such as solar street lights, solar garden lights, solar lawn lamps, solar electronic displays, solar sculptures, solar fountains, solar kiosks, solar ice bars, solar kiosks, solar mobile toilets, solar bus shelters, solar boats, *etc*. The large-scale solar energy technology applications and a variety of demonstrations enable people to experience solar energy technology to bring about changes in human life, and fully reflects that the Expo will be involved in addressing global climate change and sustainable development of human society problems, reflecting the "Better City, Better Life" theme; the promotion of solar and other new energy technologies will expand the applications.

Figure 5.25 Rooftop-mounted PV system at International Garden and Flower Expo Park.

5.8.4 The PV System at the International Garden and Flower Expo Park in Shenzhen

Figure 5.25 shows the roof-mounted PV power plant installed across the buildings of Shenzhen's International Garden and Flower Expo Park.[45] The plant has a capacity of 1 MW_p electricity output and covers around 5325 m^2 of surface area. Over 4160 BP Solar PV modules, each of 160–170 W, were connected to form the solar array, which ranges in capacity from 90 to 280 kW_p. A total of 24 central and string inverters ranging in size from 2.5 to 90 kW were installed in string and centralized configurations to maximize the scale and minimize the potential for power reduction from any shading. It is estimated to generate annually 755 MWh of energy and reduce CO_2 emission by 1050 tonnes. State-of-the-art communication systems were also integrated into the system design to allow for simple data collection and remote monitoring of the systems, which will go a long way towards minimizing maintenance costs over the long term.

5.9 Implementation in Taiwan

According to the Industrial Technical Research Institute's Photovoltaic Technology Center,[46] at the end of 2008, nationwide Taiwan has 384 completed PV system installations with a combined installed capacity of 4 MW_p. Following the passing of the "Renewable Energy Development Act", it is hoped that by 2012 the accumulated installed capacity will reach 60 MW_p.

5.10 Implementation in Australia

Australia has led the world in the development and applications of PV systems for telecommunications and remote area power supplies since the 1970s and 1980s. The then Telecom Australia installed one of the world's first and largest

PV powered microwave repeater stations in central Australia in 1978 and continued to run an important PV research, development and testing program at its Telecom Research Laboratories until Telstra was formed.[47] In 1981, Australian National Railways installed a PV powered signal relay system, one of the world's largest at the time. Australia also established solar car racing, which raised public awareness of PV technology and also stimulated international research efforts in both PV technology and a balance of system components, such as inverters and batteries.

The current market growth in Australia is for off-grid industrial and agricultural applications. These include power systems for telecommunications, signalling, cathodic protection, water pumping and lighting. Significant markets also exist for off-grid residential and commercial power supplies and increasingly for fuel saving and peak load reduction on community diesel grid systems. Some of this market is supported by government grants aimed at reducing diesel fuel use. There is a growing market for recreational PV applications, for caravans, boats and off-road vehicles. PV installations connected to central grids have increased strongly, with the majority of installations taking advantage of government grant programmes which currently contribute 20–25% of up-front capital costs. The main applications are rooftop systems for private residences, schools and community buildings, although commercial interest is also growing. This unreached grid parity is mainly due to the higher cost per kW than other power sources because of the cost of solar panels.

Feed-in tariffs and mandatory renewable energy targets are designed to assist renewable energy commercialization. In the year 2008, 22 MW_p of new setups were added, reaching a goal of 100 MW_p cumulative installations. The government programme that supports the PV market includes the PV Rebate Programme and the Renewable Remote Power Generation Programme. Some support is also provided via the Mandatory Renewable Energy Target (MRET) and Green Power programmes. From March 2009, the Australian Capital Territory (ACT) has introduced a solar gross feed-in tariff. For systems up to 10 kW the payment is 50.05 cents per kWh and for 10–30 kW the payment is 40.04 cents per kWh.[48] The payment is made quarterly, based on energy generated. The tariff rate is guaranteed for 20 years. On 8 June 2009 the Solar Credits Programme was brought into action, where an installation of a solar system would receive five Renewable Energy Certificates (REC) for the first 1.5 kW of capacity rather than the standard 1 REC.[48] There is now an estimated of 115 MW_p PV power installed, contributing only 0.1–0.2% of total electricity production despite the hot, dry and sunny climate that would make it ideal for utilization. The federal government MRET is that renewable energy will obtain a 20% share of electricity supply by 2020.

5.10.1 The BIPV System at Kogarah Town Square

Figure 5.26 shows an aerial view of the Kogarah Town Square building (33°57′50″S 151°7′60″E) with amorphous silicon solar cells integrated into the

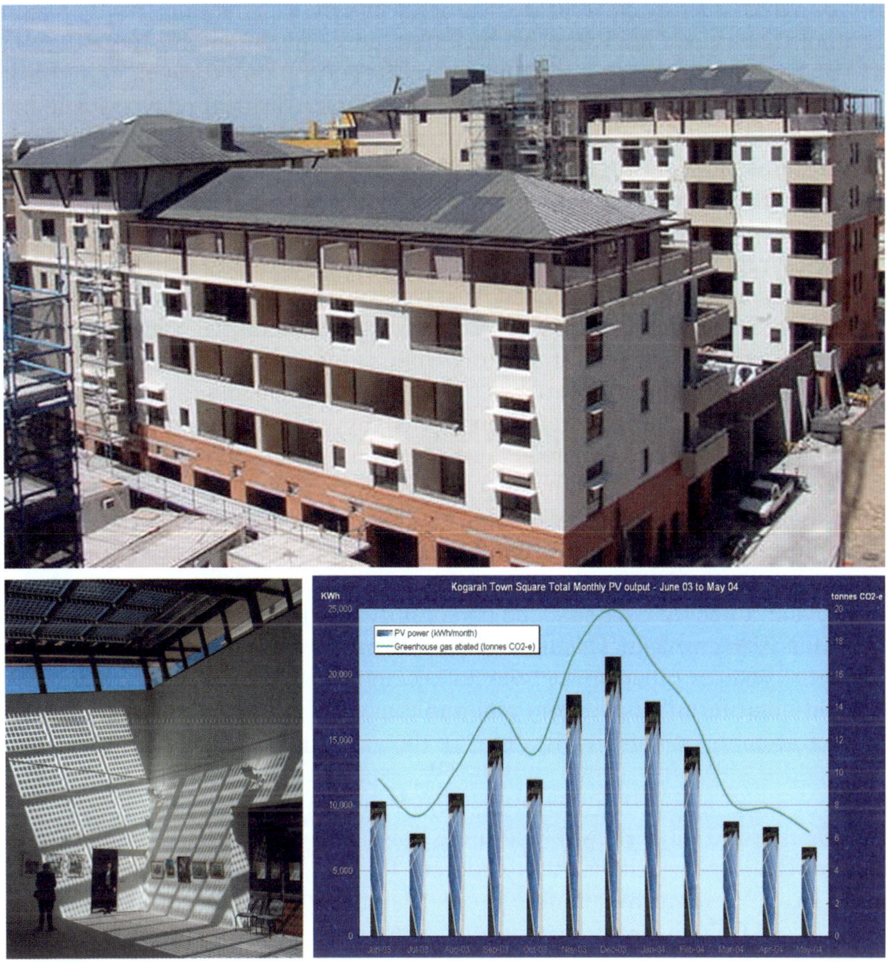

Figure 5.26 The BIPV system of Kogarah Town Square: *top*, aerial view of the building; *bottom left*, translucent PV exhibition view from inside; *bottom right*, annual performance, 2003–2004.

roof. Figure 5.26 also shows the internal view of the exhibition hall integrated with semi-transparent polycrystalline solar cells. A total of 1459 PV modules were integrated over three buildings of four to six stories and underground parking, covering 2800 m^2 of the surface area. PV modules were placed on the whole roof surfaces rather than just the north-facing roofs. Whilst the amorphous PV roof is a flexible building material that is simply laminated onto a steel spandek roof, much of the problems encountered relate to the roof penetrations, the locations of which were not clearly determined prior to construction.

The system has an overall rating of 161 kW_p (131 kW_p Unisolar amorphous cells and 30 kW_p BP Solar polycrystalline cells).[49] A total of 59 Sunpower

SP2500 inverters were installed in ventilated and easily accessible cupboards on landing areas, many in close proximity to the lifts. To reduce wiring runs the ac output powers corridor lighting, signage and contributes to the lift. Any excess power is exported to the distribution grid. Figure 5.26 shows the performance output of the BIPV system observed for a year, from June 2003 to May 2004.[49] The amount of power produced during the peak varied widely between 3.5 and 80.8 kW. The system meets 15–20% of the building electrical load, depending on the favourability of climatic characteristics. Several inverters failed during the period, indicating that regular monitoring and maintenance is essential to maintain the reliable performance. It is expected to generate 153 MWh of electricity per annum, with greenhouse gas savings of 140 tonnes per annum.

5.10.2 The Grid-connected BIPV Power Station at Adelaide Showground

Figure 5.27 shows a 1 MW_p PV power station mounted on the rooftop at the Adelaide Showground (34°56′40″S 138°35′20″E) in Goodwood Road. The station consists of 12 612 thin-film CdTe panels, each of 77.5 W_p and made by First Solar, installed over six building roofs (Goyder Pavilion, Jubilee Pavilion, Wayville Pavilion, Ridley Centre, Alpaca Pavilion and Dairy Cattle Pavilion) and 108 semi-transparent polycrystalline panels each of 200 W_p made by SunTech and installed over two building façades covering 9000 m^2 of the total surface area.

Figure 5.27 PV system mounted on the roof of buildings at Adelaide Showground.

The project is funded by a grant awarded to the Royal Agricultural & Horticultural Society by the Department of the Premier and Cabinet.[50] Built Environs in partnership with Solar Shop Australia completed the design and installation of the panels and associated systems by September 2009. The contract includes a 25 year warranty on the PV panels, a 10 year warranty on the inverter systems and 10 year maintenance/service warranty. The Showground solar system is expected to produce around 1.435 GWh of electricity annually, which is enough to meet 40% of the Showground's power needs. This would prevent the release of 1400 tonnes of CO_2 emissions into the atmosphere annually.

5.10.3 The PV System Mounted over the Tilted Roof of the Crowne Plaza Hotel in Alice Springs

Figure 5.28 shows a PV system mounted over the tilted roof of the Crowne Plaza Hotel in Alice Springs.[51] The PV power system has a capacity to generate 304.28 kW_p of electrical energy. Over 1650 m^2 of the surface area, 1326 SunPower PV modules were installed and brought into operation by the end of February 2009. It is estimated that the BIPV system will generate 531 MWh of electricity per annum, which will reduce the CO_2 emissions by 420 tonnes.[51] The installation will supply 40–80% of the instantaneous power demand during the day, depending on the time of year, and reduce the total load on the Alice Springs power grid by 0.5%. The panels are wired into the main mechanical services circuits, where the power is used by the air-conditioning units and other

Figure 5.28 The PV system mounted on the tilted roof of the Crowne Plaza Hotel.

equipment. The modules have a 25-year warranty on power output and a 10-year product warranty.

5.10.4 The PV Integrated Façade over an Eight-story Building at Melbourne University

Figure 5.29 shows a semi-transparent PV array fully integrated into the façade of a building having a curved surface. The array forms the top level of the northern façade of an eight-storey building in the campus of Melbourne University (37°48′10″S 144°57′40″E) in Parkville. The modules contain poly-crystalline solar cells, encapsulated between two layers of heat-strengthened glass. The outer layer is a 6 mm low-iron layer whereas the inner layer is 6-mm clear float glass. Permasteelisa fitted the laminated panels into frames off-site and then simply bolted them into place on-site from inside the building. The PV façade array is rated at 40 kW_p. Ten single-phase solar grid interactive inverters were used, of which six are rated at 5 kW and four at 6 kW, providing three-phase power. It is expected that the system will generate 40 MWh of electricity every year.[52]

An unusual feature of the system is that it consists of a variety of panel sizes, each with a different number of cells and therefore different output voltages. The panels are connected together (in series) into groups, and so the groups too have different nominal voltages. Each group is connected to a single inverter, so

Figure 5.29 Translucent PV system integrated into the curved façade of an eight-story building at Melbourne University.

the inverters are each designed to have an input voltage characteristic to match the relevant group. The performance of each inverter and the array can be monitored from a connected remote computer.

5.11 Implementation in the United Kingdom

While the UK may not be the most obvious location for solar energy, it does have a good solar resource, ranging from 900 to 1300 kWh m^{-2} per year. Recognizing this, the government has promoted PV technology since the late 1990s and in 2001 a joint government–industry report recommended a £150 million 10-year programme to install PV systems on 70 000 roofs and 1400 larger non-domestic buildings.[53] Although this particular plan has not yet come to fruition, three other notable policies have existed. These include the PV field trials (2000–2002, approximately 1.5 MW_p installed), the Major Photovoltaics Demonstration Programme (MDP) (2002–2006, approximately 6.7 MW_p installed) and the Low Carbon Buildings Programme (LCBP) (since April 2006).

Like the previous MDP, the current LCBP is a grant scheme that provides up to 50% of the cost of small-scale microgeneration systems ($<5\,MW_p$).[53] The programme also continues the MDP's accreditation system, which aims to build consumer confidence through the certification of microgeneration installers and technologies. By 2008, such programmes had successfully increased the installed capacity of PV systems in the UK to approximately 22 MW_p. PV systems are still too expensive for much of the general public and relatively unattractive due to the moderate level of insolation, cheap grid electricity (compared to other European countries) and low financial subsidies from government. Owing to an EU agreement to generate 15% of electricity from renewables by 2020, in June 2008 a new programme to encourage homeowners to generate their own electricity was announced, which will include a feed-in tariff.

5.11.1 The Solar Office at Doxford International Business Park

The solar office is one of the few BIPV projects designed at Doxford International Business Park, near Sunderland in the north east of England.[54] Its building location has on average 230 hours of sunshine per year. The project is supported by funding from the European Regional Development Fund and the UK Department of Trade and Industry has provided support for design development, testing and monitoring of the building performance. The construction of the building was completed in March 1998 with the exception of the internal fit-out, which will be carried out to the requirements of the tenant.

Figure 5.30 shows a photograph of the building-integrated solar façade from external and internal views. The glazed south façade has a fully integrated photovoltaic array sloped at 60° to the ground. The total array area is 646 m^2 with a nominal rating of 73.1 kW_p. The PV system (made by Kyocera) utilizes polycrystalline silicon cells in a glass-to-glass encapsulation. Bands of clear glazing have been introduced into the façade to allow views out and to ensure

Figure 5.30 The Doxford solar office with an inclined BIPV façade: *left*, external view; *right*, inside view.

good internal light levels. The risk of glare is minimized by the introduction of modules with a reduced number of cells immediately above the glazed panels and by provision for the introduction of locally controlled roller blinds. In addition, the heat accumulation behind the array is used in a positive manner to assist with the natural ventilation of the building via the stack effect. Office windows were placed on the north, north-east and north-west to avoid the need for control of solar gains from low morning and evening sunshine.

The system performance to the end of one year observation is shown in Table 5.2. Over the first two years the PV array has generated around 113 000 kWh at an average efficiency of 7.5%.[54,55] The system is sufficient to provide nearly 30% of the electricity used by the occupants over the period of a year. In summer, when the system generates more energy than required, the surplus energy is exported to the National Grid. The project achieved an "excellent" BREEAM (Building Research Establishment Environmental Assessment Method) rating. It was selected as a "Millennium Product" by the Design Council of Great Britain and won the 2000 Eurosolar award.

5.11.2 The Cladding PV Modules over the Co-operative Insurance Tower

The Co-operative Insurance Tower was built in 1962 in Manchester. As part of the buildings refurbishment in the fiscal year 2005–2006, the design team developed a weatherproof cladding which incorporated photovoltaic modules,

Table 5.2 Performance data for the BIPV façade at the Doxford solar office.

Month	*Monitoring fraction*	*Measured ac output (kWh)*	*Estimated loss due to inverter outage (kWh)*	*Estimated loss due to monitoring outage (kWh)*	*Corrected output (kWh)*
Mar	47	1530	0	1725	3225
Apr	97	4607	0	142	4749
May	85	5411	0	955	6366
Jun	100	4240	927	0	5167
Jul	100	4558	485	0	5043
Aug	100	3277	1710	0	4987
Sep	100	2740	497	0	3237
Oct	100	3267	710	0	3977
Nov	77	2330	0	695	3026
Dec	–	–	0	1343	1343
Jan	74	1987	0	698	2685
Feb	100	3687	0	0	3687

shown in Figure 5.31.[56] The cost of incorporating the technology was offset against the projected cost of replacing the traditional mosaic tiles that had previously wrapped the tower. The tower has three of its four sides completely clad with 7244 PV modules of 80 W_p (manufactured by Sharp). This allows the building to harvest the Sun's power throughout the day from dawn to dusk. The front wall is facing south, which is the main recipient of sunlight. The east and west walls receive far less light. As this structure is in the Northern Hemisphere, there is clearly no point cladding the north wall with solar panels. Curiously, only 4898 of the modules are actually functional, but they still soak up enough sunlight to generate 390 kW_p of energy. Even though the PV panels only meet 10% of the building's power needs and will probably never pay for themselves, the Co-operative Insurance Tower authorities have done as much as they possibly could to make this undoubtedly one of the greenest office buildings in the UK.[56]

5.12 Installation in India

India's photovoltaic solar energy market is emerging and set to show strong growth. India already has the world's best solar resources, over 520–630 W m^{-2}, with about 2300–3200 sunshine hours per year, depending upon location. PV systems could offer a solution to a country still struggling heavily with continuous electricity shortages. The cost of solar electricity generation is expected to fall by more than 30% within a span of 10 years (2008–2017), while grid electricity prices are expected to reach break-even point by 2015. Based on these forecast developments, India has the potential to become one of the largest markets in the world before too long.

The Ministry of New and Renewable Energy (MNRE)[57] reports indicate that India does not have silicon feedstock or significant wafer manufacturing,

Figure 5.31 The Co-operative Insurance Tower in Manchester clad with PV modules.

although recent proposals have been made to the government to build these lines. However, there are nine manufacturers of silicon-based solar cells and about twice as many module makers. There are a number of balance-of-system (BOS) manufactures who produce charge controllers, inverters and battery systems. By the end of 2009, around 560 295 solar lanterns, 55 795 street lighting systems, 342 600 home lighting systems, 6820 pumps and 1.5 MW_p power plants are based on PV technology.[58] Popular grid-connected PV plants of capacity 100 kW_p and more are at Jaitpur Kalan village, Kadmat, Kiltan,

Minicoy, Agatti, Andrott, Kavaratti, Mousuni Island and Rakalpur. The largest solar PV plant with a capacity of 239 kW_p has been set up at Steel Tubes of India in Dewas. Cumulative installed solar power in India by the end of 2008 is 110 MW_p, out of which only 3 MW_p were added in the year 2008.

In July 2009, India unveiled a $19 billion plan to produce 20 GW of solar power by 2020.[59] Under the plan, solar-powered equipment and applications would be mandatory in all government buildings, including hospitals and hotels. In November 2009 it was reported that India was ready to launch its Solar Mission under the National Action Plan on Climate Change, with plans to generate 1000 MW of power by 2013.[60] The greatest interest regarding the Solar Mission is not only that it will provide clean and safe PV electricity to millions of people, but also create tens of thousands of jobs at the same time. MNRE expects about 100 000 jobs to be created by the PV industry by 2020 and industry sources believe the figures could be higher once all job linkages are taken into consideration.[57]

5.12.1 The PV System Mounted on the Roof of the WHO South East Asia Office Building

Figure 5.32 shows the PV systems that are mounted on the roof of the World Health Organisation (WHO) South East Asia Office building. Around 568 PV

Figure 5.32 The PV system mounted on the roof of the WHO South East Asia Office building.

modules (made by Mosearbaer) containing polycrystalline solar cells are installed, titled towards the south at an angle of 30° from the horizontal. Fifteen inverters, each of 11 KVA, are used to generate 1000 V. The system has a capacity of producing 142 kW_p output. The system has been under operation from January 2010 and the observed power output for that month was 85 kW_p. It is claimed that the output was lower in January due to the winter season and partial shadows on the arrays, but it will produce higher output in the month of June (summer season). It is further expected that the system will produce about 200 MWh annually. The PV arrays have a water-pipe connection with a hose and flexible pipe for cleaning the modules once a week.

5.13 Recent Research and Conclusions

To generate PV power on a large scale, grid-connected PV power plants are installed these days. Some of the largest plants installed on farms include the 60 MW_p Olmedilla Photovoltaic Park, the 54 MW_p Strasskirchen Solar Park, the 53 MW_p Lieberose Photovoltaic Park, the 50 MW_p Puertollano Photovoltaic Park, the 46 MW_p Moura photovoltaic power station, the 45 MW_p Kothen Solar Park, the 42 MW_p Finsterwalde Solar Park and the 40 MW_p Waldpolenz Solar Park. Setting up PV power plants on a farm does not provide any advantage other than the production of electricity. Moreover, they are not cost-effective if the land cost is added to the overall cost of the plant.

Mounting PV systems over a building would help in reducing the overall cost of the plant by utilizing the unused roof and façade surface area. Recently, a 40 MW_p capacity plant has been mounted over the rooftop of a global logistics service provider with headquarters in Antwerp, Belgium. This is considered to be the largest BIPV system installed up to the end of the year 2009. A few more rooftop-mounted BIPV systems such as the 5 MW_p plant at Burstadt, the 3.84 MW_p plant at Muggensturm, the 2.9 MW_p plant at the Telefonica Business Park Complex and the 5.2 MW_p plant at the Sharp Corporation LCD plant in Kameyama have been discussed. Most of these systems are mounted directly on the flat roof without giving proper inclination to the PV modules. As the cost of PV modules is quite high, to utilize them to their highest capacity they should be inclined towards the equator at an angle equal to the latitude of the location, which is considered optimum for the best performance on an annual basis. If the PV system is to run the air-conditioner for cooling or heating purposes, depending upon the summer or winter season, then they should tilted towards equator at an angle of $\pm 15°$ of the latitude of the site.

A number of PV systems have been mounted over horizontal rooftops with inclination equal to the latitude of the site, including the BIPV system at Toyota's NAPCC manufacturing plant, the Hong Kong EMSD Headquarters and the WHO South East Asia office. Many times it is found that a shading problem occurs, especially during the winter months, which reduces the overall output of the system. To avoid any shading problem a proper distance between the two rows of the PV modules needs to be maintained by the design engineer.

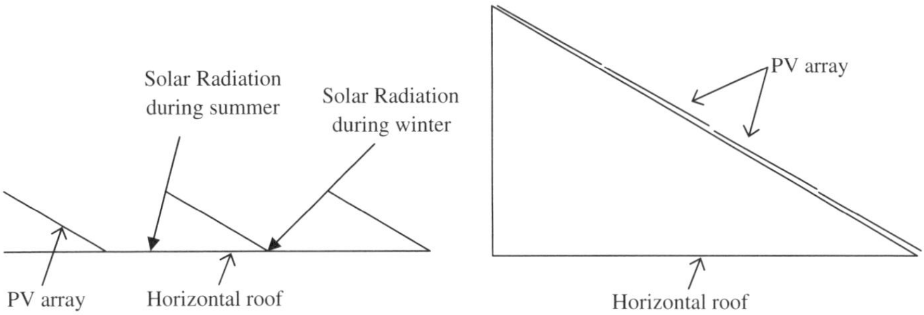

Figure 5.33 A PV system mounted in rows, end view: *left*, on a flat roof; *right*, on an inclined roof.

Let us consider a rooftop, shown by the horizontal line in Figure 5.33 (*left*), where the PV modules are mounted with an inclination. In order to avoid shading throughout the year, the number of PV rows that can be accommodated on the roof is three. Now let us consider a PV system mounted on the rooftop where the whole frame is inclined, as shown in Figure 5.33 (*right*). In this case there would be no shading problem and the rows of the array could be closer to each other. As this increases the surface area of the roof, the number of PV rows fitted on the roof is four. Thus, the mounting of the PV modules on a frame as shown are much more cost-effective. Hence, mounting the PV modules on the louvres of the roof shelter of the Itoman City Government Building is a better design. A even better option is to mount the PV modules over a tilted roof, as applied at the Crowne Plaza Hotel in Alice Springs. If the PV system mounted on the rooftop over the inclined frame is raised, it will create a living space underneath. In other words, if the PV array is integrated into the rooftop, it will be much economical as the cost of construction of the building is shared.

PV solar cells have a negative effect on the high temperature of the electrical output of PV modules. To keep the electricity efficiency close to a satisfactory level, it is essential to lower the temperature of PV modules. Generally, PV modules are combined with a thermal energy extraction unit, where a circulating fluid such as air or water is used to extract the heat, and constitute the hybrid photovoltaic thermal (PVT) system.[61] On extracting the heat, the operating temperature is reduced and leads to an increase in the electrical efficiency of the system. As PVT systems provide useful thermal energy in the form of either sensible heat or latent heat, the total exergy output from PV modules is much higher than that from plain PV modules. The effect of air mass flow rate, air channel depth, length and packing factor on a single pass have been presented.[62,63]

Hegazy[64] and Chow *et al.*[65] give interesting modelling results for air-cooled PV systems. Free convective air was used as cooling to remove heat from the back of PV modules and to keep the electrical efficiency at an acceptable level.[66,67] Free convective air cooling is a simple and low-cost method, but a forced-flow scheme provides a desirable cooling rate at all times. Chow *et al.*[68] found that forced convective cooling under a higher coolant flow velocity is better than a natural flow design. Nevertheless, the additional fan power consumption reduces the net electrical gain of the system. Tiwari *et al.*[69,70] have presented a variety of results regarding the effect of design and operation parameters on the performance of air-type PVT systems. Aste *et al.*[71] has presented experimental and theoretical results of a research and development programme carried out at the Politecnico di Milano on the design, development and performance monitoring of a hybrid PVT air collector.

PV façades and solar roofs are promising ways to integrate photovoltaics into a building structure in that they provide significant additional benefits over and above the electricity and thermal yield. They both form part of the outer structure of the building, saving the costs of conventional roof and cladding materials. Konttinenn *et al.*[72] determined the thermal stress on BIPV modules along with field testing. Bloom[73] has presented results for glass–glass PV modules as façades with forced ventilation.

Agrawal and Tiwari[74] proposed integration of PVT systems into the rooftop of a building, called the building integrated photovoltaic thermal (BIPVT) system. The BIPVT system, when used as the rooftop, reduces the construction of a conventional roof and thereby the cost. In addition to this, the system supplies the electrical power required for household purposes along with thermal energy for space heating, especially during peak load hours. Such a system is of higher use as it will produce electricity at a higher efficiency along with the required heat needed for space heating of the building to a reasonable temperature, especially in cold climatic conditions, and for industrial processing in moderate climatic conditions. A semi-transparent BIPVT system with the walls, roofs or windows of the building can provide day lighting in addition to electrical and thermal requirements. Such systems need a proper design so that the daylight received by the living space does not glare.

Apart from thin-film BIPVT systems, concentrator PV systems designed for building integration have also been gaining ground. These systems often work at low and medium concentration levels if installed on the rooftop of a building. Low concentration levels are preferred for integration into façades since direct views of the Sun are restricted and the diffuse component of light represents a larger proportion of the total irradiation available in such cases. Indeed, restrictions in the availability of direct light has led to many designs of building integrated concentrators that utilize the passive benefits of building integration, such as solar gain control, interior light distribution and collection of thermal energy for preheating of water in order to increase the total system efficiency and cost effectiveness.[1]

Even though there are at the moment a number of innovative ideas and designs for such technologies, the field of BIPVT systems still has room for improvement. Research into the complex interaction between the building envelope and BIPV modules is still required, since the inclusion of such systems can affect the solar gain of buildings as well as the thermal conductivity of the areas where they are installed.

References

1. G. Makrides, B. Zinsser, M. Norton, G. E. Georghiou, M. Schubert and J. H. Werner, Potential of photovoltaic systems in countries with high solar irradiation, *Renewable Sustainable Energy Rev.*, 2010, **14**, 754–762.
2. *Global Solar Photovoltaic Market Analysis and Forecasts to 2020*, http://www.prlog.org/10198293-global-solar-photovoltaic-market-analysis-and-forecasts-to-2020.html.
3. Akademie Mont-Cenis, http://www.sma.de/it/prodotti/referenze-impianti/inverter-solari/sezionatore/akademie-mont-cenis.html.
4. Justin, *The One Megawatt Solar Glass Complex*, http://www.metaefficient.com/renewable-power/65-million-square-feet-of-solar-rooftops-powering-162000-homes.html.
5. Mont-Cenis Academy, *Herne Sodingen Case Studies: Germany*, http://www.iea-pvps.org/cases/deu_01.htm.
6. http://www.pvdatabase.org/images_db/070608Buerstadtklitzeklein1.jpg.
7. Photovoltaikanlage Bürstadt, http://de.wikipedia.org/wiki/Photovoltaikanlage_B%C3%BCrstadt.
8. Anlage in Bürstadt wurde komplett abgeschaltet, http://ralos.de/blog/2009/06/anlage-in-burstadt-wurde-komplett-abgeschaltet.html.
9. R. Hug, *Building integrated photovoltaics (BIPV): solar electric power systems conquer large roofs and façades*, http://www.solarserver.de/solarmagazin/solar-report_0607_e.html.
10. C. F. Lange, *PV facade integration in the Lehrter station Berlin*, presented at the Proceedings of the 3rd World Conference on Photovoltaic Energy Conversion, Osoh, Japan, May 2003.
11. *Trends in Photovoltaic Applications: Survey Report of Selected IEA Countries between 1992 and 2006*, Report IEA-PVPS T1-16:2007, http://www.iea-pvps.org/products/download/rep1_16.pdf.
12. M. Drif, P. J. Pérez, J. Aguilera, G. Almonacid, P. Gomez, J. de la Casa and J. D. Aguilar, Univer project: a grid connected photovoltaic system of 200 kWp at Jaén University. Overview and performance analysis, *Sol. Energy Mater. Sol. Cells*, 2007, **91**, 670–683.
13. *UNIVER (UNIversidad VERde), Jaén Case studies: Spain*, http://www.iea-pvps.org/cases/esp_01.htm.
14. E. Hugses, *Case Study: City of Communications, Madrid*, http://www.pv-tech.org/news/_a/case_study_city_of_communications_madrid/.
15. V. Salas and E. Olias, Overview of the photovoltaic technology status and perspective in Spain, *Renewable Sustainable Energy Rev.*, 2009, **13**, 1049–1057.
16. P. Parker, Residential solar photovoltaic market stimulation: Japanese and Australian lessons for Canada, *Renewable Sustainable Energy Rev.*, 2008, **12**, 1944–1958.
17. *Photovoltaic Power System: Largest of its Kind in the World*, http://sharp-world.com/kameyama/eco/solar/mega_scale.html.

18. K. Odagaki, Practical study on 5.2 MW PV system in Sharp's Kameyama plant, *Proceedings of the Power Conversion Conference, Nagoya*, April 2007, pp. 1212–1216.
19. Itoman City Hall, http://www.ibec.or.jp/jsbd/H/tech.htm.
20. Kanazawa Station East Plaza bus terminal, http://www.pvdatabase.org/projects_view_details.php?ID = 228.
21. A. Kriscenski, *Solar Ark: World's Most Stunning Solar Building*, http://www.inhabitat.com/2008/01/14/solar-ark-worlds-most-stunning-solar-building/.
22. *PV Status Report 2009, Research, Solar Cell Production and Market Implementation of Photovoltaics*, European Commission, Joint Research Centre, Institute for Energy, August 2009.
23. *Toyota and SunPower Complete Largest Single-Roof Solar Installation in North America*, September 2008, http://solarcellsinfo.com/blog/archives/1793#more-1793.
24. M. Murphy, *Google Powers Up 1.6 MW Solar System & Hybrid Initiative*, http://www.renewableenergyworld.com/rea/news/article/2007/06/google-powers-up-1-6-mw-solar-system-hybrid-initiative-49007.
25. *California State University at Hayward Goes Solar with SunPower*, http://www.powerlight.com/business/success-stories/success-story-pdfs/education/sp_calstate_hayward_en_ltr_w_cs.pdf.
26. *Farmingdale Rooftop Photovoltaic System, Long Island, NY, USA*, http://www.power-technology.com/projects/long_island/.
27. S. Jinsoo and K. Donghwan, Trends of research and dissemination for PV systems in Korea, *Proceedings of the IEEE 4th World Conference on Photovoltaic Energy Conversion*, 2006, pp. 2268–2271.
28. Presented at EPIA's 3rd International Conference on Solar Photovoltaic Investments, Frankfurt, Germany, April 2009.
29. K. E. Park, G. H. Kang, H. I. Kim, G. J. Yu and J. T. Kim, Analysis of thermal and electrical performance of a semi-transparent photovoltaic (PV) module, *Energy*, 2009, in press.
30. S. H. Yoo and E. T. Lee, Efficiency characteristic of building integrated photovoltaics as a shading device, *Build. Environ.*, 2002, **37**, 615–623.
31. *EPURON Completes First Photovoltaic Project in South Korea*, http://www.epuron.de/en/desktopdefault.aspx/tabid-933/571_read-1628/.
32. N. Aste, R. S. Adhikari and L. C. Tagliabue, Evaluation of energy policies for promotion and dissemination of photovoltaic technology in Italy, *Prog. Photovoltaics Res. Appl.*, 2007, **15**, 449–460.
33. *Italian PV Market*, http://www.epia.org/policy/national-policies/italy/italian-pv-market.html.
34. *Italy Proposes Policy Changes: New FiT Looms*, http://www.pv-tech.org/news/_a/italy_proposes_policy_changes_new_fit_looms/.
35. N. Aste, F. Groppi and C. D. Pero, The first installation under the Italian PV Rooftop Programme: a performance analysis referred to five years of operation, *Proceedings of the International Conference on Clean Electrical Power*, 2007, pp. 360–365.

36. F. Butera, R. S. Adhikari, N. Aste and R. Bracco, Solar hybrid photovoltaic-thermal (PVT) façade for heating, cooling and electricity generation, *Proceedings of the International Conference on Clean Electrical Power*, Capri, Italy, 2007, pp. 766–769.
37. *Trina Solar Completes 4.7 MW Photovoltaic System in Italy*, http://www.solarserver.de/solarmagazin/news-ea2009m04.html.
38. *Mitsubishi Electric 3 MW Solar Panels to Power Italy's Coop Logistic Centre*, http://www.mitsubishielectric.eu/download/204.
39. *Research Report of China's Building Integrated Photovoltaic (BIPV) Industry*, China Research and Intelligence, 2009.
40. *Government Policies Help PV Market in China*, http://www.solarfeeds.com/pcs-solar-photovoltaics-blog-/10195-government-policies-help-pv-market-in-china.html.
41. H. Yang, G. Zheng, C. Lou and J. Burnett, Grid-connected building-integrated photovoltaics: a Hong Kong case study, *Sol. Energy*, 2004, **76**, 55–59.
42. Y. Hongxing and L. Yutong, *Potential of building-integrated photovoltaic applications*, www.iea-pvps.org/products/download/rep7_04.pdf.
43. Y. Hongxing, L. Chengzhi and S. Liangliang, Building-integrated photovoltaics for maximum power generation, *Proceedings of the Electronics System-Integration Technology Conference*, Greenwich, UK, 2008, pp. 39–44.
44. Shanghai World Expo will provide a stage for solar energy, http://www.tysilicon.com/english/shownews.asp?id = 36.
45. *China Grid Connect 1 MW Green Garden Solar Project – Shenzhen, China*, BP Solar Report 6038A-1 09/04, 2004.
46. Energy Report Editor Office, Ministry for Economic Affairs promote PV policy with bright results, *Energy Rep.*, 2009, **2**, 40.
47. M. Watt, *The Commercialisation of Photovoltaics Research in Australia*, Science and Innovation Mapping Department of Education Science and Training, Sydney, September 2003.
48. M. Watt, Co-operative programme on photovoltaic power systems: exchange and dissemination of information on PV power systems, *National Survey Report of PV Power Applications in Australia*, International Energy Agency, Australia, May 2007.
49. *Solar Kogarah Town Square: PV details*, http://www.re-systems.ee.unsw.edu.au/KTS_home.htm.
50. *Adelaide Showground – Australia's largest PV system*, http://www.thisissolar.com/index.php?area = infosheet&id = 55.
51. *A SunPower first – Australia's Largest Rooftop System on the Crowne Plaza Hotel in Alice Springs*, SunPower Corporation, http://www.sunpowercorp.com.au/business/success-stories/success-story-pdfs/sp_crowne_plaza_en_a4 w cs.pdf.
52. University of Melbourne Building Integrated Photovoltaic System, *Eco-Generation Mag.*, Oct/Nov 2002, p. 21.
53. J. Keirstead, Behavioural responses to photovoltaic systems in the UK domestic sector, *Energy Policy*, 2007, **35**, 4128–4141.

54. *Solar Office Doxford International Case Studies: United Kingdom*, http://www.iea-pvps.org/cases/gbr_02.htm.
55. *Photovoltaic and Architecture*, ed. R. Thomas, M. Fordham, et al., Spon Press, London, 2001, p. 17–32.
56. D. L. Jones, C. Matson and N. M. Pearsall, The solar office: a solar powered building with a comprehensive energy strategy, *Proceedings of 2nd World Conference on Photovoltaic Solar Energy Conversion*, Vienna, Austria, July 1998, p. 2559–2562.
57. *Ministry of New & Renewable Energy, Government of India*, http://mnes.nic.in/.
58. *Solar power in India*, http://en.wikipedia.org/wiki/Solar_power_in_India.
59. K. Mukherjee and D. Fogarty, India to unveil 20 GW solar target under climate plan, *Reuters*, New Delhi, 28 July 2009.
60. N. Sethi, India targets 1000 MW solar power in 2013, *The Times of India*, New Delhi, 18 November 2009.
61. J. Ji, K. Liu, T. Chow, G. Pei and H. He, Thermal analysis of PV/T evaporator of a solar-assisted heat pump, *Int. J. Energy Res.*, 2007, **31**, 525–545.
62. A. K. Bhargava, H. P. Garg and R. K. Agarwal, Study of a hybrid solar heat system – solar air heater combined with solar cells, *Energy Convers. Manage.*, 1991, **31**, 471–479.
63. J. Prakash, Transient analysis of a photovoltaic-thermal solar collector for co-generation of electricity and hot air/water, *Energy Convers. Manage.*, 1994, **35**, 972–976.
64. A. A. Hegazy, Comparative study of the performances of four photovoltaic/thermal solar air collectors, *Energy Convers. Manage.*, 2000, **41**, 861–881.
65. T. T. Chow, Performance analysis of photovoltaic-thermal collector by explicit dynamic model, *Sol. Energy*, 2003, **75**, 143–152.
66. O. V. Ekechukwu and B. Norton, Review of solar-energy drying systems II: an overview of solar drying technology, *Energy Convers. Manage.*, 1999, **40**, 615–655.
67. J. K. Tonui and Y. Tripanagnostopoulos, Performance improvement of PV/T solar collectors with natural air flow operation, *Sol. Energy*, 2008, **82**, 1–12.
68. T. T. Chow, W. He and J. Ji, An experimental study of facade-integrated photovoltaic/water-heating system, *Appl. Thermal Eng.*, 2007, **27**, 37–45.
69. A. Tiwari and M. S. Sodha, Performance evaluation of hybrid PV/thermal water/air heating system: a parametic study, *Renewable Energy*, 2006, **31**, 2460–2474.
70. A. Tiwari and M. S. Sodha, Performance evaluation of solar PV/T system: an experimental validation, *Sol. Energy*, 2006, **80**, 751–759.
71. N. Aste, G. Chiesa and F. Verri, Design, development and performance monitoring of a photovoltaic-thermal (PVT) air collector, *Renewable Energy*, 2008, **33**, 914–927.

72. P. Konttinenn, T. Carlsson, P. Lund and T. Lehtinen, Estimating thermal stress in BIPV modules, *Int. J. Energy Res.*, 2006, **30**, 1264–1277.
73. J. J. Bloem, Evaluation of a PV-integrated building application in a well-controlled outdoor test environment, *Build. Environ.*, 2008, **43**, 205–216.
74. B. Agrawal and G. N. Tiwari, *Energy and Exergy Analysis of Building Integrated Photovoltaic Thermal (BIPVT) Systems. Energy Sources, Part A: Recovery, Utilization, and Environmental Effects*, 2010, in press.

CHAPTER 6

Thermal Modelling and Performance Analysis

6.1 Introduction

A thermal model is a mathematical model that dynamically predicts the temperature of a building living space. Its accuracy depends upon an algorithm and the accuracy of the input parameters. When developing strategies to minimize energy consumption within buildings it is crucial to understand the dynamics of energy generation and loss. For a BIPVT system the sources of heat gain include the effect of sunlight and the heat generating devices which are inside the building. Heat losses are either in the form of conduction or ventilation. Conductive heat losses can be thought of in terms of heat flowing from a hot region to a cold region through windows and walls. Ventilation meanwhile represents the direct movement of hot air out of a building through cracks and gaps as well as through deliberate ventilation.

This chapter aims to investigate some of the contributions to heat generation and loss studied through the development of thermal models. The models developed will have adjustable parameters corresponding to different contributions to the heat budget and so by accurately understanding the form of temperature variation we aim to appreciate the most important factors in the energy consumption and production within a building.

6.2 Assumptions

In order to write the energy balance equation for each component of the BIPVT system, the assumptions made are as follows:

- The system is in quasi-steady-state condition.
- The specific heat of the air remains constant. It does not change with any rise in the temperature of the air.

RSC Energy Series No. 4
Building Integrated Photovoltaic Thermal Systems: For Sustainable Developments
By Basant Agrawal and G. N. Tiwari
© B. Agrawal and G. N. Tiwari 2011
Published by the Royal Society of Chemistry, www.rsc.org

- The transmissivity of poly(ethylene–vinyl acetate) (EVA) is approximately 100%.
- The temperatures of the glass cover, solar cells, Tedlar [a poly(vinyl fluoride) film], duct and insulation vary only in the direction of the flow of air.
- The side losses from the system are negligible.
- The airflow through the duct is uniform for the forced mode of operation for streamline flow.

6.3 Thermal Modelling

6.3.1 A Roof-integrated Opaque PVT System without an Air Duct

Figure 6.1 shows the cross-sectional view of a roof integrated opaque-type PVT system without a duct over which solar intensity is received.

If b is the width of the PVT system (in m) and dx the elemental length (in m), then the energy balance for the PV module for the elemental area $b\,dx$ is given by:

$$\begin{bmatrix}\text{Rate of heat}\\ \text{received by}\\ \text{the solar cell}\end{bmatrix} + \begin{bmatrix}\text{Rate of heat}\\ \text{received by the}\\ \text{non-packing area}\end{bmatrix} = \begin{bmatrix}\text{Rate of heat transfer}\\ \text{from PV module to}\\ \text{air as the top loss}\end{bmatrix} + \begin{bmatrix}\text{Rate of heat transfer}\\ \text{from PV module to}\\ \text{the back surface/Tedlar}\end{bmatrix} + \begin{bmatrix}\text{Rate of}\\ \text{electricity}\\ \text{production}\end{bmatrix} \tag{6.1}$$

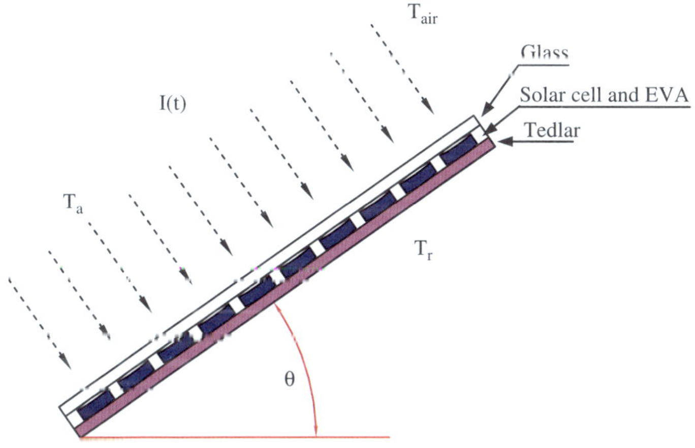

Figure 6.1 Cross-sectional view of a roof integrated PVT system.

Let τ_g be the transmissivity of the top glass, α_c the absorptivity of solar cell, β_c the packing factor, α_T the absorptivity of the Tedlar and $I(t)$ the solar intensity on the BIPVT system in W m^{-2}. Also, let T_c be the temperature of the solar cell in kelvin, T_a the temperature of the atmospheric air in kelvin, T_{bs} the temperature of the back surface of the Tedlar in kelvin and η_c the electrical conversion efficiency of the solar cell. Therefore, eqn (6.1) can be rewritten as:

$$\tau_g[\alpha_c\beta_c + (1-\beta_c)\alpha_T]I(t)b\,dx = [U_T(T_c - T_a) + h_T(T_c - T_{bs})]b\,dx + \eta_c\alpha_c\beta_c I(t)b\,dx \tag{6.2}$$

where the top heat loss coefficient from the solar cell to the ambient air is $U_T = \left(\frac{L_g}{K_g} + \frac{1}{h_o}\right)^{-1}$ W m^{-2} K^{-1} and the heat transfer coefficient for Tedlar is $h_T = \left(\frac{L_T}{K_T}\right)^{-1}$ W m^{-2} K^{-1}. L_g and L_T are the thickness (in m) of the top glass and the Tedlar, respectively, K_g and K_T are the thermal conductivity in W m^{-1} K^{-1} of the glass and Tedlar, respectively, and h_o is the heat transfer coefficient for the ambient air.

Simplifying eqn (6.2), the expression for a solar cell temperature can be obtained as:

$$T_c = \frac{h_T T_{bs} + U_T T_a + I(t)(\alpha\tau)_{eff}}{U_T + h_T} \tag{6.3}$$

where $(\alpha\tau)_{eff} = \tau_g[\alpha_c\beta_c + \alpha_T(1-\beta_c)] - \eta_c\alpha_c\beta_c$.

The energy balance for the Tedlar back surface of the PV module for elemental area $b\,dx$ is given by:

$$\begin{bmatrix}\text{Rate of heat transfer from} \\ \text{cell to Tedlar}\end{bmatrix} = \begin{bmatrix}\text{Rate of heat transfer from} \\ \text{Tedlar to room}\end{bmatrix} \tag{6.4}$$

Let h_i be the heat transfer coefficient of the air inside the room in W m^{-2} K^{-1} and T_r the average temperature of the room. Therefore, eqn (6.4) can be rewritten as:

$$h_T(T_c - T_{bs})b\,dx = h_i(T_{bs} - T_r)b\,dx \tag{6.5}$$

Substituting T_c from eqn (6.3) into eqn (6.5), the expression for the back-surface temperature is obtained as:

$$T_{bs} = \frac{h_r T_r + U_{tT} T_a + h_{p1} I(t)(\alpha\tau)_{eff}}{U_{tT} + h_r} \tag{6.6}$$

where a penalty factor due to the presence of the solar cell material, Tedlar and EVA is $h_{p1} = \frac{h_T}{U_T + h_T}$ and the overall heat transfer coefficient is $U_{tT} = \frac{U_T \times h_T}{U_T + h_T} = \left(\frac{1}{h_T} + \frac{1}{U_T}\right)^{-1}$.

The rate of useful thermal energy (in Wh) for the BIPVT system is given by:

$$\dot{Q}_{ux} = A_{roof} \times h_i (T_{bs} - T_r) \tag{6.7}$$

where A_{roof} is the area of roof in m^2 and h_i is the heat transfer coefficient of the room air.

The available useful thermal energy is used to heat the room but part of it is lost. The energy balance for the space heating of the building is given by:

$$A_{roof} \times h_i (T_{bs} - T_r) = \dot{m}_r C_{air} \left(\frac{dT_r}{dt}\right) + (UA)_t (T_r - T_a) + 0.33 N_o V (T_r - T_a) \tag{6.8}$$

where $\dot{m}_r$ is the air mass flow-rate in the room (in $kg\,s^{-1}$), dt is the elemental time (in s), N_o is the number of air changes per hour and V is the volume of the space in the room. The heat loss capacity (in $W\,K^{-1}$) through the building surfaces is the sum of the heat loss capacities through the walls, windows and doors, *i.e.* $(UA)_t = (UA)_{t\_w} + (UA)_{t\_win} + (UA)_{t\_d}$. The heat loss capacities through the walls, windows and doors are given by:

$$(UA)_{t\_wall} = \frac{A_w}{\left(\frac{1}{h_o} + \frac{L_w}{K_w} + \frac{1}{h_r}\right)},$$

$$(UA)_{t\_win} = \frac{A_{win}}{\left(\frac{1}{h_o} + \frac{L_{win}}{K_{win}} + \frac{1}{h_r}\right)}$$

and

$$(UA)_{t\_d} = \frac{A_d}{\left(\frac{1}{h_o} + \frac{L_d}{K_d} + \frac{1}{h_r}\right)},$$

respectively, where A_w, L_w and K_w are the area, thickness and thermal conductivity of the wall, respectively, A_{win}, L_{win} and K_{win} are the area, thickness and thermal conductivity of the windows, respectively, and A_d, L_d and K_d are the area, thickness and thermal conductivity of the doors, respectively. On solving eqn (6.8) we have:

$$\left(\frac{dT_r}{dt}\right)+\frac{1}{\dot{m}_r C_{air}}\left[(UA)_t+0.33N_oV+A_{roof}h_i\right]T_r = \frac{1}{\dot{m}_r C_{air}}\left[\{0.33N_oV+(UA)_t\}T_a+A_{roof}h_iT_{bs}\right] \tag{6.9}$$

On integrating eqn (6.9) and applying the initial condition at time $t=0$, $T_r=T_{ri}$, the room air temperature (T_r) is given by:

$$T_r=\frac{f(t)}{a}(1-e^{-at})+T_{ri}e^{-at}. \tag{6.10}$$

where

$$a=\frac{1}{\dot{m}_r C_{air}}\left[(UA)_t+0.33N_oV+A_{roof}h_i\right]$$

and

$$f(t)=\frac{1}{\dot{m}_r C_{air}}\left[\{0.33N_oV+(UA)_t\}T_a+A_{roof}h_iT_{bs}\right]$$

6.3.2 A Roof-integrated Semi-transparent PVT System without an Air Duct

Figure 6.2 shows the cross-sectional view of a roof-integrated semi-transparent PVT system without a duct over which solar intensity is received. If b is the width of the PVT system (in m) and dx the elemental length (in m), then the

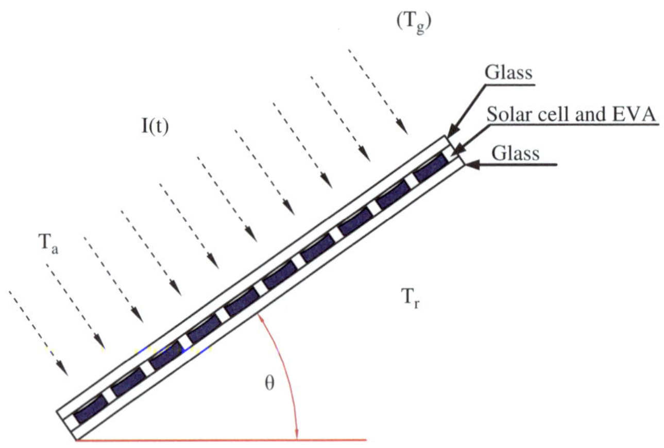

Figure 6.2 Cross-sectional view of a roof-integrated semi-transparent PVT system.

energy balance for a PV module of elemental area $b\,dx$ is given by:

$$\begin{bmatrix}\text{Rate of heat}\\ \text{received by}\\ \text{the solar cell}\end{bmatrix} = \begin{bmatrix}\text{Rate of heat transfer}\\ \text{from PV module to air}\\ \text{as the top loss}\end{bmatrix} + \begin{bmatrix}\text{Rate of heat transfer}\\ \text{from PV module to}\\ \text{the room space}\end{bmatrix} + \begin{bmatrix}\text{Rate of}\\ \text{electricity}\\ \text{production}\end{bmatrix} \quad (6.11)$$

and

$$\begin{bmatrix}\text{Daylight received by}\\ \text{the room space}\end{bmatrix} = \begin{bmatrix}\text{Rate of heat received by}\\ \text{the non-packing area}\end{bmatrix} \quad (6.12)$$

Let τ_g be the transmissivity of both of the glasses, α_c the absorptivity of the solar cell, β_c the packing factor and $I(t)$ the solar intensity on the BIPVT system in $\mathrm{W\,m^{-2}}$. Also, let T_c be the temperature of the solar cell (in K), T_a the temperature of the atmospheric air (in K), T_r the room space temperature (in K) and η_c the electrical conversion efficiency of the solar cell. Therefore eqns (6.11) and (6.12) can be rewritten as:

$$\tau_g\alpha_c\beta_c I(t) b\,dx = [U_T(T_c - T_a) + U_B(T_c - T_r)]b\,dx + \eta_c\alpha_c\beta_c I(t) b\,dx \quad (6.13)$$

$$\text{and the daylight received by the room space} = \tau_g^2(1-\beta_c)\alpha_T \times 100 \text{ lux} \quad (6.14)$$

where the top heat loss coefficient from the solar cell to the ambient air is $U_T = \left(\frac{L_g}{K_g} + \frac{1}{h_o}\right)^{-1}$ $\mathrm{W\,m^{-2}\,K^{-1}}$ and the bottom heat loss coefficient from the solar cell to the room is $U_B = \left(\frac{L_g}{K_g} + \frac{1}{h_i}\right)^{-1}$ $\mathrm{W\,m^{-2}\,K^{-1}}$, where h_o and h_i are the heat transfer coefficients of the ambient and room air, respectively, in $\mathrm{W\,m^{-2}\,K^{-1}}$.

Simplifying eqn (6.13), the expression for a solar cell temperature can be obtained as:

$$T_c = \frac{U_B T_r + U_T T_a + I(t)\{\tau_g\alpha_c\beta_c - \eta_c\}}{U_T + U_B} \quad (6.15)$$

The rate of useful thermal energy (in Wh) for the BIPVT system is given by:

$$\dot{Q}_{ux} = A_{roof} \times U_B(T_c - T_r) \quad (6.16)$$

where A_{roof} is the area of the roof.

The available useful thermal energy is used to heat the room but part of it is lost. The energy balance for space heating of the building is given by:

$$A_{\text{roof}} \times U_{\text{B}}(T_{\text{c}} - T_{\text{r}}) = \dot{m}_{\text{r}} C_{\text{air}} \left(\frac{dT_{\text{r}}}{dt} \right) + (UA)_{\text{t}}(T_{\text{r}} - T_{\text{a}}) + 0.33 N_{\text{o}} V (T_{\text{r}} - T_{\text{a}}) \tag{6.17}$$

where $\dot{m}_{\text{r}}$ is the air mass flow-rate in the room (in kg s^{-1}), dt is the elemental time (in s), N_{o} is the number of air changes per hour and V is the volume of the space in the room. The heat loss capacity (in W K^{-1}) throughout the building surfaces is the sum of the heat loss capacities through the walls, windows and doors, *i.e.* $(UA)_{\text{t}} = (UA)_{\text{t\_w}} + (UA)_{\text{t\_win}} + (UA)_{\text{t\_d}}$. On solving eqn (6.17) we have:

$$\left(\frac{dT_{\text{r}}}{dt} \right) + \frac{1}{\dot{m}_{\text{r}} C_{\text{air}}} \left[(UA)_{\text{t}} + 0.33 N_{\text{o}} V + A_{\text{roof}} U_{\text{B}} \right] T_{\text{r}} = \frac{1}{\dot{m}_{\text{r}} C_{\text{air}}} \left[\{ 0.33 N_{\text{o}} V + (UA)_{\text{t}} \} T_{\text{a}} + A_{\text{roof}} U_{\text{B}} T_{\text{c}} \right] \tag{6.18}$$

On integrating eqn (6.18) and applying the initial condition at time $t = 0$, $T_{\text{r}} = T_{\text{ri}}$, the room air temperature (T_{r}) is given by:

$$T_{\text{r}} = \frac{f(t)}{a}(1 - e^{-at}) + T_{\text{ri}} e^{-at} \tag{6.19}$$

where

$$a = \frac{1}{\dot{m}_{\text{r}} C_{\text{air}}} \left[(UA)_{\text{t}} + 0.33 N_{\text{o}} V + A_{\text{roof}} U_{\text{B}} \right]$$

and

$$f(t) = \frac{1}{\dot{m}_{\text{r}} C_{\text{air}}} \left[\{ 0.33 N_{\text{o}} V + (UA)_{\text{t}} \} T_{\text{a}} + A_{\text{roof}} U_{\text{B}} T_{\text{c}} \right]$$

6.3.3 A Roof-integrated Opaque PVT System with an Air Duct

Figure 6.3 (*left*) shows the cross-sectional view of a roof-integrated PVT system while Figure 6.3 (*right*) shows an elemental area b dx over which the solar intensity is received.

The energy balance for the PV module of the BIPVT system for elemental area b dx is given by eqn (6.2) as:

$$\tau_{\text{g}}[\alpha_{\text{c}}\beta_{\text{c}} + (1 - \beta_{\text{c}})\alpha_{\text{T}}]I(t)b\, dx = [U_{\text{T}}(T_{\text{c}} - T_{\text{a}}) + h_{\text{T}}(T_{\text{c}} - T_{\text{bs}})]b\, dx + \eta_{\text{c}}\alpha_{\text{c}}\beta_{\text{c}} I(t) b\, dx \tag{6.20}$$

Simplifying eqn (6.20), the expression for a solar cell temperature can be obtained as:

$$T_{\text{c}} = \frac{h_{\text{T}} T_{\text{bs}} + U_{\text{T}} T_{\text{a}} + I(t)(\alpha\tau)_{\text{eff}}}{U_{\text{T}} + h_{\text{T}}} \tag{6.21}$$

where $(\alpha\tau)_{\text{eff}} = \tau_{\text{g}}[\alpha_{\text{c}}\beta_{\text{c}} + \alpha_{\text{T}}(1 - \beta_{\text{c}})] - \eta_{\text{c}}\alpha_{\text{c}}\beta_{\text{c}}$.

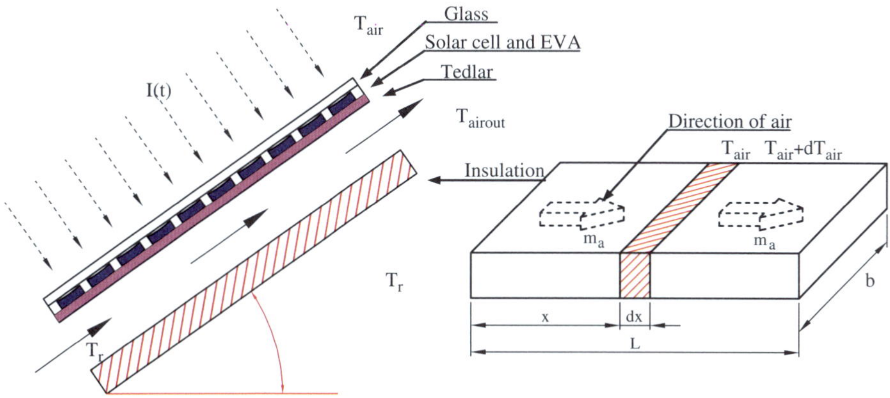

Figure 6.3 A roof-integrated PVT system: *left*, cross-sectional view; *right*, airflow pattern over elementary area b dx.

The energy balance for the back surface of the PV module for elemental area $b\,dx$ is given by:

$$\begin{bmatrix}\text{Rate of heat transfer from} \\ \text{cell to the Tedlar}\end{bmatrix} = \begin{bmatrix}\text{Rate of heat transfer from} \\ \text{the Tedlar to air side in the duct}\end{bmatrix} \quad (6.22)$$

Let h_{air} be the heat transfer coefficient of the air flowing through the duct (in $W\,m^{-2}\,K^{-1}$) and T_{air} be the average temperature of the air flowing through the duct. Therefore, eqn (6.22) can be rewritten as:

$$h_T(T_c - T_{bs})b\,dx = h_{air}(T_{bs} - T_{air})b\,dx \quad (6.23)$$

Substituting T_c from eqn (6.21) into eqn (6.23), the expression for the back-surface temperature is obtained as:

$$T_{bs} = \frac{h_{air}T_{air} + U_{tT}T_a + h_{p1}I(t)(\alpha\tau)_{eff}}{U_{tT} + h_{air}} \quad (6.24)$$

where the penalty factor due to the presence of the solar cell material, Tedlar and EVA is $h_{p1} = \frac{h_T}{U_T + h_T}$ and the overall heat transfer coefficient is $U_{tT} = \frac{U_T \times h_T}{U_T + h_T} = \left(\frac{1}{h_T} + \frac{1}{U_T}\right)^{-1}$.

The energy balance for air flowing in the duct of the BIPVT system for elemental area $b\,\mathrm{d}x$ is given by:

$$\begin{bmatrix}\text{Rate of heat transfer}\\ \text{from Tedlar to air}\\ \text{side in the duct}\end{bmatrix} = \begin{bmatrix}\text{Rate of heat gain by}\\ \text{air flowing in the duct}\\ \text{of the BIPVT system}\end{bmatrix} + \begin{bmatrix}\text{Rate of heat}\\ \text{loss from air}\\ \text{through insulation}\end{bmatrix} \tag{6.25}$$

Let $\dot{m}_{\mathrm{air}}$ be the air mass flow-rate through duct (in $\mathrm{kg\,s^{-1}}$), C_{air} the specific heat of air at constant pressure (in $\mathrm{J\,kg^{-1}\,K^{-1}}$) and T_{r} the average temperature of the room (in K). Therefore eqn (6.25) can be rewritten as:

$$h_{\mathrm{air}}(T_{\mathrm{bs}} - T_{\mathrm{air}})b\,\mathrm{d}x = \dot{m}_{\mathrm{air}}C_{\mathrm{air}}\left(\frac{\mathrm{d}T_{\mathrm{air}}}{\mathrm{d}x}\right)\mathrm{d}x + U_{\mathrm{bb}}(T_{\mathrm{air}} - T_{\mathrm{r}})b\,\mathrm{d}x \tag{6.26}$$

where the overall heat transfer for the insulation plate is $U_{\mathrm{bb}} = \left(\frac{1}{h_{\mathrm{air}}} + \frac{L_{\mathrm{i}}}{K_{\mathrm{i}}} + \frac{1}{h_{\mathrm{r}}}\right)^{-1}$.

On substituting T_{bs} from eqn (6.24) into eqn (6.26) we have:

$$h_{\mathrm{air}}\left[\frac{h_{\mathrm{p1}}I(t)(\alpha\tau)_{\mathrm{eff}} - U_{\mathrm{tT}}(T_{\mathrm{air}} - T_{\mathrm{a}})}{U_{\mathrm{tT}} + h_{\mathrm{air}}}\right]b\,\mathrm{d}x = \dot{m}_{\mathrm{air}}C_{\mathrm{air}}\left(\frac{\mathrm{d}T_{\mathrm{air}}}{\mathrm{d}x}\right)\mathrm{d}x + U_{\mathrm{bb}}(T_{\mathrm{air}} - T_{\mathrm{r}})b\,\mathrm{d}x \tag{6.27}$$

$$\text{or}\quad \left(\frac{dT_{\mathrm{air}}}{dx}\right) + \frac{bU_L}{\dot{m}_{\mathrm{air}}C_{\mathrm{air}}}T_{\mathrm{air}} = \frac{b}{\dot{m}_{\mathrm{air}}C_{\mathrm{air}}}\left[U_{bb}T_r + U_{\mathrm{tair}}T_{\mathrm{a}} + h_{\mathrm{p2}}h_{\mathrm{p1}}I(t)\,(\alpha\tau)_{\mathrm{eff}}\right]$$

where $U_{\mathrm{tair}} = \frac{U_{\mathrm{tT}} \times h_{\mathrm{air}}}{U_{\mathrm{tT}} + h_{\mathrm{air}}} = \left(\frac{1}{h_{\mathrm{air}}} + \frac{1}{U_{\mathrm{tT}}}\right)^{-1}$, $U_L = (U_{\mathrm{bb}} + U_{\mathrm{tair}})$ and the penalty factor due to the presence of interference between the Tedlar and the air through the duct is $h_{\mathrm{p2}} = \frac{h_{\mathrm{air}}}{U_{\mathrm{tT}} + h_{\mathrm{air}}}$.

On integrating eqn (6.27) with the boundary condition at $x = 0$, $T_{\mathrm{air}} = T_{\mathrm{r}}$,, and at $x = L$, $T_{\mathrm{air}} = T_{\mathrm{airout}}$, the outlet air temperature (T_{airout}) of the flowing air in the duct of the BIPVT system for length L, is given by:

$$T_{\mathrm{airout}} = \left[\frac{U_{\mathrm{bb}}T_{\mathrm{r}} + U_{\mathrm{tair}}T_{\mathrm{a}} + h_{\mathrm{p2}}h_{\mathrm{p1}}I(t)(\alpha\tau)_{\mathrm{eff}}}{U_L}\right] \times \left\{1 - \exp\left[\left(-\frac{bUL}{\dot{m}_{\mathrm{air}}C_{\mathrm{air}}}\right)L\right]\right\} + T_{\mathrm{r}}\exp\left[\left(-\frac{bUL}{\dot{m}_{\mathrm{air}}C_{\mathrm{air}}}\right)L\right] \tag{6.28}$$

and the average air temperature of the air flowing in the duct of the BIPVT system is given by:

$$\overline{T}_{air} = \left[\frac{U_{bb}T_r + U_{tair}T_a + h_{p2}h_{p1}I(t)(\alpha\tau)_{eff}}{U_L}\right] \times \left\{1 - \left[\frac{1 - \exp\left(-\frac{bU_L}{\dot{m}_{air}C_{air}}L\right)}{\frac{bU_L}{\dot{m}_{air}C_{air}}L}\right]\right\} + T_r\left[\frac{1 - \exp\left(-\frac{bU_L}{\dot{m}_{air}C_{air}}L\right)}{\frac{bU_L}{\dot{m}_{air}C_{air}}L}\right] \tag{6.29}$$

The rate of useful thermal energy (in Wh) obtained for n_{pv} rows of the BIPVT system is given by:

$$\begin{aligned}\dot{Q}_{ud} &= n_{pv} \times \dot{m}_{air}C_{air}(T_{airout} - T_r) \\ &= n_{pv} \times \dot{m}_{air}C_{air}\left[\frac{U_{bb}T_r + U_{tair}T_a + h_{p2}h_{p1}I(t)(\alpha\tau)_{eff}}{U_L} - T_r\right] \\ &\quad \times \left[1 - \exp\left(-\frac{bU_L}{\dot{m}_{air}C_{air}}L\right)\right]\end{aligned} \tag{6.30}$$

The available useful thermal energy is used to heat the room but a part of it is lost. The energy balance for space heating of the building is given by:

$$\begin{aligned}& n_{pv} \times \dot{m}_{air}C_{air}\left[\frac{U_{bb}T_r + U_{tair}T_a + h_{p2}h_{p1}I(t)(\alpha\tau)_{eff}}{U_L} - T_r\right] \\ &\quad \times \left[1 - \exp\left(-\frac{bU_L}{\dot{m}_{air}C_{air}}\right)\right] + U_{bb}(\bar{T}_{air} - T_r)A_{roof} \\ &= \dot{m}_r C_{air}\left(\frac{dT_r}{dt}\right) + (UA)_t(T_r - T_a) + 0.33N_oV(T_r - T_a)\end{aligned} \tag{6.31}$$

where $\dot{m}_r$ is the air mass flow-rate in the room (in kg s^{-1}). The heat loss capacity (in W K^{-1}) through the building surfaces is the sum of the heat loss capacities through the walls, windows and doors, *i.e.* $(UA)_t = (UA)_{t\_w} + (UA)_{t\_win} + (UA)_{t\_d}$; h_o and h_r are the heat transfer coefficients for the ambient air and the room air, respectively. On solving eqn (6.31) we have:

$$\begin{aligned}\left(\frac{dT_r}{dt}\right) &= \frac{1}{\dot{m}_r C_{air}}\left[\begin{array}{l}\{(UA)_t + 0.33N_oV\} - \left\{n_{pv} \times \dot{m}_{air}C_{air}\left(\frac{U_{bb}}{U_L} - 1\right)\left[1 - \exp\left(-\frac{bU_L}{\dot{m}_{air}C_{air}}L\right)\right]\right\} \\ - U_{bb}\left\{\frac{U_{bb}}{U_L}\left[1 - \frac{1 - \exp\left(-\frac{bU_L}{\dot{m}_{air}C_{air}}L\right)}{\frac{bU_L}{\dot{m}_{air}C_{air}}L}\right] + \left[\frac{1 - \exp\left(-\frac{bU_L}{\dot{m}_{air}C_{air}}L\right)}{\frac{bU_L}{\dot{m}_{air}C_{air}}L}\right] - 1\right\}A_{roof}\end{array}\right]T_r \\ &= \frac{1}{\dot{m}_r C_{air}}\left[\begin{array}{l}\{(UA)_t + 0.33N_oV\}T_a + \left\{n_{pv}\dot{m}_{air}C_{air}\left[\frac{U_{tair}T_a + h_{p2}h_{p1}I(t)(\alpha\tau)_{eff}}{U_L}\right]\left[1 - \exp\left(-\frac{bU_L}{\dot{m}_{air}C_{air}}L\right)\right]\right\} \\ \qquad + U_{bb}\left\{\frac{U_{tair}T_a + h_{p2}h_{p1}I(t)(\alpha\tau)_{eff}}{U_L}\right\}\left[1 - \frac{1 - \exp\left(-\frac{bU_L}{\dot{m}_{air}C_{air}}L\right)}{\frac{bU_L}{\dot{m}_{air}C_{air}}L}\right]A_{roof}\end{array}\right]\end{aligned} \tag{6.32}$$

On integrating eqn (6.32) and applying the initial condition at time $t=0$, $T_r = T_{ri}$, the room air temperature (T_r) is given by:

$$T_r = \frac{f(t)}{a}(1 - e^{-at}) + T_{ri}\, e^{-at} \tag{6.33}$$

where:

$$a = \frac{1}{\dot{m}_r C_{air}} \times \left[\begin{array}{l} \{(UA)_t + 0.33 N_o V\} - \left\{ n_{pv} \dot{m}_{air} C_{air} \left(\frac{U_{bb}}{U_L} - 1 \right) \left[1 - \exp\left(-\frac{bU_L}{\dot{m}_{air} C_{air}} L \right) \right] \right\} \\ - U_{bb} \left\{ \frac{U_{bb}}{U_L} \left[1 - \frac{1 - \exp\left(-\frac{bU_L}{\dot{m}_{air} C_{air}} L \right)}{\frac{bU_L}{\dot{m}_{air} C_{air}} L} \right] + \frac{1 - \exp\left(-\frac{bU_L}{\dot{m}_{air} C_{air}} L \right)}{\frac{bU_L}{\dot{m}_{air} C_{air}} L} - 1 \right\} A_{roof} \end{array} \right]$$

and:

$$f(t) = \frac{1}{\dot{m}_r C_{air}} \times \left[\begin{array}{l} \{(UA)_t + 0.33 N_o V\} T_a + \left\{ \begin{array}{r} n_{pv} \dot{m}_{air} C_{air} \left[\frac{U_{tair} T_a + h_{p2} h_{p1} I(t) (\alpha\tau)_{eff}}{U_L} \right] \\ \times \left[1 - \exp\left(-\frac{bU_L}{\dot{m}_{air} C_{air}} L \right) \right] \end{array} \right\} \\ + U_{bb} \left\{ \frac{U_{tair} T_a + h_{p2} h_{p1} I(t)(\alpha\tau)_{eff}}{U_L} \right\} \left[1 - \frac{1 - \exp\left(-\frac{bU_L}{\dot{m}_{air} C_{air}} L \right)}{\frac{bU_L}{\dot{m}_{air} C_{air}} L} \right] A_{roof} \end{array} \right]$$

6.3.4 A Roof-integrated Semi-transparent PVT System with an Air Duct

Figure 6.4 shows the cross-sectional view of a roof-integrated semi-transparent PVT system over which solar intensity is received.

The energy balance for a PV module of the BIPVT system for elemental area $b\,dx$ is given by eqns (6.13) and (6.14):

$$\tau_g \alpha_c \beta_c I(t) b\, dx = [U_T(T_c - T_a) + U_B(T_c - T_{air})] b\, dx + \eta_c \alpha_c \beta_c I(t) b\, dx \tag{6.34}$$

and the daylight received by the room space is:

$$\tau_g^2 (1 - \beta_c) \alpha_T \times 100 \text{ lux} \tag{6.35}$$

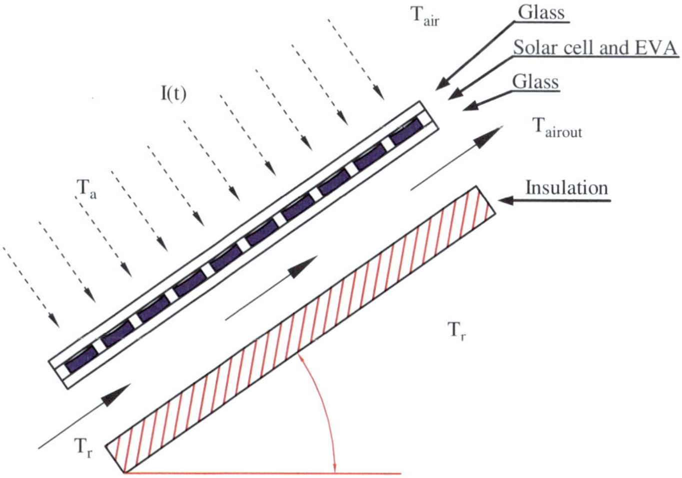

Figure 6.4 Cross-section of a semi-transparent BIPVT system with an air duct.

where the top heat-loss coefficient from the solar cell to the ambient air is $U_T = \left(\frac{L_g}{K_g} + \frac{1}{h_o}\right)^{-1}$ (in W m^{-2} K^{-1}) and the bottom heat-loss coefficient from the solar cell to the room is $U_B = \left(\frac{L_g}{K_g} + \frac{1}{h_{air}}\right)^{-1}$ (in W m^{-2} K^{-1}); h_o is the heat transfer coefficient of the ambient air and h_{air} is the heat transfer coefficient of the air flowing through duct (in W m^{-2} K^{-1}).

Simplifying eqn (6.34), the expression for a solar cell temperature can be obtained as:

$$T_c = \frac{U_B T_{air} + U_T T_a + I(t)(\alpha\tau)_{eff}}{U_T + U_B} \tag{6.36}$$

where $(\alpha\tau)_{eff} = (\tau_g - \eta_c)\alpha_c\beta_c$.

The energy balance for air flowing in the duct of the BIPVT system for elemental area $b\,dx$ is given by:

$$\begin{bmatrix}\text{Rate of heat transfer}\\ \text{from solar cell to}\\ \text{air in the duct}\end{bmatrix} = \begin{bmatrix}\text{Rate of heat gain by}\\ \text{air flowing in the duct}\\ \text{of the BIPVT system}\end{bmatrix} + \begin{bmatrix}\text{Rate of heat}\\ \text{loss from air}\\ \text{through insulation}\end{bmatrix} \tag{6.37}$$

Let $\dot{m}_{air}$ be the air mass flow-rate through the duct (in kg s^{-1}), C_{air} the specific heat of air at a constant pressure (in J kg^{-1} K^{-1}) and T_r the average temperature of the room (in K). Therefore eqn (6.37) can be rewritten as:

$$U_B(T_c - T_{air})b\,dx = \dot{m}_{air}C_{air}\left(\frac{dT_{air}}{dx}\right)dx + U_{bb}(T_{air} - T_r)b\,dx \tag{6.38}$$

where the overall heat transfer for a semi-transparent insulation plate is:

$$U_{bb} = \left(\frac{1}{h_{air}} + \frac{L_i}{K_i} + \frac{1}{h_r}\right)^{-1}$$

On substituting T_{bs} from eqn (6.29) into eqn (6.38) we have:

$$U_B\left[\frac{I(t)(\alpha\tau)_{eff} - U_T(T_{air} - T_a)}{U_T + U_B}\right] b\,dx = \dot{m}_{air}C_{air}\left(\frac{dT_{air}}{dx}\right)dx + U_{bb}(T_{air} - T_r)b\,dx$$

or

$$\left(\frac{dT_{air}}{dx}\right) + \frac{bU_L}{\dot{m}_{air}C_{air}}T_{air} = \frac{b}{\dot{m}_{air}C_{air}}\left[U_{bb}T_r + U_{tair}T_a + h_p I(t)(\alpha\tau)_{eff}\right] \tag{6.39}$$

where $U_{tair} = \frac{U_T \times U_B}{U_T + U_B} = \left(\frac{1}{U_T} + \frac{1}{U_B}\right)^{-1}$, $U_L = (U_{bb} + U_{tair})$ and the penalty factor due to air flow through the duct is $h_p = \frac{U_B}{U_T + U_B}$.

On integrating eqn (6.39) with the boundary condition at $x = 0$, $T_{air} = T_{r,}$, at $x = L$, $T_{air} = T_{airout}$, the outlet air temperature (T_{airout}) of the flowing air in the duct of the BIPVT system for length L is given by:

$$T_{airout} = \left[\frac{U_{bb}T_r + U_{tair}T_a + h_p I(t)(\alpha\tau)_{eff}}{U_L}\right] \times \left[1 - \exp\left(-\frac{bU_L}{\dot{m}_{air}C_{air}}L\right)\right] + T_r \exp\left(-\frac{bU_L}{\dot{m}_{air}C_{air}}L\right) \tag{6.40}$$

and the average air temperature of the air flowing in the duct of the BIPVT system is given by:

$$\bar{T}_{air} = \left[\frac{U_{bb}T_r + U_{tair}T_a + h_p I(t)(\alpha\tau)_{eff}}{U_L}\right]\left[1 - \frac{1 - \exp\left(-\frac{bUL}{\dot{m}_{air}C_{air}}L\right)}{\frac{bU_L}{\dot{m}_{air}C_{air}}L}\right] + T_r \frac{1 - \exp\left(-\frac{bUL}{\dot{m}_{air}C_{air}}L\right)}{\frac{bU_L}{\dot{m}_{air}C_{air}}L} \tag{6.41}$$

The rate of useful thermal energy (in Wh) obtained for n_{pv} rows of the BIPVT system is given by:

$$\dot{Q}_{ud} = n_{pv} \times \dot{m}_{air}C_{air}(T_{airout} - T_r) = n_{pv} \times \dot{m}_{air}C_{air}\left[\frac{U_{bb}T_r + U_{tair}T_a + h_p I(t)(\alpha\tau)_{eff}}{U_L} - T_r\right] \times \left[1 - \exp\left(-\frac{bU_L}{\dot{m}_{air}C_{air}}L\right)\right] \tag{6.42}$$

The available useful thermal energy is used to heat the room but part of it is lost. The energy balance for the space heating of the building is given by:

$$\begin{aligned} & n_{pv}\times \dot{m}_{air}C_{air}\left[\frac{U_{bb}T_r+U_{tair}T_a+h_pI(t)(\alpha\tau)_{eff}}{U_L}-T_r\right]\left[1-\exp\left(-\frac{bU_L}{\dot{m}_{air}C_{air}}L\right)\right] \\ & \quad + U_{bb}(\bar{T}_{air}-T_r)A_{roof} = \dot{m}_r C_{air}\left(\frac{\mathrm{d}T_r}{\mathrm{d}t}\right) \\ & \quad + (UA)_t(T_r-T_a)+0.33N_oV(T_r-T_a) \end{aligned} \tag{6.43}$$

where $\dot{m}_r$ is the air mass flow-rate in the room (in kg s^{-1}). The heat loss capacity (in W K^{-1}) through the building surfaces is the sum of the heat loss capacities through the walls, windows and doors, *i.e.* $(UA)_t=(UA)_{t\_w}+(UA)_{t\_win}+(UA)_{t\_d}$; h_o and h_r are the heat transfer coefficients for the ambient air and the room air, respectively. On solving eqn (6.43) we have:

$$\begin{aligned} \left(\frac{\mathrm{d}T_r}{\mathrm{d}t}\right) = & \frac{1}{\dot{m}_rC_{air}}\left[\begin{array}{l}\{(UA)_t+0.33N_oV\}-\left\{n_{pv}\times\dot{m}_{air}C_{air}\left(\frac{U_{bb}}{U_L}-1\right)\left[1-\exp\left(-\frac{bU_L}{\dot{m}_{air}C_{air}}L\right)\right]\right\} \\ -U_{bb}\left\{\frac{U_{bb}}{U_L}\left[1-\frac{1-\exp\left(-\frac{bUL}{\dot{m}_{air}C_{air}}L\right)}{\frac{bU_L}{\dot{m}_{air}C_{air}}L}\right]+\left[\frac{1-\exp\left(-\frac{bUL}{\dot{m}_{air}C_{air}}L\right)}{\frac{bU_L}{\dot{m}_{air}C_{air}}L}\right]-1\right\}A_{roof}\end{array}\right]T_r \\ = & \frac{1}{\dot{m}_rC_{air}}\left[\begin{array}{l}\{(UA)_t+0.33N_oV\}T_a+\left\{n_{pv}\dot{m}_{air}C_{air}\left[\frac{U_{tair}T_a+h_pI(t)(\alpha\tau)_{eff}}{U_L}\right]\left[1-\exp\left(-\frac{bU_L}{\dot{m}_{air}C_{air}}L\right)\right]\right\} \\ +U_{bb}\left\{\frac{U_{tair}T_a+h_pI(t)(\alpha\tau)_{eff}}{U_L}\right\}\left[1-\frac{1-\exp\left(-\frac{bUL}{\dot{m}_{air}C_{air}}L\right)}{\frac{bU_L}{\dot{m}_{air}C_{air}}L}\right]A_{roof}\end{array}\right] \end{aligned} \tag{6.44}$$

On integrating eqn (6.44) and applying the initial condition at time $t=0$, $T_r=T_{ri}$, the room air temperature (T_r) is given by:

$$T_r=\frac{f(t)}{a}(1-\mathrm{e}^{-at})+T_{ri}\mathrm{e}^{-at} \tag{6.45}$$

where:

$$\begin{aligned} a = & \frac{1}{\dot{m}_rC_{air}} \\ & \times\left[\begin{array}{l}\{(UA)_t+0.33N_oV\}-\left\{n_{pv}\dot{m}_{air}C_{air}\left(\frac{U_{bb}}{U_L}-1\right)\left[1-\exp\left(-\frac{bU_L}{\dot{m}_{air}C_{air}}L\right)\right]\right\} \\ -U_{bb}\left\{\frac{U_{bb}}{U_L}\left[1-\frac{1-\exp\left(-\frac{bU_L}{\dot{m}_{air}C_{air}}L\right)}{\frac{bU_L}{\dot{m}_{air}C_{air}}L}\right]+\frac{1-\exp\left(-\frac{bUL}{\dot{m}_{air}C_{air}}L\right)}{\frac{bU_L}{\dot{m}_{air}C_{air}}L}-1\right\}A_{roof}\end{array}\right] \end{aligned}$$

and:

$$f(t) = \frac{1}{\dot{m}_r C_{air}} \times \left[\begin{array}{l} \{(UA)_t + 0.33 N_o V\} T_a + \left\{ \begin{array}{l} n_{pv} \dot{m}_{air} C_{air} \left[\frac{U_{tair} T_a + h_p I(t)(\alpha\tau)_{eff}}{U_L} \right] \\ \times \left[1 - \exp\left(-\frac{b U_L}{\dot{m}_{air} C_{air}} L \right) \right] \end{array} \right\} \\ + U_{bb} \left\{ \frac{U_{tair} T_a + h_p I(t)(\alpha\tau)_{eff}}{U_L} \right\} \left[1 - \frac{1 - \exp\left(-\frac{b U_L}{\dot{m}_{air} C_{air}} L \right)}{\frac{b U_L}{\dot{m}_{air} C_{air}} L} \right] A_{roof} \end{array} \right]$$

6.3.5 Façade-integrated PVT Systems with and without a Duct

The thermal modelling for façade-integrated PVT systems is similar to that of the roof-integrated PVT systems. However, the solar radiation falling on them is of reduced intensity and the natural convection current helps the air to flow faster through the duct. Hence, the blower required by the system is of reduced capacity and the PVT façade generates lower power than the inclined roof-integrated PVT system.

6.4 Overall Performance

As the BIPVT system provides electrical energy along with thermal energy, the overall performance of the system may be presented either in the form of net thermal output or net exergy output.

6.4.1 Net Thermal Output

For the net thermal output the electrical output from the system is converted into the equivalent thermal energy required by the thermal power plant and added to the thermal output of the system. Using empirical relations given by Evans,[1] the instantaneous electrical efficiency (η_{ca}) of the BIPVT system at the solar cell temperature T_c is given by:

$$\eta_{ca} = \eta_{ref}[1 - \phi_{ref}(T_c - T_{ref})] \tag{6.46}$$

where η_{ref} is the efficiency at the reference temperature T_{ref} and ϕ_{ref} is the efficiency correction coefficient. The quantities η_{ref}, T_{ref} and ϕ_{ref} are usually given by the PV manufacturers, but they can be also obtained from flash tests.

Thus, the hourly electrical output of a BIPVT system is given by:

$$\begin{aligned} E_{\text{out}} &= \text{electrical efficiency} \times \text{solar insolation} \\ &= \eta_{\text{ca}} \times I(t) \times bL \times n_{\text{s}} \times n_{\text{p}} \end{aligned} \tag{6.47}$$

The conventional electrical output can be converted into an equivalent thermal output by dividing the electrical output with the conversion factor of the thermal power plant as follows:

$$E_{\text{ETH}} = \frac{E_{\text{out}}}{C_{\text{f}}} \tag{6.48}$$

The value of the conversion factor C_{f} depends on the quality of the coal; it varies between 0.2 and 0.4. For coal with a low ash content, its value is maximum is 0.4; for thermal power plants, C_{f} is about 0.38.

The hourly overall thermal output of a BIPVT system can be calculated by adding the hourly thermal gain ($\dot{Q}_{\text{u}}$) of the BIPVT system to eqn (6.48). Therefore:

$$\dot{Q}_{\text{hourly}} = \frac{E_{\text{out}}}{C_{\text{f}}} + \dot{Q}_{\text{u}} \tag{6.49}$$

The daily overall thermal output of the BIPVT system can be calculated by:

$$\dot{Q}_{\text{daily}} = \sum_{j=1}^{N} \frac{(\eta_{\text{ca}})_j [I(t)]_j \times n_{\text{s}} n_{\text{p}} \times bL}{C_{\text{f}}} + \sum_{j=1}^{N} (\dot{Q}_{\text{u}})_j \tag{6.50}$$

where j is an integer and N is the number of sunshine hours in a day.

Adding the daily overall thermal output for a year will give the annual overall thermal output. The overall thermal efficiency of the BIPVT system can be calculated by dividing the overall thermal output by the overall thermal gain from the solar energy. Thus:

$$\begin{aligned} \eta_{\text{th}} &= \frac{\text{net thermal output}}{\text{net solar insolation}} \\ &= \frac{\sum_{j=1}^{N} \frac{(\eta_{\text{ca}})_j [I(t)]_j \times n_{\text{s}} n_{\text{p}} \times bL}{C_{\text{f}}} + \sum_{j=1}^{N} (\dot{Q}_{\text{u}})_j}{\sum_{j=1}^{N} [I(t)]_j bL \times n_{\text{pv}}} \end{aligned} \tag{6.51}$$

6.4.2 Net Exergy Output

For the net exergy output the thermal output from the system is converted into the equivalent electrical energy, using the second law of thermodynamics, and added to the electrical output of the system. As there is a temperature difference between the hot air coming out of the BIPVT system as a heat source and the atmospheric air as a heat sink, the thermal energy can be transformed into work. The magnitude of the transformable thermal energy to work is restricted by the Carnot efficiency. Thus, from the instantaneous quantity of heat, $\dot{Q}_u$, produced by the BIPVT system, the instantaneous thermal exergy is:

$$\dot{E}x_{th} = \dot{Q}_u \times \left(1 - \frac{T_a}{T_{airout}}\right) \tag{6.52}$$

Therefore the net exergy gain from the BIPVT system is the sum of the electrical gain and the instantaneous thermal exergy:

$$\begin{aligned}\dot{E}x_{out} &= \dot{E}x_{el} + \dot{E}x_{th}\\ &= \left[\eta_{ca} \times I(t) \times bL \times n_s n_p\right] + \left[\dot{Q}_u \times \left(1 - \frac{T_a}{T_{airout}}\right)\right]\end{aligned} \tag{6.53}$$

Taking into account the total exergy inflow and outflow of the system, the overall exergy efficiency for a BIPVT system is given by:

$$\eta_{\dot{E}x} = \frac{\sum_{j=1}^{N} \dot{E}x_{out}}{\sum_{j=1}^{N} \dot{E}x_{in}} = \frac{\sum_{j=1}^{N} (\eta_{ca})_j \times [I(t)]_j \times bL \times n_s n_p + \sum_{j=1}^{N} (\dot{Q}_u)_j \times \left\{1 - \frac{(T_a)_j}{(T_{airout})_j}\right\}}{\sum_{j=1}^{N} [I(t)]_j \times bL} \tag{6.54}$$

where T_a and T_{airout} are in kelvin.

6.5 A Case Study of a Roof-integrated PVT System with an Air Duct

Agrawal and Tiwari[2] have made a detail study of a roof-integrated opaque-type PVT system with an air duct. This section describes their system configuration and findings.

6.5.1 System Description

The BIPVT system has been considered as a rooftop for Srinagar, India, situated at 34°01′N and 74°51′E, having cold climatic conditions. Each module of the BIPVT system has 72 monocrystalline silicon solar cells connected in series with two redundantly interconnected strings. The cells are laminated in UV-stabilized polymer and mounted behind a high-transmission toughened glass surface. A lamination process using Tedlar–polyester–Tedlar, EVA polymer, scrim and edge sealing of the laminate by aluminium sealant tape with an acrylic adhesive provides complete sealing against extreme temperature and weather conditions. The PV module is framed on anodized aluminium. The PV modules are manufactured to stringent quality specifications and tested to withstand adverse environmental conditions. The model conforms to the CEI/IEC 61215 standard.

For fitting the BIPVT system an experimental laboratory at Srinagar is under consideration, whose orthographic views are shown Figure 6.5. Tables 6.1 and 6.2 provide the detailed specification of the experimental room and the BIPVT system. On the roof are 48 BIPVT systems, eight in series and six in parallel. The electrical output from the BIPVT panel junction box is used to charge a 24 V battery efficiently in virtually any climate and is also connected to the power grid. Thus, solar electric generation can replace the highest cost electricity during times of peak demand (in most climatic regions). It can also

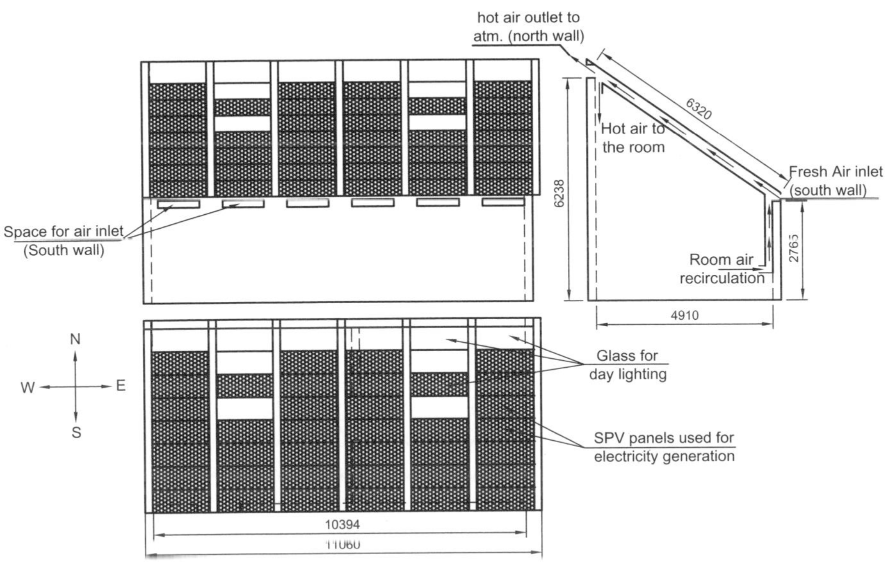

Figure 6.5 Orthographic projection of the experimental room (in mm).

Table 6.1 Specification of the building.

Maximum panel voltage	420 V
Efficiency	16%
Size of room 1	5580 × 4910 mm
Size of room 2	5480 × 4910 mm
Side wall height	south wall 2765 mm, north wall 6238 mm
Roof area	11 060 × 6144 mm
Roof inclination	35°
Windows	1 on east wall, 1 on west wall and 4 on north wall

Table 6.2 Design parameters of the BIPVT system.

Parameter	*Value*
Length of each BIPVT system	1600 mm (1594 mm)
Width of each BIPVT system	790 mm (789 mm)
Nominal output of a BIPVT system	150 W
Duct depth	250 mm
Air flow	single pass below Tedlar
C_{air} (J kg^{-1} K^{-1})	1005
C_f	0.38
h_o (W m^{-2})	$5.7 + 3.8 \times v_a$
h_i (W m^{-2})	2.8
h_T (W m^{-2})	$2.8 + 3 \times v_{air}$
K_c (W m^{-2} K^{-1})	0.039
K_G (W m^{-2} K^{-1})	0.8
K_i (W m^{-2} K^{-1})	0.035
K_T (W m^{-2} K^{-1})	0.38
L_c (mm)	0.3
L_G (mm)	32
L_i (mm)	10
L_T (mm)	3
α_c	0.7
α_t	0.7
β_c	0.9
η_c	0.16
τ_g	0.85
ρ_a (kg m^{-3})	1.29

reduce the grid loading and eliminate the need for a huge number of local batteries for use in times of darkness and high local demand. Six fans, each of 12 W, taking power from the battery bank are used to flow the air through the BIPVT system. The air is heated in the system by the transfer of heat from the back surface of the PV module. The heated air enters the room for space heating.

The opening of the duct in the south wall of the BIPVT system is the fresh air inlet. The other end of the duct is opened near the north wall to bring hot air into the room for space heating. Also, a provision is made in the north wall for the heated air to escape into the atmosphere when space heating is not required, especially during the summer season. The model is implemented to predict the temperature profile of the BIPVT roof for the given solar irradiance and ambient temperature. The approach uses individual BIPVT systems to compute electrical and outlet air temperatures using heat transfer relations. The outlet air from one BIPVT system in a column is used as the inlet air for the next BIPVT system. For the first BIPVT system the inlet air is a mixture of the circulated air of the room and fresh ambient air.

6.5.2 Observations and Discussion

Matlab 7 was used to evaluate the theoretical performance and computation of useful exergy of the BIPVT system. The solar radiation and climatic conditions of Srinagar city were collected for 11 years and classified into four climatic conditions, depending upon the ratio of daily diffuse to daily global radiation and the number of sunshine hours, namely:

- Type a: clear days (blue sky); the ratio of daily diffuse to daily global radiation is less than or equal to 0.25 and the number of sunshine hours is greater than or equal to 9 h.
- Type b: hazy days (fully); the ratio of daily diffuse to daily global radiation is between 0.25 and 0.50 and the number of sunshine hours is between 7 and 9 h.
- Type c: hazy and partially cloudy days; the ratio of daily diffuse to daily global radiation is between 0.50 and 0.75 and the number of sunshine hours is between 5 and 7 h.
- Type d: fully cloudy days; the ratio of daily diffuse to daily global radiation is more than or equal to 0.75 and the number of sunshine hours is less than or equal to 5 h.

The summarized data are presented in Appendix A6. The hourly variations of diffuse and beam radiation were used to calculate the hourly variation of the solar intensity on the inclined roof, using the Liu and Jordan formula.[3] Figure 6.6 shows the hourly variation of solar intensity over the surface of the BIPVT system and the ambient temperature for a typical day in January. The figure clearly shows that the solar intensity increases from early in the morning

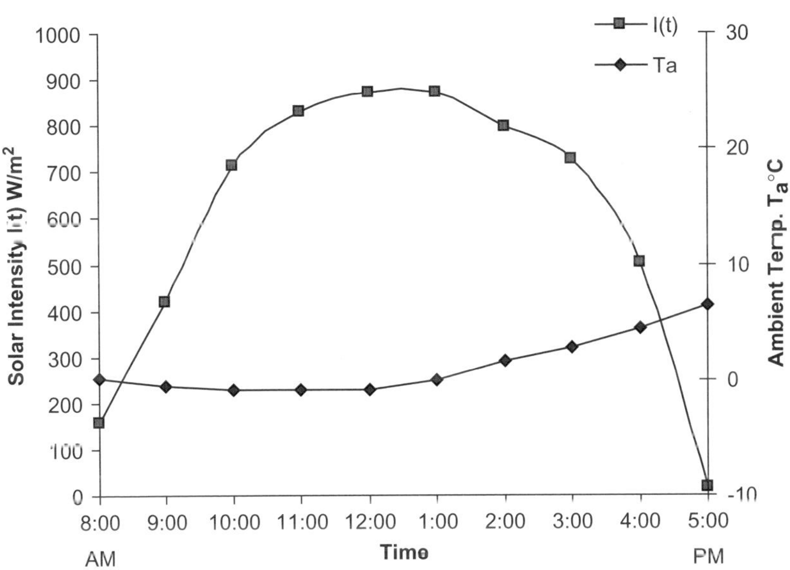

Figure 6.6 Hourly variation of the solar intensity on the BIPVT system and the ambient temperature (T_a) for a typical January day.

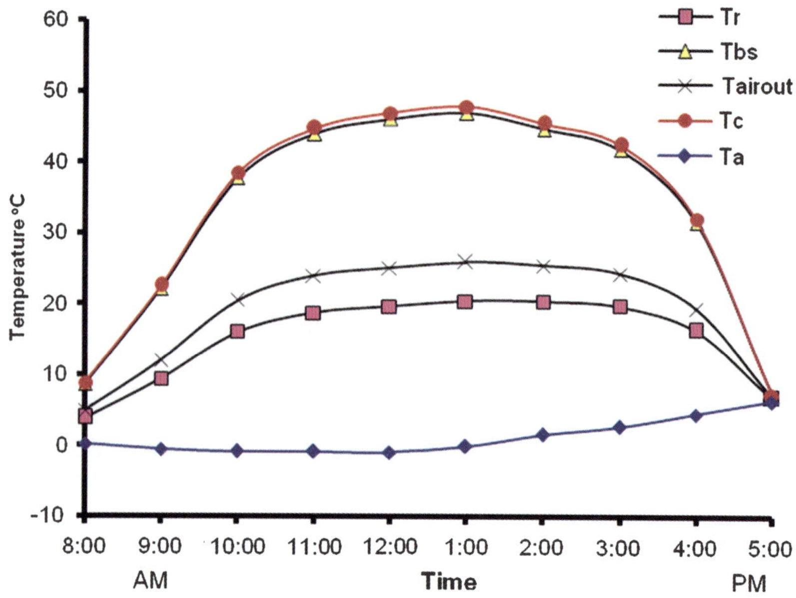

Figure 6.7 Hourly variation of solar cell temperature (T_c), back-surface temperature (T_{bs}), air outlet temperaturer (T_{airout}), room temperature (T_r) and ambient temperature (T_a) for a typical day.

and reaches maximum at 1.00 pm and thereafter decreases, whereas the ambient temperature increases and is a maximum at 5.00 pm.

Part of the solar intensity received by the system is converted into useful electricity and the remaining part is responsible for the increase in temperature of the solar cells. Equations (6.21) and (6.24) are used to determine the hourly variation of the solar cell temperature and the back-surface temperature of a module of the system. Figure 6.7 shows the hourly variation of solar cell temperature, back-surface temperature, air outlet temperature and room temperature. The temperatures of the solar cell and back surface of the module are almost same because the solar cells are very thin (0.3 mm) and the thermal conductivity of Tedlar is low (0.38 W m^{-1} K^{-1}). The heat from the back surface is carried away by the air flowing in the duct. Equation (6.28) is used to determine the hourly variation of the outlet temperature of the air from the duct. The hot air entering the room increases the room temperature. Equation (6.33) is used to calculate the hourly variation of the room temperature. The increase in the room temperature depends upon the thermal conductivities of the walls, doors and windows, the infiltration of the air, and the heat capacity of the room. To maintain a higher space temperature there should be lower infiltration. The figure shows that the maximum value of the room temperature is 20.4 °C on a typical clear day in January. Moreover, there must be lower infiltration of the air to keep the living space hotter during cold winters days and higher infiltration to keep the living space cooler during the hot summer days.

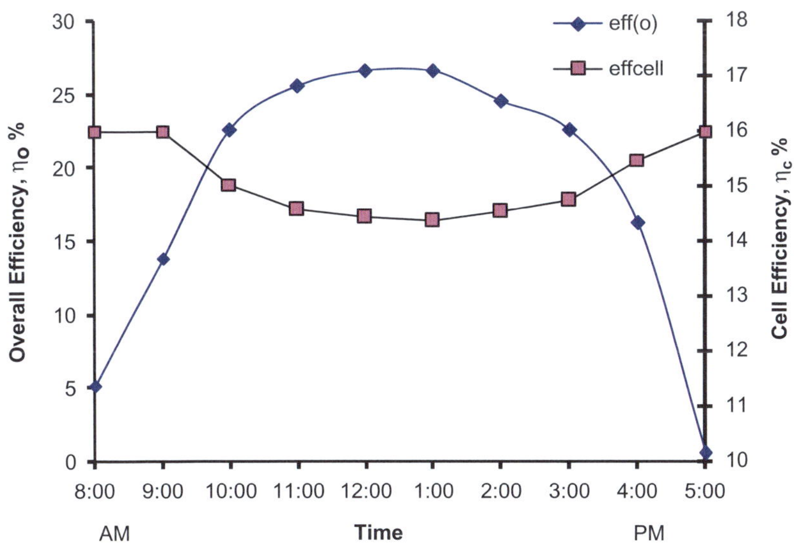

Figure 6.8 Hourly variation of solar cell efficiency (η_c) and overall thermal efficiency (η_o).

The cell efficiency and the overall thermal efficiency are determined using eqns (6.46) and (6.51), respectively. Figure 6.8 shows that the minimum value of the cell efficiency is 14.3% and the maximum value of the overall thermal efficiency is 51.3%. The overall thermal efficiency increases with the increase in solar intensity, while the cell efficiency increases with the decrease in solar intensity. This is because at higher intensity the solar cell temperature increases, leading to a reduction in cell efficiency and an increase in thermal loss. As this thermal loss is used to heat the air in the duct, the thermal efficiency of the system improves.

The useful thermal energy, the electrical output and the net exergy are obtained from eqns (6.30), (6.47) and (6.53), respectively, and plotted in Figure 6.9. For the clear days of January the maximum electrical output, thermal output and net exergy are 7.6, 6.6 and 8.2 kWh, respectively. Figure 6.10 shows the effect of the number of air changes per hour on the air temperature of the room. It can be concluded that an increase in the number of air changes per hour from 0 to 8, commonly called infiltration, reduces the maximum room air temperature from 29.5 °C to 10.4 °C, because more room air is replaced by ambient air. The rate of decrease in room air temperature decreases with the increase in the changes of air per hour. The range between the maximum and minimum values of the room air temperature in a day is also reduced from 24 °C to 9.3 °C.

Figure 6.11 shows the effect of an increase in the air mass flow-rate in the duct from 0 to 0.8 kg s^{-1}. The increase in air mass flow-rate increases the maximum room air temperature from 9 °C to 25.8 °C. It can be seen that at zero

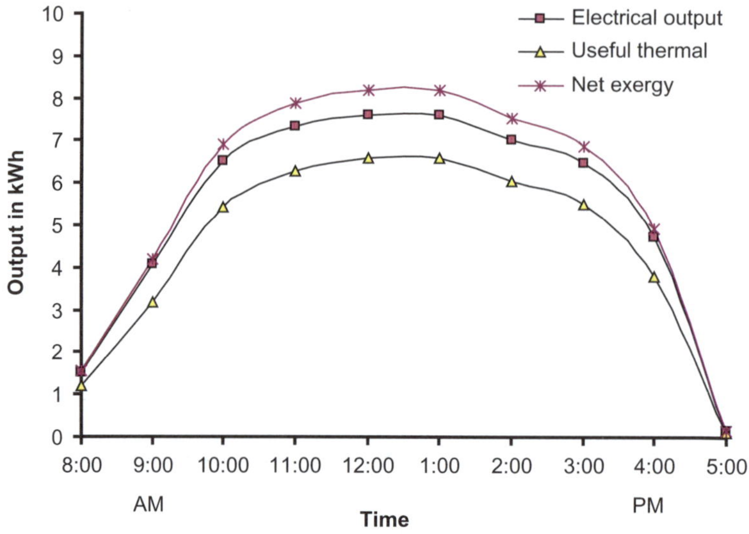

Figure 6.9 Hourly variation of electrical output (E_{out}), useful thermal (Q_u) and net exergy.

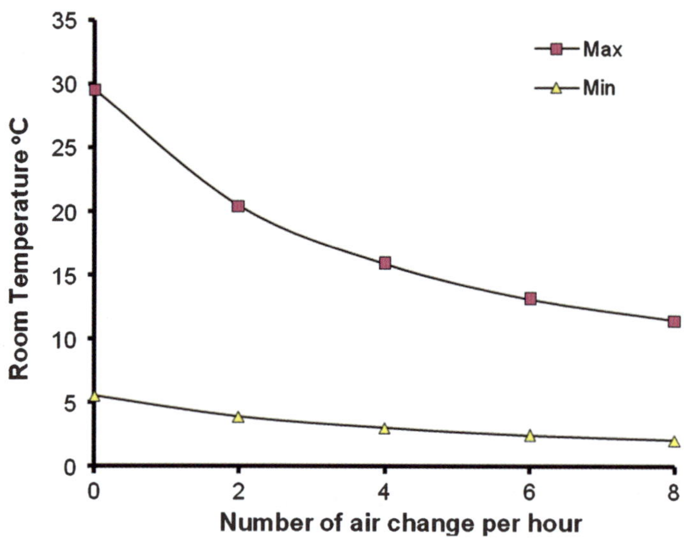

Figure 6.10 Minimum and maximum values of the room temperature for different air changes per hour.

mass flow-rate the rise in room temperature is much less because the hot air from the duct is not entering the room and at high mass flow-rate the room temperature is higher because the room air is circulated a greater number of times through the duct, raising the duct air outlet temperature. The rate of

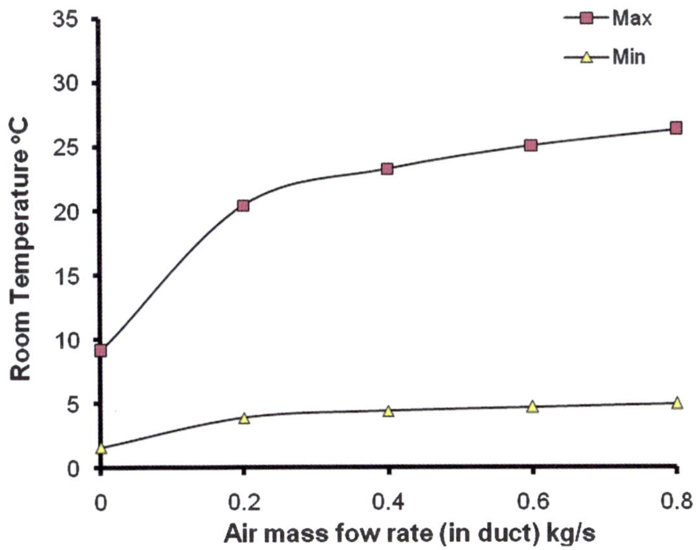

Figure 6.11 Minimum and maximum values of the room temperature for different air mass flow-rates.

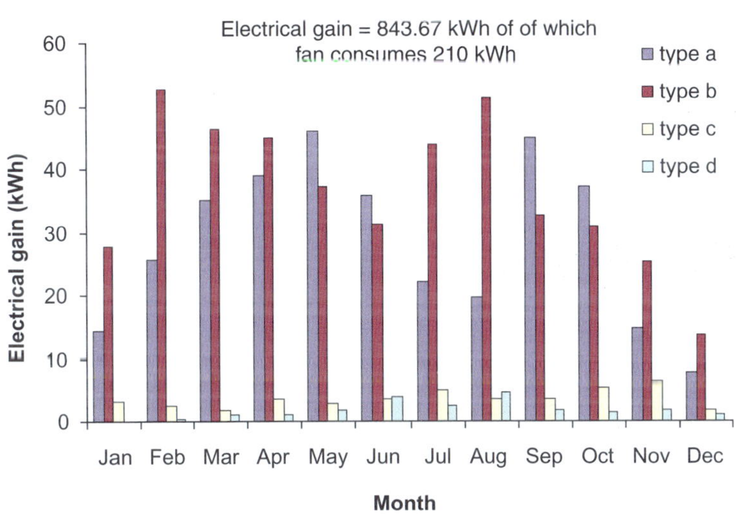

Figure 6.12 Electrical gain through using a BIPVT system in lieu of a BIPV system for type a, b, c and d climatic conditions.

increase in room air temperature decreases with the increase in change of mass flow-rate because the capacity to gain heat by the hot circulated air decreases. The daily temperature range of the room air varies from 7.5 °C to 21.3 °C.

Figure 6.12 shows the electrical gain, if a BIPVT system is used in place of a BIPV system, for type a, b, c and d climatic conditions. The analysis shows that there is an annum electrical gain of 342 kWh during clear days, 437.6 kWh during fully hazy days, 42.8 kWh during partially hazy and cloudy days and 21.1 kWh during totally cloudy days. Out of this electrical gain, 210 kWh is utilized to run the fan. Thus, the overall electrical gain for a year is nearly 633 kWh. Figure 6.13 shows the thermal gain from the BIPVT system. The annum thermal gains for type a, b, c and d are 4828, 7053, 1442 and 627 kWh, respectively. The net thermal gain is 13 950.5 kWh. Figure 6.14 shows the net

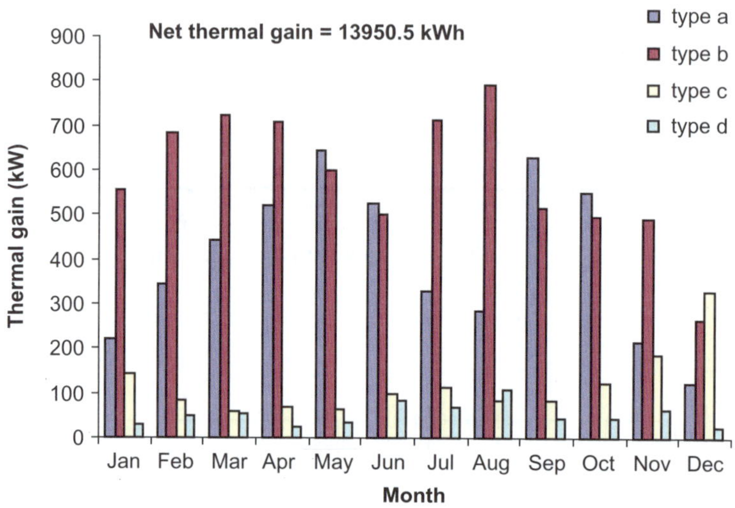

Figure 6.13 Net thermal gain by using BIPVT a system in lieu of a BIPV system for type a, b, c and d climatic conditions.

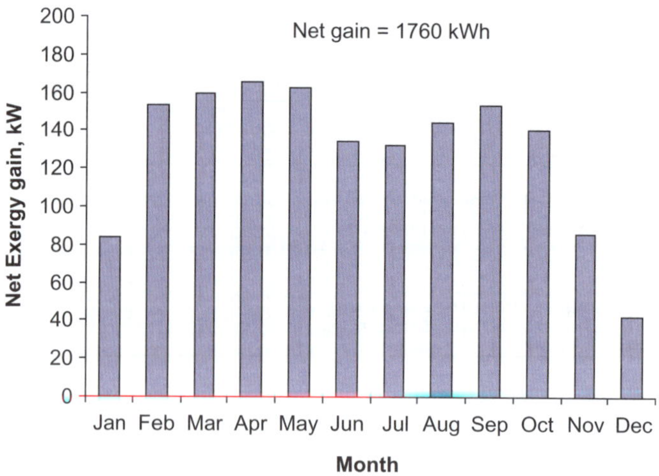

Figure 6.14 Annual exergy gain.

exergy gain for different months of the year. The net exergy gain for a year is 1560 kWh.

6.6 Optimization by Analyzing Different Configurations

For optimization of the roof-integrated opaque-type PVT system, Agrawal and Tiwari[4] connected the ducts in various manners to form configurations as follows:

- Case 1: all six rows of the BIPVT system are connected in parallel, as shown in Figure 6.15 (*top left*).
- Case 2: three rows of the BIPVT system are connected in parallel, each having two rows in series, as shown in Figure 6.15 (*top right*).
- Case 3: two rows of the BIPVT system are connected in parallel, each having three rows in series, as shown in Figure 6.15 (*bottom left*).
- Case 4: all rows of the BIPVT system are connected in series, as shown in Figure 6.15 (*bottom tight*).

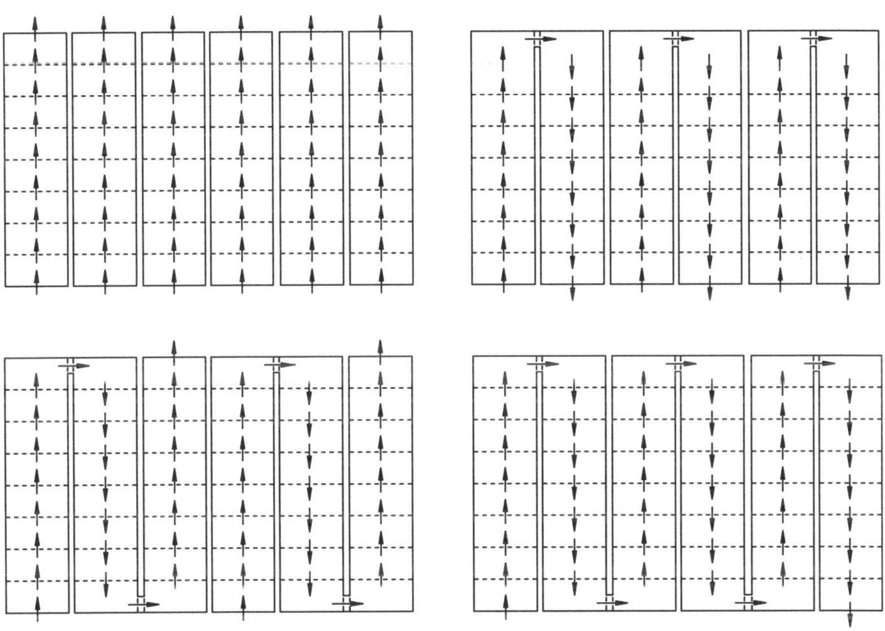

Figure 6.15 *Top left*, case 1, all rows of the BIPVT system are connected in parallel; *top right*, case 2, three rows of the BIPVT system are connected in parallel, each having two rows in series; *bottom left*, case 3, two rows of the BIPVT system are connected in parallel, each having three rows in series; *bottom right*, case 4, all rows of the BIPVT system are connected in series.

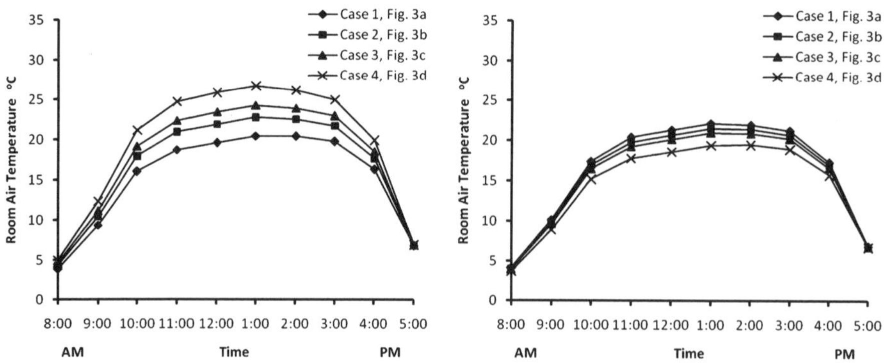

Figure 6.16 Hourly variation of room air temperature for all combinations of the BIPVT systems (cases 1 to 4) on a typical day at: *left*, constant mass flow-rate of air; *right*, constant velocity of air flow.

6.6.1 Methodology for Analysis

Owing to the different configurations, the output will vary. The best configuration is one which gives the higher energy and exergy output for the same solar intensity. The air in case 1 entering at a constant mass flow-rate of 1.2 kg s^{-1} is distributed equally in six rows; therefore the resulting velocity of air inside each duct is 0.65 m s^{-1}. Similarly for cases 2, 3 and 4, the air velocities in the duct are 1.3, 1.95 and 3.9 m s^{-1}, respectively. Note the following points.

1. The hourly variation of the ambient temperature and the solar intensity for a typical day is presented in Figure 6.16.
2. The room air temperature (T_r) is calculated with the help of eqn (6.33), taking the building specifications and design parameters from Tables 6.1 and 6.2, respectively. The results are presented in Figure 6.16, which show the hourly variation of the room temperature at a constant mass flow-rate and a constant velocity of air.
3. The outlet air temperature (T_{airout}) of the air flowing through the duct is calculated using eqn (6.28). The results are presented in Figure. 6.17 to show the hourly variation of air outlet temperature for all combinations of the BIPVT systems. Also the monthly variation of maximum and minimum room air temperature is presented in Figure 6.18.
4. The average temperature ($\bar{T}_{air}$) of the air flowing in the duct and the back-surface temperature of the photovoltaic panel (T_{bs}) are calculated using eqns (6.29) and (6.24), respectively. The solar cell temperature (T_c) is obtained by substituting the value of T_{bs} in eqn (6.21). The actual cell efficiency is obtained by substituting the solar cell temperature (T_c) from eqn (6.46). The monthly variation of electrical efficiency is presented in Figure 6.19.
5. The electrical output is obtained by substituting the solar cell efficiency in eqn (6.47). The useful thermal energy is calculated by substituting the

values of the air outlet and room air temperatures in eqn (6.30). The net exergy gain is determined by substituting the values of the electrical output and useful thermal energy in eqn (6.53). The monthly variation of net exergy gain is presented in Figure 6.20.

6. The annual overall thermal gain is obtained by summing up $\dot{Q}_{\text{daily}}$ from eqn (6.50) for a year. Finally, the overall thermal efficiency is determined from eqn (6.51), where the annual overall thermal gain is divided by the net solar insolation. The monthly variation of the overall thermal efficiency is shown in Figure 6.21. Tables 6.3 and 6.4 summarize the annual electrical output, thermal output, overall exergy gain and overall thermal efficiency of the systems at constant mass flow-rate and constant velocity of air, respectively.

6.6.2 Results and Discussion

Figure 6.6 shows that the solar intensity is a maximum for a clear day at noon. Figure 6.16 (*left*) shows that for a constant mass flow-rate of air, the BIPVT system connected in series [case 4, Figure 6.15 (*bottom right*)] provides a higher room temperature. For this case, the room air temperature is higher by up to 8 °C than the other combinations of the systems, such as cases 1, 2 and 3. This indicates that in the case of a series combination the air entering the room has a higher heat content. Figure 6.16 (*right*) shows that for a constant velocity of air flow, the BIPVT system connected in parallel [case 1, Figure 6.15 (*top left*)] provides a higher room air temperature by up to 3.5 °C than the other combinations of the systems, such as cases 2, 3 and 4, and hence a higher heat content.

Figure 6.17 (*left*) shows that for a constant mass flow-rate, the combination of the BIPVT systems connected in series provides the maximum air outlet temperature. This helps in maintaining a higher temperature of the room. It is further observed that for case 4, the air outlet temperature is about 3–8 °C higher than the other cases. Since in the case of a series combination the mass flow-rate and specific heat of air are constant, the air flows in the duct at a

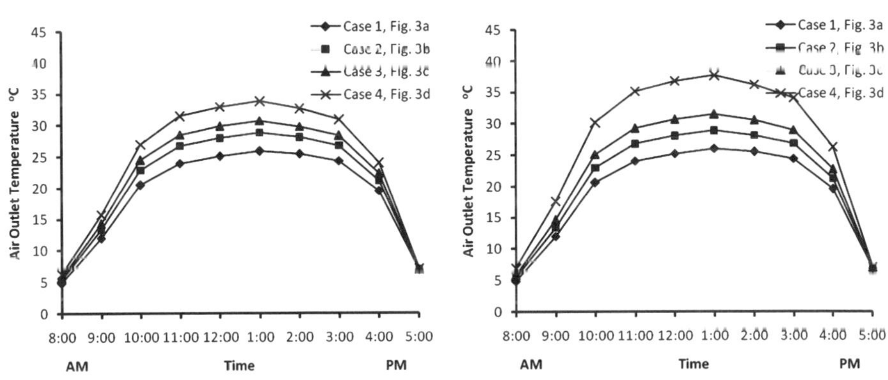

Figure 6.17 Hourly variation of air outlet temperature for all combinations of the BIPVT systems (cases 1 to 4) on a typical day at: *left*, constant mass flow-rate of air; *right*, a constant velocity of air flow

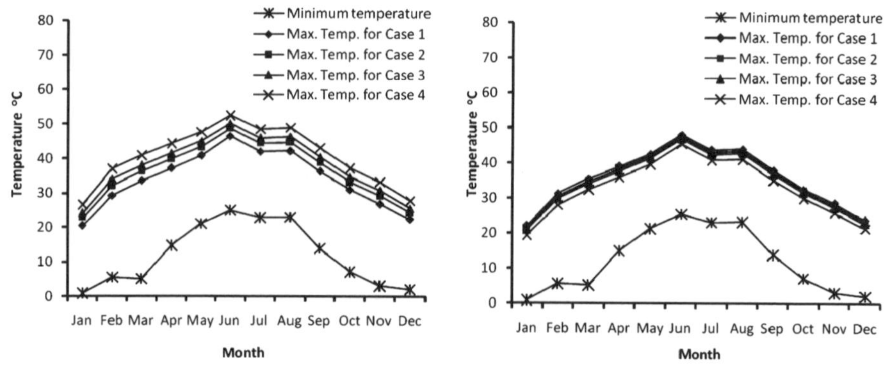

Figure 6.18 Monthly maximum and minimum room air temperature variation range for all combinations of the BIPVT systems (cases 1 to 4) at: *left*, a constant mass flow-rate of air; *right*, a constant velocity of air flow.

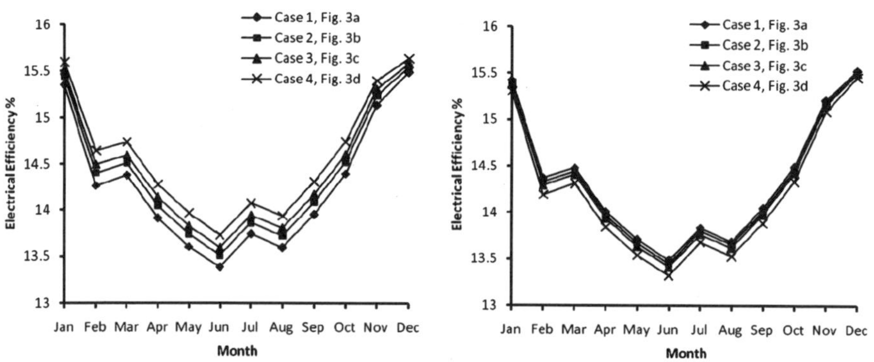

Figure 6.19 Monthly variation of electrical efficiency for all combinations of the BIPVT systems (cases 1 to 4) at: *left*, a constant mass flow-rate of air; *right*, a constant velocity of air flow.

higher velocity and collects a higher amount of heat; thus it will maintain a lower temperature of the solar cells. Figure 6.17 (*right*) shows that for a constant velocity of air flow, the combination of the BIPVT systems connected in parallel has a higher mass flow-rate of air and hence provides a lower air outlet temperature as well as a lower temperature of the solar cells. Both parts of Figure 6.18 show the variation of the temperature of the air in the room for all combinations of the BIPVT systems. It is observed that a lower value for the number of air changes per hour makes the room temperature pleasant in cold days, whereas a higher value is required in hot days.

Figure 6.19 (*left*) shows the variation of the electrical efficiency for all combinations of the BIPVT systems for a constant mass flow-rate of air. As the series combination has a lower cell temperature, the electrical efficiency is higher

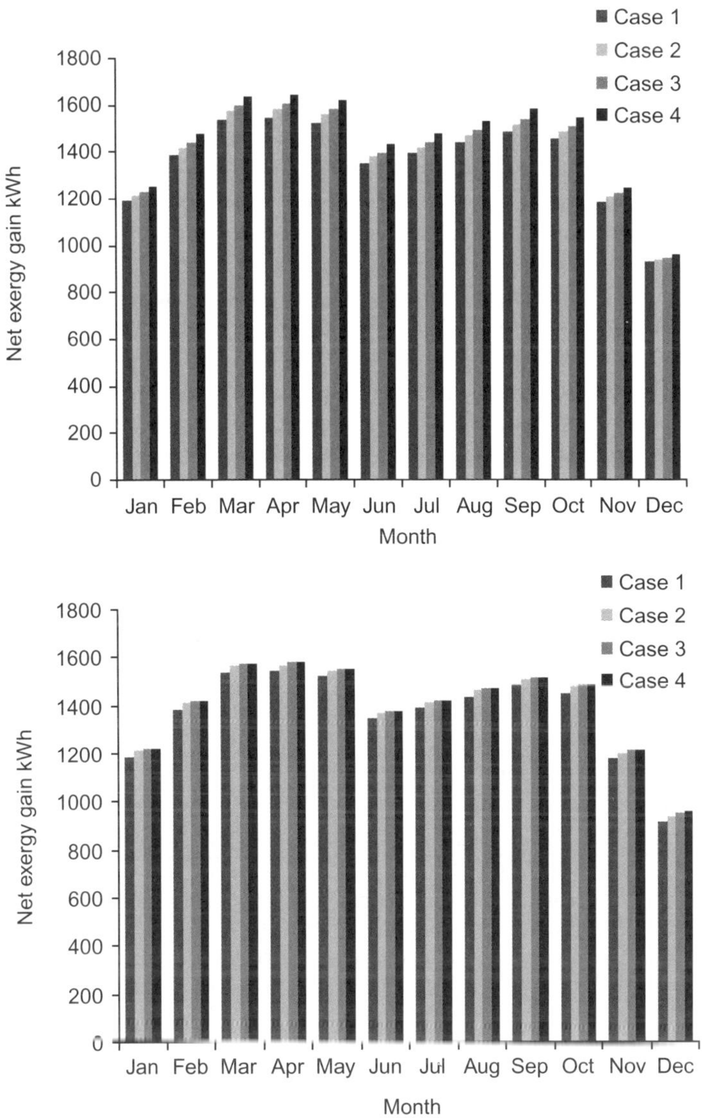

Figure 6.20 Net exergy gain through the BIPVT systems at: *top*, a constant mass flow-rate of air; *bottom*, a constant velocity of air flow.

in this case. The series combination produces an annual electrical energy of 16 209 kWh, which is 371 kWh higher than the parallel combination of the BIPVT systems and 1182 kWh higher than the similar BIPV system. The net thermal gain for the series combination is 18 062 kWh, which is 4112 kWh higher than the parallel combination of systems. Figure 6.19 (*right*) shows that for the constant velocity of air flow the electrical efficiency for parallel combinations of

BIPVT systems is higher owing to a lower cell temperature. The parallel combination produces annual electrical energy of 15 942 kWh, which is 170 kWh higher than the series combination of the BIPVT systems and 916 kWh higher than the similar BIPV system. The net thermal gain for the parallel combination is 15 285 kWh, which is 3300 kWh higher than the series combination of systems.

Figure 6.20 (*top*) shows that for a constant mass flow-rate of air, the annual overall exergy gain is 2363 kWh in the case of a series combination of the BIPVT systems, which is up to 975 kWh higher than the other combinations. Thus, from the electrical point of view, if the mass flow-rate is made constant the system connected in series is capable of producing power at a cheaper rate and hence is more economical. Figure 6.20 (*bottom*) shows that for a constant velocity of air flow, the annual overall exergy gain is 1765 kWh in the case of a series combination, which is up to 428 kWh higher than in the other combinations, because the blower consumes less power. Thus, from the electrical point of view, if the velocity of the air flow is made constant, the system connected in series is capable of producing power at a cheaper rate and hence is more economical.

The analytical results presented in Figure 6.21 are in agreement with the experimental results of Joshi *et al.*[5] Figure 6.21 (*left*) shows that for a constant mass flow-rate of air, the series combination provides an overall thermal efficiency of 53.7%, which is relatively higher than the other combinations of the BIPVT systems. Thus from the thermal point of view, if the mass of the air flow is made constant, the series combination of the BIPVT systems is better. Figure 6.21 (*right*) shows that for a constant velocity of air flow the parallel combination provides an overall thermal efficiency of 50%, which is relatively higher than the other combinations of the BIPVT systems. Thus from the thermal point of view, if the velocity of the air flow is made constant, the parallel combination of the BIPVT systems is better.

Thus, it is concluded that for a constant mass flow-rate of air, the series combination is more suitable for buildings fitted with BIPVT systems on the rooftop. The system produces annual electrical and thermal exergies of

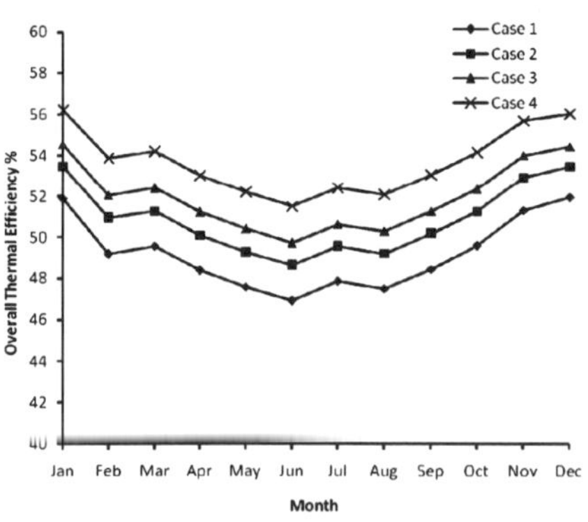

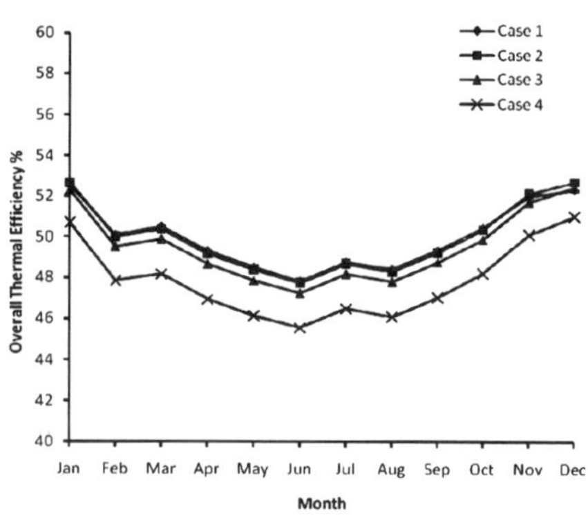

Figure 6.21 Overall efficiency of the BIPVT systems at: *left*, a constant mass flow-rate of air; *right*, a constant velocity of air flow.

Table 6.3 Annual exergies and efficiencies of the BIPVT systems for different combinations (cases 1 to 4) at a constant mass flow-rate of 1.2 kg s^{-1}.

Case	*Annual electrical output from BIPVT system (kWh)*	*Annual electrical gain from BIPVT system (kWh)*	*Annual thermal energy from BIPVT system (kWh)*	*Annual thermal exergy from BIPVT system (kWh)*	*Annual power consumed by air blower (kWh)*	*Annual overall exergy gain (kWh)*	*Thermal equivalent (net electrical/0.38) (kWh)*	*Overall thermal energy (kWh)*	*Overall thermal efficiency (%)*
1	15 838	811	13 950	927	350	1388	40 756	54 706	48.2
2	15 975	949	15 473	1134	350	1733	41 118	56 591	50.9
3	16 064	1038	16 451	1279	350	1967	41 353	57 814	51.9
4	16 209	1182	18 052	1531	350	2363	41 732	59 794	53.7

Table 6.4 Annual exergies and efficiencies of the BIPVT systems for different combinations (cases 1 to 4) at a constant air flow velocity of $1\,\mathrm{m\,s^{-1}}$.

Case	*Annual electrical output from BIPVT system (kWh)*	*Annual electrical gain from BIPVT system (kWh)*	*Annual thermal energy from BIPVT system (kWh)*	*Annual thermal exergy from BIPVT system (kWh)*	*Annual power consumed by air blower (kWh)*	*Annual overall exergy gain (kWh)*	*Thermal equivalent (net electrical/0.38) (kWh)*	*Overall thermal energy (kWh)*	*Overall thermal efficiency (%)*
1	15 942	916	15 285	1021	600	1337	40 374	55 659	50.01
2	15 904	877	14 544	1070	300	1647	41 063	55 607	49.99
3	15 858	841	13 845	1101	200	1742	41 231	55 076	49.52
4	15 771	745	11 984	1120	100	1765	41 240	53 224	47.88

16 209 kWh and 1531 kWh, with an average overall thermal efficiency of 53.7%. The net exergy output is up to 975 kWh higher than any other combination of the BIPVT systems and up to 2713 kWh higher than a BIPV system. It is further concluded that, for a constant velocity of air flow, the parallel combination provides an overall thermal efficiency of 50%, which is higher than any other combinations. Moreover, the result for a constant velocity helps for a better understanding of the functioning strategies. The thermal gain may be used either for space heating or for drying purposes. As the system is producing energy only during the daytime, it is helpful in reducing the peak demand on the grid to a greater extent.

6.7 Case Study of a Façade-integrated PVT System with an Air Duct

Dubey *et al.*[6] derived analytical expressions for N hybrid photovoltaic thermal (PVT) air collectors connected in series. The performance of the air collectors fully covered with a semi-transparent PV module is evaluated by considering the air flow in the following manner:

- Case I: above the absorber plate, as shown in Figure 6.22 (*left*).
- Case II: below the absorber plate, as shown in Figure 6.22 (*right*).

For solar cells the energy balance is given by eqn (6.34) as:

$$\tau_g \alpha_c \beta_c I(t) b\,dx = [U_T(T_c - T_a) + U_B(T_c - T_{air})] b\,dx + \eta_c \alpha_c \beta_c I(t) b\,dx \quad (6.55)$$

6.7.1 Thermal Modelling

6.7.1.1 Case I: Air Flow above the Absorber Plate

The energy balance equations for other components of the air collector for an elemental area $b\,dx$ in the flow direction are as follows. For a blackened

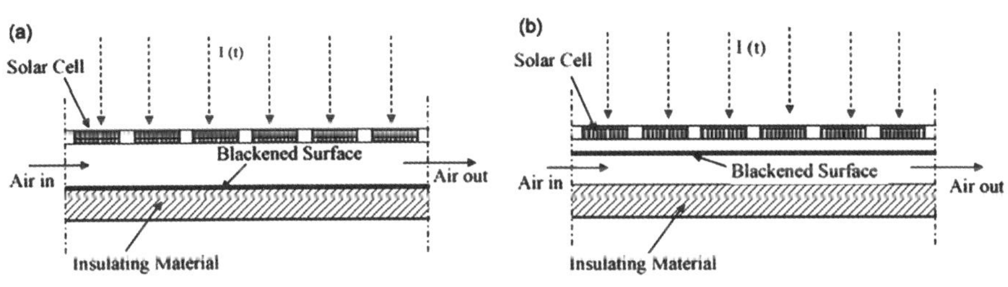

Figure 6.22 Cross-sectional view of the air collector considering the flow of air: *left*, above the blackened surface; *right*, below the blackened surface.

absorber plate the energy balance is given by:

$$\begin{bmatrix}\text{Rate of solar energy available}\\ \text{on blackened surface from}\\ \text{non-packing area of PVT}\end{bmatrix} = \begin{bmatrix}\text{Rate of heat transfer}\\ \text{from blackened plate}\\ \text{to flowing fluid}\end{bmatrix} + \begin{bmatrix}\text{An overall heat loss}\\ \text{from flowing fluid}\\ \text{to ambient}\end{bmatrix}$$

If α_p is the absorptivity of the absorber plate, $h_{p,f}$ is the heat transfer coefficient from the plate to the duct air, $U_{bp,a}$ is the overall heat transfer coefficient from the duct air to the ambient air, and T_p and T_f are the plate temperature and the flowing fluid temperature, respectively, then:

$$\left[\alpha_p(1-\beta_c)\tau_g^2 I(t)\right] b\,dx = \left[h_{p,f}(T_p - T_f) + U_{b\,p,a}(T_p - T_a)\right] b\,dx \tag{6.56}$$

For air flowing through the duct the energy balance is given by:

$$\begin{bmatrix}\text{Rate of heat withdrawal}\\ \text{by the flowing fluid}\end{bmatrix} = \begin{bmatrix}\text{Rate of heat transfer from}\\ \text{blackened plate to flowing fluid}\end{bmatrix} + \begin{bmatrix}\text{Rate of heat transfer from}\\ \text{cell to flowing fluid}\end{bmatrix}$$

If $\dot{m}_f$ is the mass flow-rate of the fluid and C_f is the specific heat, then:

$$\dot{m}_f C_f \frac{dT_f}{dx} dx = \left[h_{p,f}(T_p - T_f) + U_B(T_c - T_f)\right] b\,dx \tag{6.57}$$

6.7.1.2 Case II: Air Flow below the Absorber Plate

For a blackened absorber plate the energy balance is given by:

$$\begin{bmatrix}\text{Rate of solar energy available}\\ \text{on blackened surface from}\\ \text{non-packing area of PVT}\end{bmatrix} = \begin{bmatrix}\text{Rate of heat transfer}\\ \text{from back side of module}\\ \text{to blackened plate}\end{bmatrix} + \begin{bmatrix}\text{An overall heat loss}\\ \text{from blackened plate}\\ \text{to flowing fluid}\end{bmatrix}$$

or

$$\left[\alpha_p(1-\beta_c)\tau_g^2 I(t)\right] b\,dx = \left[h_{p,f}(T_p - T_f) + h_{c,p}(T_p - T_c)\right] b\,dx \tag{6.58}$$

For air flowing through the duct the energy balance is given by:

$$\begin{bmatrix}\text{Rate of heat withdrawal}\\ \text{by the flowing fluid}\end{bmatrix} = \begin{bmatrix}\text{Rate of heat transfer from}\\ \text{blackened plate to flowing fluid}\end{bmatrix}$$

If $\dot{m}_f$ is the mass flow-rate of the fluid and C_f is the specific heat, then:

$$\dot{m}_f C_f \frac{dT_f}{dx} dx = h_{p,f}(T_p - T_f) b\,dx \tag{6.59}$$

6.7.2 Analysis of the System

The room shown in Figure 6.23 having a surface area of 16 m^2 has been considered for evaluating the variation of room air temperature. PVT air collectors are integrated on the south wall (12 m^2) of the room and connected in series. All the remaining three walls and the roof are insulated with wooden structures to minimize heat loss. The rate of thermal energy available at the outlet of the

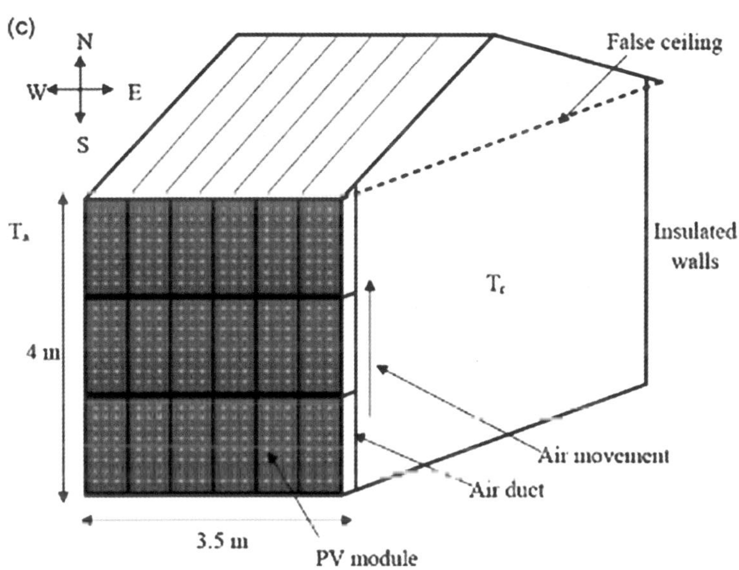

Figure 6.23 Schematic diagram of a façade-integrated PVT system.

PVT air collector is circulated into the insulated room; then the energy balance of whole system is written as:

$$\dot{Q}_{u,N} = M_a C_a \frac{dT_r}{dt} + \sum U_i Ai(T_r - T_a) \quad (6.60)$$

The above equation can be solved by assuming $T_{fi} = T_r$ due to the perfectly insulating connecting pipes. Equation (12a) can be written as:

$$(\alpha\tau)_{eff} I(t) - (UA)_{eff}(T_r - T_a) = M_a C_a \frac{dT_r}{dt} + \sum U_i A_i (T_r - T_a) \quad (6.61)$$

or

$$\frac{dT_r}{dt} + aT_r = f(t) \quad (6.62)$$

where $a = \frac{\left[(UA)_{eff} + \sum U_i A_i\right]}{M_a C_a}$ and $f(t) = \frac{(\alpha\tau)_{eff} I(t) + \left[(UA)_{eff} + \sum U_i A_i\right] T_a}{M_a C_a}$.

6.7.3 Results and Discussion

The values of the design parameters of an air collector for both cases are given in Table 6.5. The hourly variation of solar intensity and ambient temperature for a typical day in a summer month (May) is shown in Figure 6.24. The air velocity varies from 0.02 to 0.5 m s^{-1} and the number of collectors considered are from 1 to 8.

The effect of mass flow-rate on hourly outlet air temperature and useful heat gain for Cases I and II at $N = 1$ is shown in Figures 6.25 and 6.26, respectively. The results show that the hourly variation in outlet air temperature and useful heat gain for both cases are nearly the same at high air velocity (0.5 m s^{-1}). The outlet air temperature increases and the heat gain decreases with a decrease in air velocity (0.1 m s^{-1} and 0.02 m s^{-1}) for both cases. At a lower air velocity, a higher temperature is obtained due to an increase in the contact time period between the air and absorber and simultaneously a higher heat transfer. However, the thermal losses increase with an increase in the outlet temperature, which decreases useful heat gain. It is also seen that at lower air velocity a higher temperature and a higher heat gain is obtained in Case II compared with Case I. With decreases in air velocity the outlet air temperature increases from 51.3 °C to 104.7 °C and from 51.1 °C to 117.3 °C and the useful heat gain decreases from 0.182 kWh to 0.032 kWh and from 0.179 kWh to 0.038 kWh for Cases I and II, respectively. A lower air velocity (0.02 m s^{-1}) can be considered as a natural flow.

Figures 6.27 and 6.28 show the effect of the number of collectors on the hourly outlet air temperature and useful heat gain at an air velocity of 0.5 m s^{-1} for Cases I and II. The results show that as the number of collectors

Table 6.5 Design parameters for a photovoltaic thermal air collector.

Parameter	*Value*	*Parameter*	*Value*
PV module area, A (m^2)	0.675	PV module length, L (m)	1.5
PV width, b (m)	0.45	Heat transfer coefficient from blackened surface to ambient, $U_{L,I}$ (Case I) (W m^{-2})	3.58
Specific heat of fluid, C_f, (J kg^{-1} K^{-1})	1005	Heat transfer coefficient from blackened surface to ambient, $U_{L,II}$ (Case II) (W m^{-2})	3.44
Flow-rate factor, F_R (Case I)	0.88	$U_{t\ c,a}$ (W m^{-2} K^{-1})	9.5
Flow-rate factor, F_R (Case II)	0.90	Velocity of air flow, V (m s^{-1})	0.5
Heat transfer coeff. from cell to fluid, U_B (W m^{-2} K^{-1})	4.3	Thickness of collector, w (m)	0.05
Heat transfer coefficient from cell to absorber plate, $h_{c,p}$ (W m^{-2} K^{-1})	5.7	Absorptivity of solar cell, α_c	0.9
Heat transfer coefficient from plate to fluid, $h_{p,f}$ (W m^{-2} K^{-1})	100	Transmissivity of solar cell, τ_c	0.95
Heat transfer coefficient from fluid to ambient, $U_{f,a}$ (W m^{-2})	0.62	Packing factor, β_c	0.83
Mass flow-rate in duct, $\dot{m}$ (kg s^{-1})	0.0113	Efficiency of module, η_c	0.12
Penalty factor, PF_1 (Case I)	0.312	Absorptivity of plate, α_p	0.8
Penalty factor, PF_2 (Case II)	0.965	Transmissivity of glass, τ_g	0.95

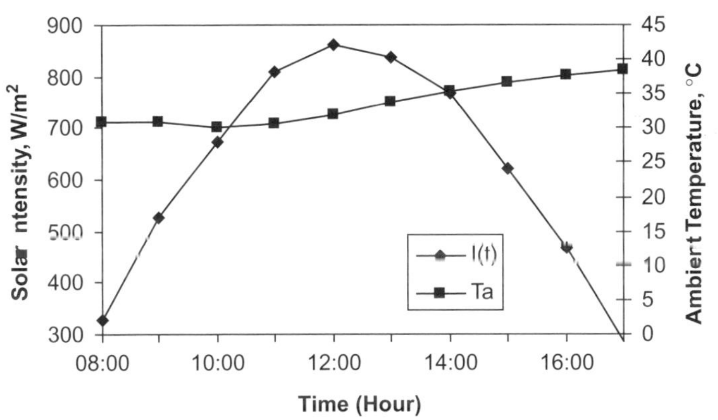

Figure 6.24 Hourly variation of solar intensity and ambient temperature of a typical day in the month of May.

increases, the outlet air temperature increases from 51.3 °C to 95.5 °C and from 51.1 °C to 101.5 °C, and the useful heat gain increases from 0.179 kWh to 0.698 kWh and from 0.176 kWh to 0.768 kWh for Cases I and II, respectively. It is also found that the higher temperature and higher heat gain is

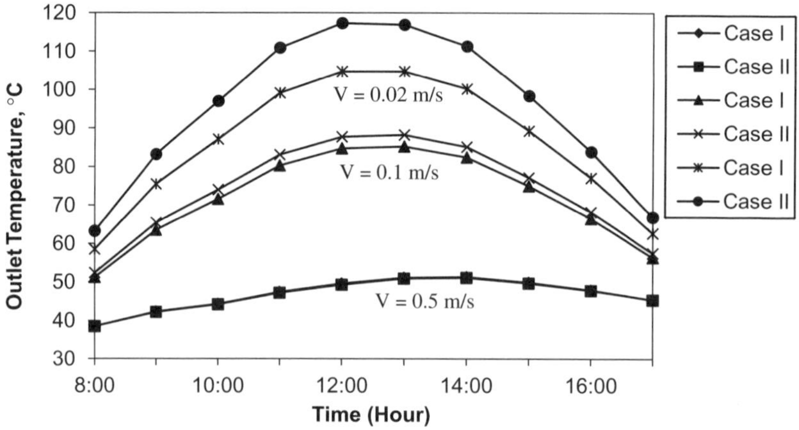

Figure 6.25 Hourly variation in outlet air temperature by varying the air velocity for $N = 1$.

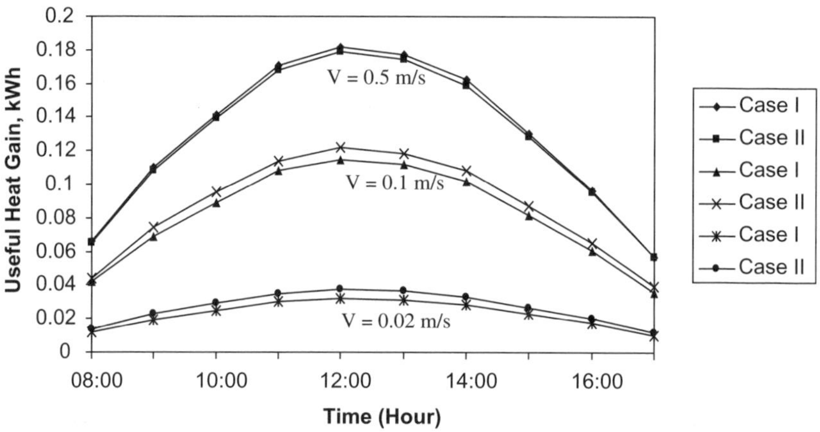

Figure 6.26 Hourly variation in useful heat gain by varying the air velocity for $N = 1$.

obtained for Case II when four or more collectors are connected in series compared to Case I.

The comparison chart of annual gain in thermal energy, exergy and electrical energy considering the four types of weather conditions for five different cities of India (New Delhi, Bangalore, Mumbai, Srinagar and Jodhpur) is shown in Figure 6.29. The detailed analysis shows that the maximum annual gain is obtained for Jodhpur and the minimum for Srinagar. The percentage variation between the maximum and minimum is 23.1%. The percentage variation between New Delhi, Mumbai and Bangalore with Srinagar are 9.7%, 14.5% and 8.8%, respectively.

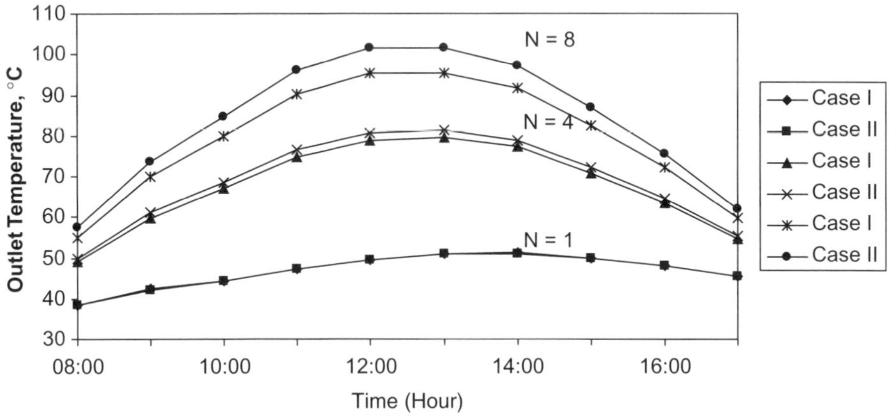

Figure 6.27 Hourly variation in outlet air temperature by varying the number of collectors for $V = 0.5\,\mathrm{m\,s^{-1}}$.

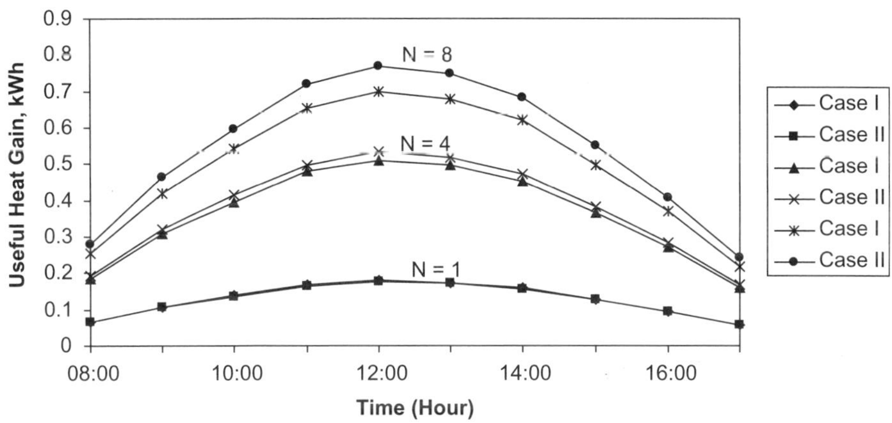

Figure 6.28 Hourly variation in useful heat gain by varying the number of collectors for $V = 0.5\,\mathrm{m\,s^{-1}}$.

With the help of above analysis it is concluded that:

1. The design of Case I is efficient at a higher air velocity and for one collector. However, the design of Case II is efficient for a lower air velocity and a higher number of collectors connected in series.
2. The design of Case II is beneficial in terms of hot air production and simultaneous electricity generation.
3. This type of system integrated with the building envelope will simultaneously fulfil the electrical power needs for domestic applications and the hot air requirements for space heating or drying.

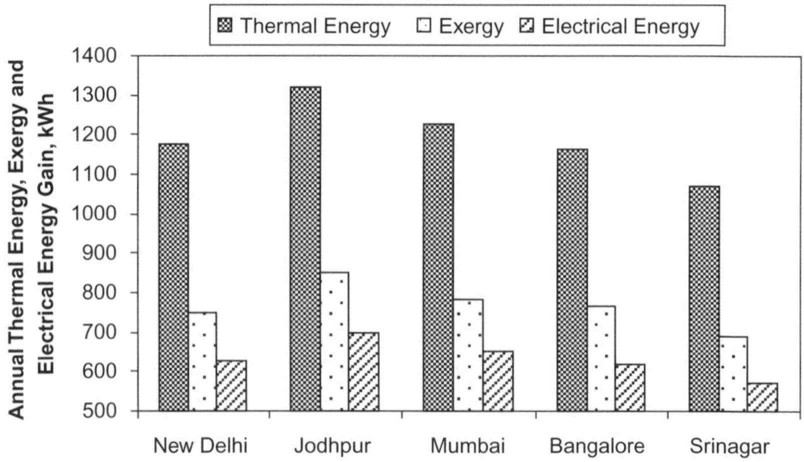

Figure 6.29 Annual gain in thermal energy, exergy and electrical energy for five different cities of India.

4. It is also found that Jodhpur is the best place for installation of such a type of system.

6.8 Case Study with a Greenhouse-integrated PVT System

Heating of a greenhouse is one of the most important and essential requirements for proper growth of summer growing crops in winter. Greenhouse heating can be carried out either by passive or active methods. Studies of greenhouse heating by passive methods have been made by many scientists, *e.g.* Tiwari *et al.*[7] and Santamouris *et al.*[8] Nayak and Tiwari[9] have performed an energy and exergy analysis of a greenhouse-integrated PVT system. This section describes Nayak's greenhouse[9] and the performance analysis.

6.8.1 System Description

Figure 6.30 shows the perspectve view and the orthographic front view of the greenhouse-integrated PVT system. There are two PV arrays, each consisting of 8 PV modules. Each module (made by Siemens solar industries) are rated at 75 W covering 0.605 m^2 area; the short circuit current (I_{sc}) = 4.8 A, the rated current = 4.4 A, the open circuit voltage (V_{oc}) = 21.7 V and the rated voltage = 17.0 V. The modules are connected in series and mounted on a wooden structure. The PV module with a wooden structure is placed on a steel frame. The inclination of the frame is maintained at around 45° for receiving the maximum solar radiation during the winter season. An IR thermometer has

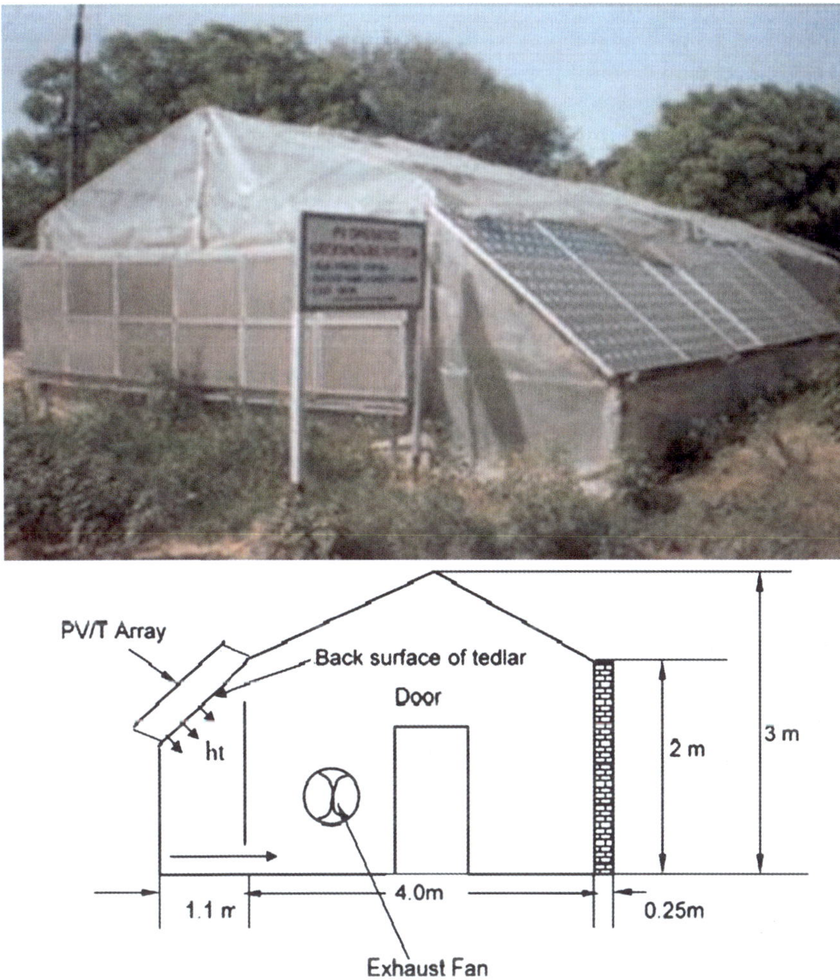

Figure 6.30 Greenhouse-integrated PVT system: *top*, pictorial view; *bottom*, schematic front view.

also been provided to measure the surface temperature of the solar cell and the Tedlar. A fan of capacity 12 W has been provided at the outlet to induce the flow of air inside the greenhouse. The power generated by the solar modules is stored in a set of 12 dc batteries each rated at 6 V and 180 Ah-10 h. An inverter of rating 2.1 KVA is employed for running the ac equipment.

On a clear summer day the temperature is sufficient high (40–45 °C) but during the winter season the temperature goes below 2 °C, which reduces the growth of plants significantly. The integration of the PVT system on the south wall helps in increasing the temperature of the greenhouse by convection. When the solar radiation falls on the greenhouse-integrated PVT system, it generates electrical power for the operation of greenhouse accessories and increases the temperature inside the greenhouse for the proper growth of plants.

The experiments were performed under a forced convection mode of operation during the year 2006–2007. Hourly observations of solar radiation, various temperatures (ambient air, greenhouse room air, solar cell, Tedlar back surface), the battery current, the short circuit current of arrays and battery voltage, the open circuit voltage of arrays, the inverter current and voltage were measured during the experimentation with the help of a calibrated solarimeter, a mercury thermometer, an IR thermometer and a tong meter, respectively.

6.8.2 Thermal Modelling

For the PV module the energy balance is given by eqn (6.2) as:

$$\tau_g[\alpha_c\beta_c + (1-\beta_c)\alpha_T]I(t)b\,dx = [U_T(T_c - T_a) + h_T(T_c - T_{bs})]b\,dx + \eta_c\alpha_c\beta_c I(t)b\,dx \tag{6.63}$$

For the back surface of the Tedlar the energy balance is given by eqn (6.5) as:

$$h_T(T_c - T_{bs})b\,dx = h_i(T_{bs} - T_r)b\,dx \tag{6.64}$$

For a plant (having mass m_p, specific heat c_p, temperature T_p, absorptivity α_p, heat transfer coefficient h_p and surface area A_p) the energy balance is given by:

$$\alpha_p F_p \sum A_i I_i \tau_i = m_p c_p \frac{dT_p}{dt} + h_p A_p (T_p - T_r) \tag{6.65}$$

The floor energy balance is given by:

$$\alpha_g(1 - F_p)\sum A_i I_i \tau_i = h_{gr} A_g (T_{g|x=0} - T_r) + h_{g\infty} A_g (T_{g|x=0} - T_\infty) \tag{6.66}$$

where h_{gr} is the heat transfer coefficient from the greenhouse floor to the room, $h_{g\infty}$ is the heat transfer coefficient from the floor to a larger depth of ground, A_g is the floor area, $T_{g|x=0}$ is the temperature of the ground surface and T_∞ is the temperature at the larger depth of ground.

The energy balance for the greenhouse air is given by:

$$\begin{aligned}(1-\alpha_p)F_p\sum A_i I_i\tau_i &+ (1-\alpha_g)(1-\alpha_p)\sum A_i I_i \tau_i + h_p A_p(T_p - T_r)\\ &+ h_{grr} A_g (T_{g|x=0} - T_r) + (\dot{Q}_u)_{daily}\\ &= 0.33 N_o V (T_r - T_a) + \sum A_i U_i (T_r - T_a)\end{aligned} \tag{6.67}$$

where $\dot{Q}_u = \frac{\dot{m}_a c_a}{U_L}\left[h_{p1}h_{p2}(\alpha\tau)_{eff}I(t) - U_L(T_r - T_a)\right][1 - \exp(\frac{-bU_L}{\dot{m}_a c_a}L)]$ with air flow and $\dot{Q}_u = b \times L\left[h_{p1}h_{p2}(\alpha\tau)_{eff}I(t) - U_L(T_r - T_a)\right]$ without air flow.

Solving the above equations, the air temperature inside the greenhouse is given by:

$$T_r = \frac{F_2 + h_p T_p A_p + (UA)_{eff} T_a + F_R h_{p_1} h_{p_2} (\alpha\tau)_{eff} I(t)}{(UA)_{eff} + h_p A_p} \tag{6.68}$$

The coefficient of correlation (r) and the root mean square percentage deviation (e) can be calculated using the following expressions:

$$r = \frac{n\left(\sum_{i-1}^{N} X_i Y_i\right) - \left(\sum_{i-1}^{N} X_i\right)\left(\sum_{i-1}^{N} Y_i\right)}{\sqrt{n\left(\sum_{i-1}^{N} X_i^2\right)\left(\sum_{i-1}^{N} X_i\right)^2 - n\left(\sum_{i-1}^{N} Y_i^2\right)\left(\sum_{i-1}^{N} Y_i\right)^2}}$$

$$\text{and } e = \left(\sqrt{\frac{\sum_{i-1}^{N} e_i^2}{N}}\right)$$

Table 6.6 Design parameters of greenhouse-integrated PVT system.

Parameter	*Value*	*Parameter*	*Value*
Transmissivity of greenhouse, τ_i	0.5	Surface area, A_1 (m^2)	10
Transmissivity of glass, τ_g	0.95	Surface area, A_2 (m^2)	10
Packing factor of module, β_c	0.83	Surface area, A_3 (m^2)	24
Efficiency of module, η_c	0.12	Surface area, A_4 (m^2)	24
Penalty factor, h_p (W m^{-2} K^{-1})	30.25	Surface area, A_5 (m^2)	15
Absorptivity of Tedlar, α_T	0.5	Surface area, A_6 (m^2)	26.83
Absorptivity of solar cell, α_c	0.9	Absorptivity of plant, α_p	0.5
Thermal conductivity of floor, k_g (W m^{-1} K^{-1})	0.52	Absorptivity of floor, α_g	0.4
Thickness of glass, L_g (m)	0.003	Heat transfer coefficient from greenhouse floor to ambient, h_{ga} (W m^{-2} K^{-1})	1.3
Thermal conductivity of glass, K_g (W m^{-1} K^{-1})	1.0	Heat transfer coefficient from floor to greenhouse space, h_{gr} (W m^{-2} K^{-1})	5.7
Overall heat transfer coefficient for greenhouse, U (W m^{-2} K^{-1})	3.5	Constant for plant, F_p	0.3
Number of air changes per minute, N_o	50	Volume of greenhouse space, V (m^3)	144
Duct air velocity, u (m s^{-1})	2.5	Mass of plant, m_p (kg)	50–250

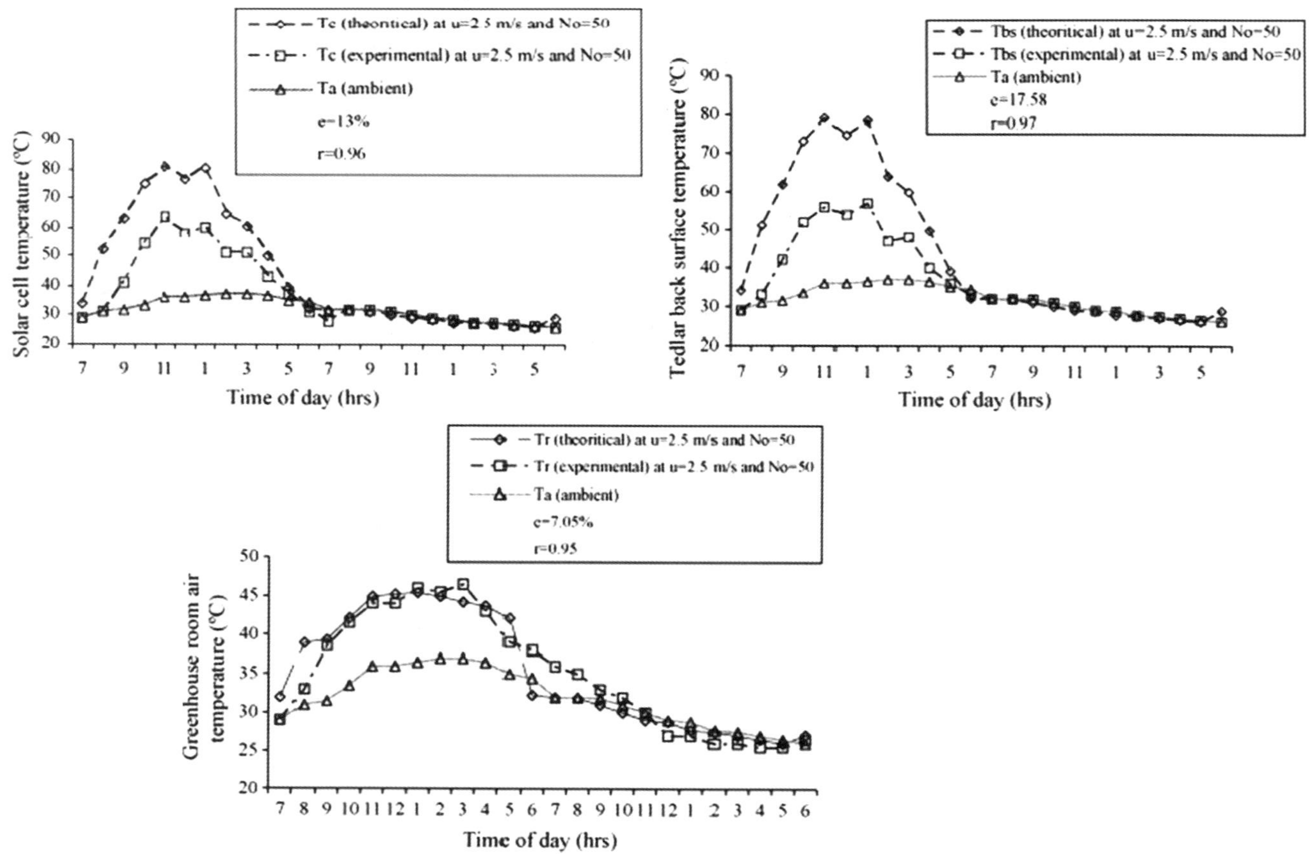

Figure 6.31 For a typical August day, the hourly variation of: *top left*, solar cell temperature; *top right*, Tedlar back-surface temperature; *bottom*, greenhouse room space air temperature.

6.8.3 Results and Discussion

The design parameters considered for numerical computation of a greenhouse-integrated PVT system is given in Table 6.6. For the number of air changes (N_o) as 50 and an air flow velocity (u) of 2.5 m s^{-1}, the hourly variations in the solar cell temperature, the back surface of the Tedlar temperature and the greenhouse space air temperature are shown in Figure 6.31. It has been observed that the solar cell temperature is around 3–4 °C higher than the Tedlar back-surface temperature and around 15–16 °C higher than the greenhouse room air temperature. The experimental values for each case are shown in the figure along with the correlation coefficient (r) and root mean square percent deviation (e). There is a fair agreement between the experimental and theoretical results for each case, with a correlation coefficient of 0.95–0.97 and a root mean square percent deviation of 7.05–17.58%. The root mean square percent deviation is due to the gap between the 16 modules in a panel of length of 1.62 m.

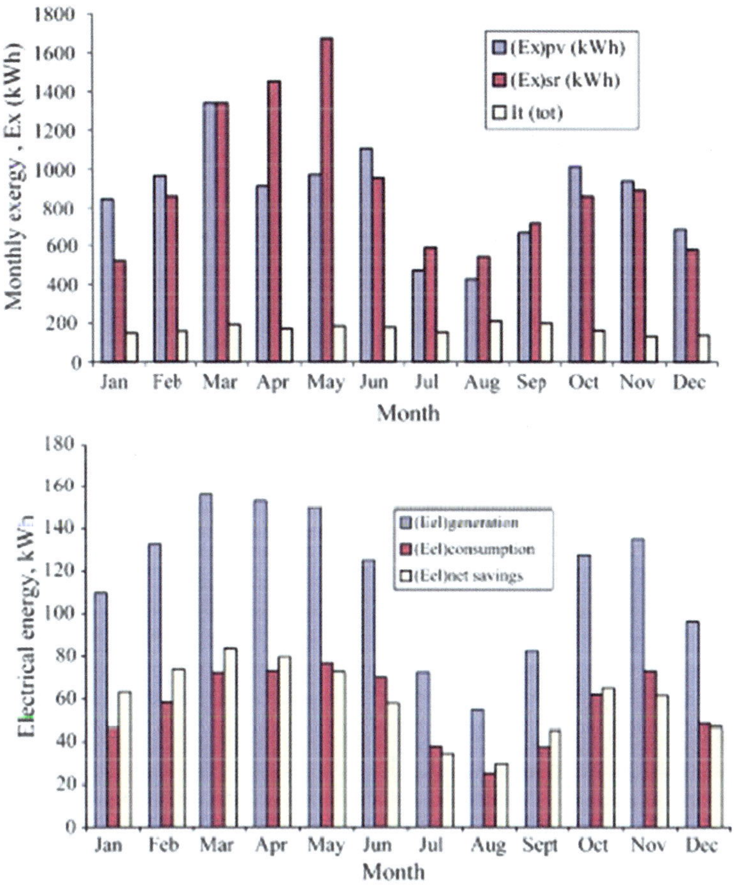

Figure 6.32 *Top*, annual exergy output; *bottom*, annual electrical gain.

Figure 6.32 shows the annual exergy input from the inclined PVT surface and south roof of the greenhouse, which in total was 21 291 kWh for the year 2006–2007. The figure also shows the annual net electrical energy savings as around 716 kWh. The exergy efficiency was 4%.

On the basis of the experimental and theoretical results it is concluded that:

1. There is fair agreement between the experimental and theoretical results for the temperatures of the solar cell, the Tedlar back surface and the greenhouse room air, with a correlation coefficient of 0.95–0.97 and the root mean square percent deviation of 7.05–17.58%.
2. The yearly exergy thermal energy is 12.8 kWh. The yearly net electrical energy gain is 716 kWh and the total exergy output is 728.8 kWh.
3. The exergy efficiency for the greenhouse-integrated PVT system is 4%.

6.9 Conclusions

The results help to develop empirical relations to understand better the dynamics of energy usage. The thermal models presented attempt to do this by considering the basic concepts of solar radiation availability and heat transfer. It helps to predict the temperature variation of the BIPVT system and the room space. This has been extended to determine the energy and the exergy efficiencies. The analysis shows that the roof and the façade-integrated PVT systems are far superior to any horizontal or inclined roof-mounted PV system.

References

1. D. L. Evans, Simplified method for predicting photovoltaic array output, *Sol. Energy*, 1981, **27**, 555–560.
2. B. Agrawal and G. N. Tiwari, Performance evaluation of building integrated photovoltaic thermal (BIPVT) systems, *Energy Sources, Part A*, 2010, in press.
3. B. Y. H. Liu and R. C. Jordan, The interrelationship and characteristic distribution of direct, diffuse and total solar radiation, *Sol. Energy*, 1960, **4**, 1–19.
4. B. Agrawal and G. N. Tiwari, Optimizing the energy and exergy of building integrated photovoltaic thermal (BIPVT) systems under cold climatic conditions, *Appl. Energy*, 2010, **87**, 417–426.
5. A. S. Joshi and A. Tiwari A, Energy and exergy efficiencies of a hybrid photovoltaic–thermal (PV/T) air collector, *Renewable Energy*, 2007, **32**, 2223–2241.
6. S. Dubey, S. C. Solanki and A. Tiwari, Energy and exergy analysis of PV/T air collectors connected in series, *Energy Build.*, 2009, **41**, 863–870.

7. G. N. Tiwari and N. K. Dhiman, Design and optimization of a winter greenhouse for the Leh-type climate, *Energy Conserv. Manage.*, 1986, **26**, 71–78.
8. M. Santamouris, A. Argiriou and M. Vallindras, Design and operation of a low energy consumption passive solar agricultural greenhouse, *Sol. Energy*, 1994, **52**, 371–378.
9. S. Nayak and G. N. Tiwari, Energy and exergy analysis of photovoltaic/thermal integrated with a solar greenhouse, *Energy Build.*, 2008, **40**, 2015–2021.

CHAPTER 7

Life Cycle Energy Analysis

7.1 Introduction

The electrical and thermal energies obtained from BIPVT systems are produced by the Sun and are considered free in environmental terms. However, most of the components of BIPVT systems are manufactured using fossil fuel intensive materials and processes. The energy consumed in the manufacture of such systems is commonly referred as embodied energy. The embodied energy includes the energy for assembly and the energy embodied in the input of goods and services to the manufacturing process, including transportation in all mining and manufacturing phases.[1]

The extent to which BIPVT systems can save energy can be shown through a life cycle energy analysis. This method determines the time it takes for annual operational savings to overtake the energy embodied in a particular product, such as a BIPVT system, *i.e.* the "energy payback time". Previous studies have shown that the energy embodied in PV systems may be quite significant, depending on which embodied energy analysis method is used.[2] The extent of this significance has an impact on the amount of time required to pay this energy back. This chapter presents the results of a life cycle energy analysis of BIPVT systems.

7.2 Embodied Energy

The concept of embodied energy is derived from the field of thermodynamics, initially involving the development of steam engines. The problem then was to balance heat gains and losses so that fuels could be consumed efficiently. More recently, energy conservation has become a global issue due to the rapid depletion of economically viable fossil fuel reserves, the potential for an enhanced greenhouse effect, problems in managing nuclear sources and wastes,

RSC Energy Series No. 4
Building Integrated Photovoltaic Thermal Systems: For Sustainable Developments
By Basant Agrawal and G. N. Tiwari
© B. Agrawal and G. N. Tiwari 2011
Published by the Royal Society of Chemistry, www.rsc.org

public resistance to hydroelectric schemes and the mining of uranium and the currently high cost of alternative energy sources, such as photovoltaics.[3]

For conventional energy sources the energy consumed at the point of use is lower than the energy required to supply this energy to the consumer, owing to conversion and transmission losses and the energy embodied in the fuels and derivatives. The energy used by the consumer is known as delivered energy, while the base form of energy required is known as primary energy. For systems involving photovoltaics, the energy supplied requires no fossil fuels and thus disconnects an entire sector of the economy and environment related to the production and distribution of power.

Traditionally, operational energy has been the focus of many studies dealing with life cycle energy. This may be partly due to the conceptual failure in quantifying the life cycle energy requirements of products through underestimating the possible importance of embodied energy. Although the operational energy consumption of buildings accounts for the highest proportion of the total energy consumed in the life cycle of a building, there is still a considerable amount of energy that is consumed in the other phases of a building's life. One of the most significant of these phases, incorporating the extraction and processing of raw materials, manufacturing of building materials and products and construction of the building, includes the embodied energy of the building and its fittings and finishes, but this varies significantly between building design, materials, systems and products.[2] Embodied energy is particularly important due to the complexity of the supply chain. This complexity means that the supply chain has to be modelled for each product and process upstream of the raw materials.[2]

For BIPVT systems, the embodied energy comprises indirect and direct energy. The direct energy may include that used on-site for the operation of power tools, while the indirect energy may include that used directly in the manufacture of PV modules and the structural steel of the building. The indirect energy of the solar cells and steel would in turn comprise energy embodied directly in the extraction and transport of silicon and iron ore. The accuracy and extent of an embodied energy analysis is dependent on which of the three main methods is chosen: process analysis, input–output analysis or hybrid analysis.

7.2.1 Process Analysis

Common definitions of process analysis identify the task of studying the inputs to and outputs from systems and the task of assigning energy values to the product flows. The total embodied energy comprises the energy required directly for the main manufacturing process and the indirect energy embodied in the material inputs to the process. Embodied energy thus comprises a chain of direct energy requirements.

This method of embodied energy analysis is seen to have major limitations, most significantly, system boundary incompleteness. The most important stage of this method is the quantification of the inputs to the product or system. Traditionally, a boundary has been drawn around the quantification of inputs to the product(s) being assessed mainly due to difficulties in obtaining necessary data and the understanding of these data. Many inputs are therefore neglected in the quantification of inputs to a product and thus the system boundary is incomplete. The magnitude of the incompleteness varies with the type of product or process and depth of study but can be 50% or more.[2–5]

7.2.2 Input–Output Analysis

National average statistics that model the financial flows between sectors of the economy, referred to as *I–O* data, can be used to fill the gaps that are caused by system boundary incompleteness.[2,3] The use of *I–O* data in an *I–O* analysis is generally treated as a black box with no understanding of the composition values being assumed in the model for each process. Also, because they are based on many inherent assumptions appropriate for national modelling, even a perfect *I–O* model may not lead to valid results for a particular product.[2,6] While *I–O* analysis is systemically complete, some *I–O* systems are inappropriately constructed and may leave out significant aspects of the economy. Some of the other main limitations of *I–O* analysis are detailed by Miller and Blair[7] and by Lenzen[5] and include homogeneity and proportionality assumptions, sector classification and aggregation.

7.2.3 Hybrid Analysis

Owing to the inherent problems with process analysis and *I–O* analysis, hybrid methods of embodied energy analysis have been developed in an attempt to minimize the limitations and errors of these traditional methods.[7] Direct energy and quantities of goods and services are obtained for critical aspects of the process under consideration by process analysis. This could, for example, mean that for materials where the manufacture represents the main bulk of the overall environmental impact, the production processes are examined and quantified in detail by the process analysis method. The energy intensities of goods and services further upstream are then obtained using *I–O* analysis. With this approach the errors associated with *I–O* analysis are thus removed from a large proportion of the results, but the energy intensities derived only apply to materials and products manufactured by the specific process(es) audited and cannot be applied globally.

In any embodied energy calculation it is important to utilize realistic conversion factors when comparing one form of energy with another. Many studies only consider the electrical energy directly required for a particular process. However, when losses related to electrical generation, conversion and transmission are considered, the energy actually required for the process might be considerably greater. In order for any embodied energy calculation to be complete, all inputs must be accounted for. Care must be taken to ensure the user is comparing like with like within different studies and that the boundaries of the study are the same.

7.3 Life Cycle Metrics

Life cycle metrics are also referred as energy metrics. Recent work has demonstrated that PV modules are net energy producers. Over their lifetime they generate 10–20 times the energy required to produce them. Empirical investigations of the energy requirements and net energy production are characterized by the parameter of energy payback time.

7.3.1 Energy Payback Time

Energy payback time (EPBT) is one metric adopted by several analysts in characterizing the energy sustainability of various technologies. It is defined as the time necessary for a system to generate the energy equivalent to that used to produce, install and decommission the system. Mathematically:

$$\begin{aligned}\text{EPBT} &= \frac{\text{Total energy input in manufacturing, installation and decommission}}{\text{Annual energy output from the system}}\\ &= \frac{E_{\text{in}}}{E_{\text{aout}}}\end{aligned} \tag{7.1}$$

For BIPVT systems the EPBT depends on a number of factors: solar cell technology, inverter, charge controller, battery, building materials, *etc.* The performance of a BIPVT system is determined by irradiation and the performance ratio. However, mathematically, the EPBT for a BIPVT system can be expressed as:

$$\text{EPBT} = \frac{E_{\text{building}} + E_{\text{support}} + E_{\text{BIPVT}} + E_{\text{cc}} + E_{\text{inv}} + E_{\text{bat}} + E_{\text{inst+MO}} + E_{\text{dec}} - E_{\text{rec}}}{E_{\text{aout}}} \tag{7.2}$$

where E_{building}, E_{support}, E_{BIPVT}, E_{cc}, E_{inv}, E_{bat}, $E_{\text{inst+M\&O}}$, E_{dec} and E_{rec} are the embodied energy of the building, support for the BIPVT system, BIPVT system, charge controller, inverter, battery, installation maintenance and operation, decommissioning and recycling, respectively; E_{aout} is the annual exergy energy produced from the system, which is given by:

$$E_{\text{aout}} = \dot{E}x_{\text{el}} + \dot{E}x_{\text{th}} \tag{7.3}$$

where $\dot{E}x_{\text{el}}$ and $\dot{E}x_{\text{th}}$ are the annual electrical and the thermal exergies equivalent. Substituting, we have the annual exergy output as:

$$E_{\text{aout}} = \left[\eta_{\text{ca}} \times I(t) \times bL \times n_{\text{s}} n_{\text{p}}\right] + \left[\dot{Q}_{\text{u}} \times \left(1 - \frac{T_{\text{a}}}{T_{\text{airout}}}\right)\right] \qquad (7.4)$$

For a sustainable energy character, the EPBT should be less than the entire installation service period:

$$\text{EPBT} \leq n_{\text{sys}} \qquad (7.5)$$

The energy payback time as an indicator of energy performance has an appeal because of the similarity with economic payback times. A drawback of EPBT is that it does not account for the energy gain during the rest of the economic lifetime. Indicators that fulfil this requirement are the electricity production factor and the life cycle conversion efficiency.

7.3.2 Electricity Production Factor

The electricity production factor (EPF) is commonly used in the power sector by engineers and scientists to predict the overall performance of the system. On a lifetime basis, it is defined as the ratio of the total energy output during the service time of the system to the total energy input to the system. Mathematically:

$$\chi_{\text{LT}} = \frac{\text{Total energy output during the service time of the system}}{\text{Total energy input to the system during the service time}} = \frac{E_{\text{out}}}{E_{\text{tot}}} \qquad (7.6)$$

The total energy output during the service time may be considered equivalent to the product of annual energy produced (E_{aout}) and the lifetime of the system (n_{sys}). Therefore eqn (7.6) can be rewritten as:

$$\chi_{\text{LT}} = \frac{E_{\text{aout}} \times n_{\text{sys}}}{E_{\text{in}}} \qquad (7.7)$$

or

$$\chi_{\text{LT}} = \frac{n_{\text{sys}}}{\text{EPBT}} \qquad (7.8)$$

Thus, the electricity production factor (EPF) is inversely proportional to the EPBT. For sustainable energy, the EPF the EPBT should be greater than unity:

$$\chi_{\text{LT}} \geq 1 \qquad (7.9)$$

7.3.3 Life Cycle Conversion Efficiency

The term life cycle conversion efficiency (LCCE) is mostly used by energy planners. It is defined as the ratio of the net energy productivity over the lifetime of the system to the solar input during the same period. Mathematically:

$$\varphi = \frac{\text{Total energy over the lifetime of the system}}{\text{Total energy input during the same time}} = \frac{E_{out} - E_{in}}{E_{sol} \times n_{sys}} \tag{7.10}$$

or

$$\varphi = \frac{E_{aout} \times n_{sys} - E_{in}}{E_{sol} \times n_{sys}} = \frac{E_{aout}}{E_{sol}}\left(1 - \frac{E_{in}}{E_{aout} \times n_{sys}}\right) \tag{7.11}$$

The ratio of the annual energy output to the solar energy input is called the efficiency (η) of the system. Therefore, eqn (7.11) can be written in the terms of efficiency as:

$$\varphi = \eta\left(1 - \frac{\text{EPBT}}{n_{sys}}\right) = \eta\left(1 - \frac{1}{\chi_{LT}}\right) \tag{7.12}$$

Thus, for a higher LCCE of the system, in addition to the improvement in the efficiency, the system should have a lower EPBT and a higher installation service period.

7.4 Greenhouse Gas Emissions

An approach similar to EPBT can be used to determine CO_2 payback times as a measure of the climate change mitigation potential associated with PV power systems. Alternatively, cumulative CO_2 emissions are recorded per kWh in order to compare them with CO_2 emissions from alternative power production technologies. For a large part the CO_2 emissions originate from the use of fossil energy carriers in the life cycle of the PV power systems. In addition to these energy-related emissions, however, other CO_2 emissions occur. Examples are the CO_2 emissions caused by the silica reduction process and the CO_2 emissions from the consumption of carbon electrodes in aluminium production.

Greenhouse gas emissions other than CO_2 should also be considered since some of them have a large global warming potential relative to CO_2. Thus, small emissions of these gases can have a significant contribution to the total global warming equivalent as expressed in equivalent CO_2 emissions. Examples of such substances are SF_6 and CF_4, gases which may be used in plasma etching processes or in the cleaning of reactor chambers. A release to the atmosphere of 1 kg of these gases will cause a greenhouse effect equivalent to 23 800 and

6500 kg of CO_2, respectively. Alternative cleaning methods and other techniques under development within the semiconductor industry will help to avoid these emissions.

7.5 Energy Payback Time Studies for Photovoltaic Systems: A Literature Review[8,9]

Previous life cycle studies reported a wide range of primary energy consumptions for Si-PV modules. Alsema[10] reviewed such analyzes from the 1990s and found considerable variance between investigators in their estimates of primary energy consumption. Normalized per m^2, the researchers reported 2400–7600 MJ of primary energy consumption for mc-Si and 5300–16 500 MJ for mono-Si modules. Besides uncertainties in the data, he attributed these differences mainly to the assumptions and allocation rules that each author adopted for modelling the purification and crystallization stages of silicon. In those days, solar cells were mostly tailored from off-spec products of electronic-grade silicon and not directly from solar-grade silicon, so that multiple allocation rules might well be applied to the energy and material inputs for each grade of silicon. Currently, only 5% of solar cells are from off-spec electronic-grade silicon. Selecting only those process steps needed to produce solar-grade silicon, Alsema's own estimates were 4200 and 5700 MJ m^{-2} for mc-Si and sc-Si modules, respectively.[10] These values correspond to an energy payback time (EPBT) of 2.5 and 3.1 years and life cycle greenhouse-gas emissions of 46 and 63 g CO_2-equivalent per kWh for mc-Si PV with 13% efficiency and for sc-Si with 14% efficiency, respectively, under southern European conditions: insolation of 1700 kWh m^{-2} a^{-1} and a performance ratio of 0.75. The balance of system (BOS) components, such as a mounting support, a frame, and electrical components, account for an additional 0.7 years of EPBT and 15 g CO_2-equivalent per kWh of greenhouse gas emissions.

Meijer *et al.*[11] more recently assessed a slightly higher energy expenditure of 4900 MJ m^{-2} to produce mc-Si modules. They assumed that the 270 μm thick Si PV module with 14.5% cell efficiency was fabricated from electronic-grade high-purity silicon, which entails greater energy consumption. Their corresponding EPBT estimate for the module was 3.5 years, excluding BOS components, which is higher than Alsema's earlier determination of 2.5 years. The increase stems mainly from the low level of insolation in the Netherlands (1000 kWh m^{-2} a^{-1}) compared with the average for southern Europe (1700 kWh m^{-2} a^{-1}) and, to a lesser degree, from the higher energy estimation for silicon. Jungbluth[12] reported the life cycle metrics of various PV systems under environmental conditions in Switzerland in 2000. He considered the environmental impacts for 300-μm thick mc-Si and sc-Si PV modules with 13.2% and 14.8% conversion efficiencies, respectively. Depending on which of the two materials he evaluated and their applications (*i.e.*, façade, slanted roof or flat roof), he arrived at figures of 39–110 g CO_2-equivalent per kWh of greenhouse gas emissions and 3–6 years of EPBT for the average insolation of

$1100\,kWh\,m^{-2}a^{-1}$ in that country. He assumed that the source of silicon material was 50% from off-grade silicon and 50% from electronic-grade silicon.

There are fewer life cycle studies of thin-film PV technologies; evaluations of the life cycle primary energy consumption of amorphous silicon range between 710 and $1980\,MJ\,m^{-2}$. The differences are largely attributed to the choice of substrate and encapsulation materials. The lowest estimate, made by Palz and Zibetta,[13] considered a single glass structure, while the highest one by Hagedorn[14] was based on a double-glass configuration to protect the active layer. For CdTe PV, Hynes *et al.*[15] based their energy analysis on two alternative technologies current at that time (1994). The first employed non-vacuum electrodeposition of a 1.5-μm absorber layer (CdTe), in conjunction with a chemical-bath deposition of the 0.2-μm window layer (CdS); the second method deposited both of these layers, *i.e.* a 5-μm thick absorber layer and a 1.7-μm thick window layer, by thermal evaporation. Their primary energy estimate for the first technology was $993\,MJ\,m^{-2}$, and that for the second was $1188\,MJ\,m^{-2}$. The energy estimates from Kato *et al.*[16] were pertinent to the scale of annual production; they suggested that energy consumption will decline as the scale of production rises. They cited values of 1523, 1234 and $992\,MJ\,m^{-2}$ for frameless modules with annual capacities of 10, 30 and $100\,MW_p$, respectively. However, these earlier estimates fall far short of describing present-day commercial-scale CdTe PV production that, unlike previously, now encompasses many large-scale production plants. Table 7.1 summarizes the results of the most cited EPBT studies made by Azzopardi and Mutale.[9]

7.6 Energy Content Coefficient for Building Materials

The concept of sustainable buildings and the use of environmentally friendly construction materials like stones, timber, thatch, mud, *etc.*, have been practiced since ancient times. However, the perception of people about strong and durable buildings has changed with the advent and lavish use of the present modern materials like steel, cement, aluminium and glass. A large amount of fuel energy is consumed in producing such materials. These materials, being industrial products, further need to be transported long distances before they are used in buildings, thus making them energy intensive. The most common building materials used in construction activity today are cement, steel, bricks, stones, glass, aluminium and timber. The estimates of the energy consumed in the manufacture and extraction of a few major building materials chosen from the study made by Venkatarama Reddy and Jagadish[17] are discussed below.

Portland cement represents one of the major materials consumed in bulk quantities for building construction. The energy of the cement arises from the use of coal in the rotary kilns and the energy needed for crushing and grinding the clinker. It is manufactured by employing both wet and dry processes. The wet process used in earlier cement plants leads to an energy consumption of $7.5\,MJ\,kg^{-1}$ of cement, whereas modern plants employing precalcination and

Table 7.1 Summary of EPBTs for roof-mounted PV systems.

			Some assumptions					
Author(s)	*Year*	*PV technology*	*PR*	*η (%)*	*Lifetime*[a]	*BOS*	*EPBT*	*g CO_2 per kWh*
Hunt	1976						12	
Hyne *et al.*	1991	CIS		10	20		0.83–1.42	
Palz *et al.*	1991	mc-Si and a-Si	0.7	12 and 6			2.1 and 1.2	
Alsema	1996	mc-Si and a-Si					4.0 and 4.5	
Keoleian *et al.*	1997	a-Si		5	10	Y	7.4	
Kato *et al.*	1997	mc-Si	0.8			Y	15.5/11/4	91/65/21
Dones *et al.*	1998	mc-Si and sc-Si		14–16.5	30	Y		189/114
Kato *et al*	1998	sc-Si/mc-Si/a-Si	0.8	12.2/11.6/8	20	Y	11.8/2.4/2.1	83/20/17
Frankl *et al.*	1998	sc-Si	0.9	11.2	25	Y	9	200
Alsema *et al.*	2000	mc-Si and a-Si	0.8	13.0 and 7.0	30	Y	3–4 and 2.5–4	60 and 50
Bossert *et al.*	2000	mc-Si		16–13.5			4.1–2.3	
		a-Si a-(Si, Ge):H		10			1.9–3.0	
		thin-film Si		14			4.7	
		CIGS		12			1.8–1.3	
		CdTe		10			0.9–0.5	
		dye sensitized		8			1	

Knapp *et al.*	2001	sc-Si	0.8		30		4.1	
		CIS					12.1–2.2	
Greijer *et al.*	2001	dye sensitized		7.0/12/9.0	20			19/22/25
Meijer *et al.*	2003	InGaP/mc-Si		25			5.3	39
		InGaP					6.3	
		mc-Si					3.5	
Jungbluth *et al.*	2004	c-Si (sc/mc)		14.8–17.5	30		3.0–6.0	39–110
Peharz *et al.*	2005	Con III-V multi-jun		26		Y	1.3–0.67	
Alsema *et al.*	2006	mc-Si	0.8	13.2	30	Y	1.8	32.5
Fthenakis *et al.*	2006	mc-Si	0.8	13.2	30	Y	2.2	37
		CdTe	0.8	8.0/9.0	30		1.0/1.1	21/25–18
Veltkamp *et al.*	2006	dye sensitized	0.8	8			1.3–0.8-0.6	20–120
SENSE project, Shibasaki	2006	a-Si	0.8				1.25	
		CIGS					1.3	
Raugei *et al.*	2007	CdTe	0.8	9	20	Y	1.5	48
		CIS		11			2.8	95
		mc-Si		14			5.5–2.4–2.5	167–72–57
Pacca *et al.*	2007	PVL 136 a-Si	1	6.3		Y	3.27.5	34.3
		KC 120 mc-Si		12.92				72.4
Roes *et al.*	2007	polymer	0.8	5	25		0.93	727

[a]years.

Table 7.2 Specific energy content of basic building materials.

Type of material	*Specific energy content*	
	kWh kg⁻¹	*MJ kg⁻¹*
Cement	1.625	5.85
Lime	1.564	5.63
LP	0.647	2.33
Steel	11.667	42.0
Aluminium	65.778	236.8
Glass	7.167	25.8

the dry process consume 4.2 MJ kg^{-1} of cement. The average value of 5.85 MJ kg^{-1} of cement has been used in the computation of energy in various components and systems.

Hydrated lime consumes 5.63 MJ kg^{-1} of thermal energy, which is about the same as that for cement. High-energy consumption of lime can be attributed to the low thermal efficiency of small-scale kilns employed for lime burning. Lime–pozzolana (LP) cements can provide very effective alternatives to Portland cement, mainly for secondary applications such as masonry mortar, plastering, base/sub-base for flooring, *etc.* A typical LP cement consists of 30% lime, 60% pozzolana and 10% calcined gypsum, all three being interground in a ball-mill. Such cement will have an energy content of 2.33 MJ kg^{-1}.

Aluminium and steel are the two high-energy metals commonly used in building construction. Even though aluminium is much lighter than steel, it consumes six times more than the energy of steel per unit weight. The modern tendency to use aluminium doors and windows can contribute significantly to the energy input of a building. Consumption of these metals should be kept to a minimum, in order to keep the energy in a building low. Glass is another energy-intensive material used in buildings. Its energy consumption is next to steel, but its density is much lower than steel.

Table 7.2 summarizes the energy consumed during the production of basic building materials which have been used in the computations of energy in building materials/systems and buildings.

7.7 Energy for Masonry Materials

Masonry walls constitute one of the major energy-consuming components of a building, especially in the case of load-bearing masonry structures. A variety of materials are used for the construction of masonry walls. Five types of building blocks, *viz.* stone, burnt clay brick, soil–cement block, hollow concrete block and steam-cured mud block have been considered for the analysis.

7.7.1 Stone Blocks

Natural building stones have been extensively used for building construction. They are generally produced by splitting the hard natural stone into convenient sizes. Stone blocks of size about 180 × 180 × 180 mm are generally termed size stones in practice. Only manual labour is employed in the bulk of the sizing operations. Occasionally, very hard and large stones are reduced to smaller sizes (for the convenience of handling for further sizing by manual processes) using detonators. Hence, hardly any thermal energy goes into the production of size stones. However, some energy is spent in transporting the stones to the construction site.

7.7.2 Burnt Clay Bricks

These are very commonly used for building construction. The common brick size is 230 × 110 × 60–75 mm. They require considerable amounts of thermal energy during the burning process. Coal, coal cinder and firewood are the most commonly used fuels for brick burning. In general, each brick needs either 0.20 kg of coal or 0.25–0.30 kg of firewood for the burning process. This translates into a thermal energy of 3.75–4.75 MJ per brick.

7.7.3 Hollow Concrete Blocks

These are lightweight/low-density blocks very commonly used for the construction of non-load-bearing filler walls in multi-storeyed buildings. They are also used for the construction of load-bearing masonry walls to a limited extent. The basic composition of the blocks consists of cement, sand and coarse aggregate (~6 mm size). The energy content of the block will mainly depend upon the cement percentage. The energy spent on crushing the coarse aggregate will also contribute to the block energy. The cement percentage generally varies between 7% and 10% by weight. The quality of the block, particularly compressive strength, is the deciding factor for cement percentage. The energy content of a hollow concrete block of size 400 × 200 × 200 mm will be in the range of 12.3–15.0 MJ.

7.7.4 Soil–Cement Blocks

These are produced by pressing a wetted soil–cement mixture into a solid block using a manually operated or mechanized machine, and then cured. Soil–cement blocks are produced by employing manually operated machines in a highly decentralized fashion. The energy content of the blocks is mainly dependent upon the cement content. Soil–cement blocks used for load-bearing masonry buildings have a cement content of about 6–8%. Such blocks will have an energy content of 2.75–3.75 MJ per block of size 230 × 190 × 100 mm.

7.7.5 Steam Cured Mud Blocks

These are lime stabilized blocks using expansive and high clay soils. They are produced by mixing suitable proportions of lime, clayey soil and sand and then

Table 7.3 Specific energy content of masonry materials.

Type of material	*Size (mm)*	*Ingredients*	*Specific energy content in one brick or block*	
			kWh	*MJ*
Stone block	180 × 180 × 180	–	0	0
Burnt clay brick	230 × 105 × 70	–	1.18	4.25
Soil–cement block	230 × 190 × 100	6% cement	0.72	2.60
		8% cement	0.97	3.50
Hollow concrete block	400 × 200 × 200	7% cement	3.42	12.30
		10% cement	4.17	15.00
Steam cured block	230 × 190 × 100	10% lime	1.86	6.70

pressing it into a block of convenient size. The blocks produced in this manner are cured in a steam chamber at about 80 °C for 10–12 h. The steaming process accelerates lime–clay reactions and the block is ready for construction within two days after moulding. Steam curing operations involve extra energy expenditure. The total energy required will be about 6.70 MJ per block of size 230 × 190 × 100 mm using 10% lime.

The energy content of different types of blocks used for masonry construction is given in Table 7.3. The table gives details of block type, size, energy content, energy per brick equivalent and percentage of energy with respect to brick energy. The blocks are of different size; hence they have been normalized by referring to an equivalent brick size.

7.8 Energy in the Transportation of Building Materials

The transportation of materials is a major factor in the cost and energy of a building. The transportation distance may vary depending upon the location of the construction activity. Materials such as sand are transported from a distance of 70–100 km in cities like Bangalore, India. Similarly, bricks/blocks, crushed aggregate, *etc.*, travel about 40–60 km before reaching a construction site in urban and semi-urban centres. Cement and steel travel even longer distances, of the order of 500 km or more. Fancy building materials such as marble, paints, *etc.*, are sometimes transported distances greater than 1500 km in India.

Natural sand and crushed stone aggregate consume about 1.75 MJ m^{-3} for every one km of transportation distance. Similarly, bricks require about 2.0 MJ m^{-3} per km of travel. Assuming steel and cement are also transported using trucks, diesel energy of 1 MJ t^{-1} km^{-1} is spent during transportation. The thermal energy spent for natural sand production is nil, but it requires about 175 MJ of diesel energy per m^{-3} for transporting it over a 100 km distance. Crushed aggregate consumes about 20 MJ m^{-3} during its production and an additional 400–800% more during transportation for distances of 50–100 km. The energy spent during transportation of bricks is about 4–8% of its energy in production, for distances of 50–100 km. The transportation energy required for

Table 7.4 Specific energy spent in transportation of building materials.

Type of material	Energy in production	Specific energy spent in transportation: 50 km	100 km
Sand	0	87.5 MJ m^{-3}	175 MJ m^{-3}
Crushed aggregate	20.5 MJ m^{-3}	87.5 MJ m^{-3}	175 MJ m^{-3}
Burnt clay bricks	2550 MJ m^{-3}	100 MJ m^{-3}	200 MJ m^{-3}
Portland cement	5858 MJ t^{-1}	50 MJ t^{-1}	100 MJ t^{-1}
Steel	42000 MJ t^{-1}	50 MJ t^{-1}	100 MJ t^{-1}

Table 7.5 Specific energy content of mortars.[a]

Type of mortar	Proportion of materials: Cement	Soil	Sand	Specific energy content: kWh m^{-3}	MJ m^{-3}
Cement mortar	1	0	6	352.22	1268
			8	279.44	1006
Cement–pozzolana mortar	cement/pozzolana = 4:1	0	6	255.00	918
			8	204.44	736
Cement–soil mortar	1	2	6	235.83	849
			8	214.72	773
Lime–pozzolana (LP) mortar	lime/pozzolana = 1:2	0	3	203.33	732

[a]Considering the specific energy content for Portland cement as 5.85 MJ kg^{-1}, sand as 175 MJ m^{-3} and pozzolana as 1.5 MJ m^{-3}.

hauling high-energy materials such as steel and cement is marginal when compared to the energy spent during production. Table 7.4 gives the diesel energy spent during transportation of various building materials, along with the energy consumed in production.

7.9 Energy in Mortars

Mortar is a mixture of cementitious material and sand. It is used for the construction of masonry as well as plastering. Cement mortar, cement–soil mortar, cement–pozzolana mortar and lime–pozzolana (LP) mortar are used for masonry construction and plastering. Cement mortar is a common choice for masonry and rendering works. Cement–soil mortar is used for the construction of stabilized mud block (SMB) masonry. Cement–pozzolana and LP mortars can also be used for masonry construction and other applications. The total energy content of these four types of mortars is given in Table 7.5. Details of mortar type, their proportions and energy content per m^3 of mortar are given in this table.

7.10 Energy in Different Types of Masonry

Masonry is an assemblage of masonry units (such as bricks or blocks) and mortar. The individual volumes of these two components in masonry will depend

Table 7.6 Specific energy content for different types of masonry.

Type of masonry	*Constituents*	*Specific energy content of masonary* ($MJ\,m^{-3}$)
Burnt clay brick masonry	cement mortar (1:6)	2141
Hollow concrete block masonry	cement mortar (1:6)	819 (7% cement blocks) 971 (10% cement blocks)
Soil–cement block masonry	cement–soil mortar (1:2:6)	646 (6% cement blocks) 810 (8% cement blocks)
Steam-cured mud block masonry	cement–soil mortar (1:2:6)	1396 (10% lime blocks)

mainly upon the size of the masonry unit. The energy content of masonry should include the energy content of the masonry units as well as the mortar. Table 7.6 gives the energy content of four types of masonry. Cement mortar (1:6) for brick masonry and hollow concrete block masonry and cement–soil mortar (1:2:6) for soil–cement block masonry and steam-cured mud block masonry have been considered for calculating the energy content of masonry. Soil–cement block masonry is the most energy efficient among the alternatives listed in the table.

7.11 Energy in Different Types of Floor and Roofing Systems

Varieties of alternatives are available for the construction of the roof or floor of a building. The energy content and construction details of some roof and floor systems are discussed in the following sections. These alternative systems have been used for the construction of buildings in India.

7.11.1 Stabilized Mud Block Filler Slab Roof

A reinforced concrete (RC) solid slab is very commonly used for floor as well as roof construction. A portion of the material below the neutral axis in a solid slab can be replaced by filler material such as SMB. Use of such filler material can result in a reduction in dead weight of a RC slab, saving in cost as well as energy of the roof or floor system. The total energy content of the materials constituting a SMB filler slab is $590\,MJ\,m^{-2}$ of plan area of the slab. This is a floor slab designed as per IS456 code for a span of 3.6 m. There will be variations in energy content for different spans of slab.

7.11.2 Composite Brick Panel Roof or Floor Slab

This roof consists of a reinforced brickwork panel supported on RC beams. The size of the RC beams will depend upon their spacing and roof span. Both the panels as well as the RC beam can be precast and assembled into a roof slab. The energy content of such a slab for 3.6 m span is about 560 MJ m^{-2} of projected plan area of the slab.

7.11.3 Reinforced Concrete Ribbed Slab Roof

A thin RC slab of thickness 50–60 mm is supported on small RC beams. The spacing of the RC beams will be in the range of 0.75–1.0 m and the size of the beams will depend upon the roof or floor span. This type of roof or floor slab can have an energy content of 491 MJ m^{-2} of slab area, for a 3.6 m span.

7.11.4 Masonry Vault Roof

An unreinforced masonry vault roof consists of a thin masonry vault supported on ring beams with tie rods. The vault can be constructed using burnt clay bricks or SMBs. The total energy of the roof will be 575 MJ m^{-2} and 418 MJ m^{-2} for brick masonry and SMB masonry vault roofs, respectively.

7.11.5 Ferroconcrete Tile Roof

A ferroconcrete tile roof can be made locally using thin galvanized iron (GI) wires and microconcrete. The tile size is about 1.25 × 0.55 m. These tiles can be supported on wooden rafters. The energy of such a roof is 158 MJ m^{-2}.

The energy values of different types of floor and roofing systems are given in Table 7.7. The table gives energy per m^2 of plan area of roof or floor. It can be used conveniently for making a selection of the roofing system based on energy content of the roof or floor system.

7.12 Energy for Different Types of PV Modules

The life cycle stages of PV systems, as shown in Figure 7.1, involve (1) the production of raw materials, (2) their processing and purification, (3) the manufacture

Table 7.7 Specific energy content of different roof and floor systems .[a]

	Specific energy content	
Type of floor or roof	*kWh m^{-2} of plan area*	*MJ m^{-2} of plan area*
RC slab	202.78	730
SMB filler slab roof	163.89	590
RC ribbed slab roof	136.39	491
Composite brick panel roof	138.89	500
Burnt clay brick masonry vault roof	159.72	575
SMB masonry vault roof	116.11	418
Mangalore tile roof	63.05	227
Ferroconcrete roof	43.88	158

[a]Span – 3.6 m.

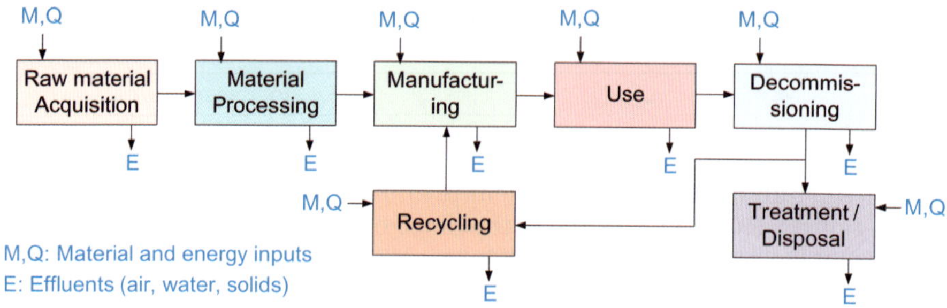

Figure 7.1 Flow of the life cycle stages, energy, materials and wastes for PV systems.

of modules and balance of system (BOS) components, (4) the installation and use of the system, and (5) their decommissioning and disposal or recycling.[8]

7.12.1 Energy for Silicon PV Modules[13]

According to Tiwari and Ghosal,[18] the material (silicon) required for preparing a PV module is 0.724 kg m^{-2}. The energy required to produce 1 kg of metallurgical grade silicon (MG-Si) by carbothermic reduction of silicon dioxide (SiO_2) is 20 kWh. The energy required to produce electronic grade silicon (EG-Si) from MG-Si is 100 kWh kg^{-1} and there is a 90% yield. The EG-Si is melted in a Czochralski crystal puller at 1400 °C and the silicon slowly crystallizes to form a single crystal ingot of silicon. The energy required is 210 kWh kg^{-1}, with a total yield of 72%. The ingot is typically sliced with a thickness of 0.2–0.5 mm. Assuming the silicon wafer thickness is 0.350 mm and the losses are 0.300 mm, the total yield is 54%. After trimming a 156.25 cm^2 PV wafer, a pseudosquare cell with an effective area of 142 cm^2 is prepared. Thus the mass of each PV cell is 11.43 g (142 × 0.035 × 2.3) with 91% yield and a density 2.3 g cm^{-3}. The energy required to prepare 1 m^2 of silicon cell is 120 kWh. These solar cells are used to make a PV module with a packing factor of 0.82 (*i.e.* 82%) silicon and 18% open space between cells. The energy requirement to prepare a module is 190 kWh m^{-2}.

For an open-field installation, the concrete, cement and steel are the main components used for the foundation and frame, which require the maximum energy. The energy requirement for an open-field installation is 500 kWh m^{-2} of panel. For a rooftop-integrated PV system the energy requirement is 200 kWh m^{-2} due to the absence of foundations and structure for the frame. The requirements for the BOS (that is, all component parts of modules) will depend largely on the desired application. Table 7.8 summarizes the energy requirement in different processes for the production of a PV module.

So the required energy for different processes for a unit panel area are as follows:

(1) Silicon purification and processing: (a) production of 2.334 kg of MG-Si = 2.334 × 20 = 46.68 kWh; (b) production of 2.011 kg of

Table 7.8 Energy requirement in different processes for the production of a PV module.

Process	*Energy requirement*
Silicon purification and processing	
(a) MG-Si production from silicon dioxide (quartz, sand)	20 kWh kg^{-1} of MG-Si
(b) EG-Si production from MG-Si	100 kWh kg^{-1} of EG-Si
(c) Czochralski silicon (Cz-Si) production from EG-Si	290 kWh kg^{-1} of EG-Si
Solar cell fabrication	120 kWh m^{-2} of silicon cell
PV module assembly	190 kWh m^{-2} of PV module
Stand-alone PV system	500 kWh m^{-2} of PV module
Rooftop integrated PV system	200 kWh m^{-2} of PV module

EG-Si = $2.011 \times 100 = 201.10$ kWh; (c) production of 1.448 kg of EG-Si for Cz-Si = $1.448 \times 210 = 304.08$ kWh.

(2) Solar cell fabrication = $120 \times (0.60534 \times 0.83) = 60.29$ kWh.

(3) PV module assembly of size $1.2 \times 0.55 \times 0.01 = 190 \times 0.66 =$ 125.40 kWh.

(4) Installation or integration = $200 \times 0.66 = 132$ kWh.

Hence, the total embodied energy required for installation or integration of a PV module (glass-to-glass) with a PVT system is $869.55 \approx 870$ kWh. The total embodied energy with BOS for an open-field installation is $870 + 500 =$ 1370 kWh and for a rooftop is $870 + 200 = 1070$ kWh. The EPBT for an open-field installation in India is 12.45 years. The EPBT in future is likely to be much lower than the current value owing to the following reasons:

1. Reduction of the thickness of the crystalline silicon solar cell to 150 μm or less.
2. Reduction in the cell processing energy to 75% of the current value.
3. Increase in cell efficiency from 14% to 23%.
4. Use of a square wafer instead of pseudosquare-shaped wafers.
5. Use of a low-cost material at the back to support the solar cells.
6. Use of fibre material for framing the modules instead of an aluminium frame.

7.12.2 Energy for Non-silicon PV Modules

In thin film technologies the photoactive P/N junction is made up of two semiconductor compounds, CdTe and CIGS, which are directly deposited in extremely thin layers, ~10 and ~0.1 μm, respectively, on a treated transparent glass pane by means of a vacuum vaporization process. Series connection of adjacent P/N junctions is achieved by means of a series of automated laser and mechanical scribing processes, and then a second protective glass pane is added

Table 7.9 Inventory of main input flows to the CdTe module manufacturing process.

Inventory	*Quantity*
Glass	24 960 g m^{-2}
Water	1250 g m^{-2}
EVA	630 g m^{-2}
{CdTe + CdS + $CdCl_2$ + Sn + Ni/V + ITO + Sb_2Te_3}	230 g m^{-2}
Electricity	236 kWh m^{-2}

Table 7.10 Inventory of main input flows to the CIGS module manufacturing process.

Inventory	*Quantity*
Glass	24 960 g m^{-2}
Water	1250 g m^{-2}
EVA	880 g m^{-2}
{Mo + Cu + In + Ga + Se + CdS + ZnO + CuSn}	70 g m^{-2}
Electricity	24.3 kWh m^{-2}

Table 7.11 Module efficiency, service period and specific energy content for PV modules.

Solar cell technology	*Module effciency, η_{PV} (%)*	*Expected life (n_{PV}) (years)*	*Specific energy content, ε_{PV}*	
			kWh m^{-2}	*MJ m^{-2}*
sc-Si	14–18	30	1120–1260	4032–4556
mc-Si	12–16	30	840–980	3024–3528
rc-Si	10–14	25	570–650	2052–2340
a-Si	5–7	20	308–448	1110–1615
CdTe	7–9	15	196–336	705–1210
CIGS	9–11	5	20–29	72–105

on top to form the finished module. Tables 7.9 and 7.10 provide the quantities of inventory used in manufacturing the CdTe and CIGS modules.

For the purposes of the present analysis, the system boundaries were drawn around the module production facility, and thorough inventories of the necessary inputs for the production process of 1 m^2 of frameless modules were made. Table 7.11 gives the typical values concerning the energy content coefficient (ε_{PV}) for manufacturing popular silicon-based and non-silicon-based photovoltaic modules.

7.13 Energy for Balance of System

Depending on the application, BIPVT systems are either rooftop or façade integrated, both operating with a proper balance of system (BOS). For a BIPVT application, the BOS typically includes charge controllers, inverters, battery banks, mounting structures and frames, cables and connectors. Large-scale ground-mounted PV installations require additional equipment and facilities, such as grid connections, office facilities and concrete.

7.13.1 Charge Controller

In solar applications, charge controllers may also be called solar regulators. Their function is to regulate the rate of flow of electricity from the PV array to the battery and the load. The charge controller keeps the battery fully charged without overcharging it. When the load is drawing power the controller allows charge to flow from the modules into the battery, the load or both. When the controller senses that the battery is fully charged, it stops the flow of charge from the modules. Many controllers will also sense when loads have taken too much electricity from the batteries and will stop the flow until sufficient charge is restored to the batteries. This last feature can greatly extend a battery's lifetime. Circuitry that functions as a charge regulator controller should be matched for the batteries used in the system. The range of specific heat content varies from 468 to 1044 MJ kW^{-1}, depending on the circuitry and the ampere capacity.

7.13.2 Inverter

An inverter is an electrical device that converts direct current to alternating current, functioning opposite to that of a rectifier. The converted ac can be at any required voltage and frequency with the use of appropriate transformers, switching and control circuits. Micro-inverters convert dc from individual PV panels into ac for the electric grid. The specific heat content and the cost is affected by several factors, including the quality of the electricity it needs to produce, whether the input dc source is supplying 12, 24, 36 or 48 V, the number of ac watts required by loads during normal operation, the amount of extra surge power ac loads need for short periods, and any additional features such as meters and indicator lights.

7.13.3 Batteries

Batteries are the energy storing devices generally employed for solar applications. They store the electricity to meet load demand during off-sunshine hours when the modules are generating insufficient or no power. To provide electricity over longer periods, PV systems require deep-cycle batteries. Lead–acid batteries are usually designed to gradually discharge and recharge 80% of their capacity hundreds of times. The energy content ranges from 0.87 to 1.19 MJ Wh^{-1}, depending on the capacity (ampere-hours) and the types of chemicals used to store and release the electricity.

The National Electric Code (NEC) battery companies and PV system designers recommend that lead–acid and wet cell batteries, which give off

Table 7.12 Energy content coefficients and service period for the BOS components, the system installation and M&O stages, and the diesel electric generator.

BIPVT system components	*Life in years*	*Range for specific heat content*	
Charge controller	10	130–290 kWh kW^{-1}	468–1044 MJ kW^{-1}
Inverter, safety factor 0.3	10	140–280 kWh kW^{-1}	504–1008 MJ kW^{-1}
PbA battery, depth of discharge 80%	5	0.24–0.33 kWh Wh^{-1}	0.87–1.19 MJ Wh^{-1}
PV frame	30	84–140 kWh m^{-2}	302–504 MJ m^{-2}
Installation	–	23.6–35 kWh m^{-2}	85–126 MJ m^{-2}
Junction box, cabling, *etc.*	30		
M&O	–	1–2 kWh m^{-2} a^{-1}	3.6–7.2 MJ m^{-2} a^{-1}
Foundation, array support, *etc.*	30	168–252 kWh m^{-2}	604–907 MJ m^{-2}
Diesel including installation	5	400–800 kWh kW^{-1}	1440–2880 MJ kW^{-1}

explosive hydrogen gas when recharging, be located in a well-ventilated space isolated from the other electrical components of the system and away from living spaces. Enough space should also be allowed for easy access during maintenance, repair and replacement. Automotive batteries are shallow-cycle batteries and should not be used in PV systems because they are designed to discharge only about 20% of their capacity.

Table 7.12 gives the typical values concerning the energy content coefficients and service periods considered for the BOS components, the system installation and maintenance and operation (M&O) stages, and the diesel electric generator.

7.14 Guidelines for Reducing Embodied Energy

The choice of materials and construction methods can significantly change the amount of energy embodied in the structure of a building. True low-energy building design will consider this important aspect and take a broader life cycle approach to energy assessment. Merely looking at the energy used to operate the building is not really acceptable. Operational energy consumption is dependent on the occupants. Embodied energy content is incurred once (apart from maintenance and renovation), whereas operational energy accumulates over time and can be influenced throughout the life of the building. Reuse of building materials commonly saves about 95% of the embodied energy that would otherwise be wasted. The building should be selected for the best combination for its application based on climate, transport distances, availability of materials and budget, balanced against known embodied energy content. General guidelines for reducing the embodied energy are as follows:[19]

1. Design for long life and adaptability, using durable low-maintenance materials.

2. Ensure materials can be easily separated.
3. Avoid building a bigger house than needed; this will save materials.
4. Modify or refurbish instead of demolishing or adding.
5. Ensure materials from demolition of existing buildings and construction wastes are reused or recycled.
6. Use locally sourced materials (including materials salvaged on site) to reduce transport.
7. Select low embodied energy materials (which may include materials with a high recycled content), preferably based on supplier-specific data.
8. Avoid wasteful material use.
9. Specify standard sizes and do not use energy-intensive materials as fillers.
10. Ensure off-cuts are recycled and avoid redundant structure, *etc*. Some very energy intensive finishes, such as paints, often have high wastage levels.
11. Select materials that can be re-used or recycled easily at the end of their lives, using existing recycling systems.
12. Give preference to materials manufactured using renewable energy sources.
13. Use efficient building envelope design and fittings to minimize materials (*e.g*. an energy efficient building envelope can downsize or eliminate the need for heaters and coolers, water-efficient taps allow downsizing of water pipes).
14. Ask suppliers for information on their products and share this information.

7.15 Modelling of Embodied Energy for BIPVT Systems

The energy content of a BIPVT system involves the stages of manufacturing, installation, maintenance and operation (M&O), final decommissioning and recycling. Taking into consideration the building and all the components, the total embodied energy of the configuration is:

$$E_{tot} = E_{in} = E_{building} + E_{support} + E_{BIPVT} + E_{cc} + E_{inv} + E_{bat} + E_{inst+M\&O} + E_{dec} - E_{rec} \quad (7.13)$$

where $E_{building}$, $E_{support}$, E_{BIPVT}, E_{cc}, E_{inv}, E_{bat}, $E_{inst+M\&O}$, E_{dec} and E_{rec} are the embodied energy of the building, the support for the BIPVT system, the BIPVT system, the charge controller, the inverter, the battery, the installation, maintenance & operation, the decommissioning and the recycling, respectively.

7.15.1 Masonry Buildings

The embodied energy of the building is the sum of the embodied energy of all the construction material used. Mathematically:

$$E_{\text{building}} = E_{\text{bricks}} + E_{\text{cement}} + E_{\text{sand}} + E_{\text{concrete}} + E_{\text{lime}} + E_{\text{MS}} + E_{\text{glass}} + E_{\text{paint}} + E_{\text{plywood}} + E_{\text{floor}} \tag{7.14}$$

$$E_{\text{building}} = m_{\text{bricks}} \times \xi_{\text{bricks}} + m_{\text{cement}} \times \xi_{\text{cement}} + m_{\text{sand}} \times \xi_{\text{sand}} + m_{\text{concrete}} \times \xi_{\text{concrete}} + m_{\text{lime}} \times \xi_{\text{lime}} + m_{\text{MS}} \times \xi_{\text{MS}} + m_{\text{glass}} \times \xi_{\text{glass}} + m_{\text{paint}} \times \xi_{\text{paint}} + m_{\text{plywood}} \times \xi_{\text{plywood}} + A_{\text{floor}} \times \varepsilon_{\text{floor}} \tag{7.15}$$

where m is the mass and ξ is the embodied energy per unit mass.

The support for mounting a BIPVT system is made of aluminium and has a service time equal to n_{sys}. Its embodied energy is:

$$E_{\text{support}} = m_{\text{Al}} \times \xi_{\text{Al}} \tag{7.16}$$

where m_{Al} is the mass of aluminium used in construction of the support and ξ_{Al} is the embodied energy of the aluminium per unit mass.

7.15.2 PVT Systems

The embodied energy of the PVT system is the sum of the embodied energy of the PV array, frame and duct. Therefore:

$$E_{\text{BIPVT}} = E_{\text{PV}} + E_{\text{frame}} + E_{\text{duct}} \tag{7.17}$$

Usually the frame is made of aluminium alloy and the duct is of plywood. Thus, for a PVT system having a PV array area of A_{PV} and a service time of n_{sys}, the embodied energy is:

$$E_{\text{BIPVT}} = A_{\text{PV}}\varepsilon_{\text{PV}}\left[1 + \text{int}\left(\frac{n_{\text{sys}} - 1}{n_{\text{PV}}}\right)\right] + A_{\text{frame}}\varepsilon_{\text{frame}}\left[1 + \text{int}\left(\frac{n_{\text{sys}} - 1}{n_{\text{frame}}}\right)\right] + A_{\text{PV}}\varepsilon_{\text{plywood}}\left[1 + \text{int}\left(\frac{n_{\text{sys}} - 1}{n_{\text{plywood}}}\right)\right] \tag{7.18}$$

where ε_{PV}, $\varepsilon_{\text{frame}}$ and $\varepsilon_{\text{plywood}}$ are the embodied energy per unit area for the PV module, the PV frame and the plywood used for duct manufacture, respectively; n_{PV}, n_{frame} and n_{plywood} are the service times for the PV module, the PV frame and the duct plywood, respectively.

The panel area A_{PV} results from the peak power of the PV generator, N_{PV}, and the corresponding efficiency, η_{PV}:

$$A_{PV} = \frac{N_{PV}}{\eta_{PV} G} \tag{7.19}$$

where G is the solar radiation at standard test conditions and its value is considered as $1000\,\text{W}\,\text{m}^{-2}$.

7.15.3 Balance of System

The embodied energy of a charge controller is:

$$E_{cc} = \zeta_{cc} P_{cc}\left[1 + \text{int}\left(\frac{n_{sys} - 1}{n_{cc}}\right)\right] \tag{7.20}$$

where P_{cc} is the rated power of the charge controller, which is slightly higher than the peak power generated by the PV array:

$$P_{cc} = P_{PV-peak} + \delta P_{cc} \approx P_{PV-peak} \tag{7.21}$$

The inverter should have its rated capacity (P_{inv}) such that it fulfils the peak load demand (P_{LD}) of the consumer. Its embodied energy during n_{sys} is:

$$E_{inv} = \zeta_{inv} P_{inv}\left[1 + \text{int}\left(\frac{n_{sys} - 1}{n_{inv}}\right)\right] \tag{7.22}$$

The rated capacity of the inverter is given by:

$$P_{inv} = P_{LD}(1 + SF) \tag{7.23}$$

where SF is the safety factor, which is usually considered as 0.3.

The rated capacity of the battery C_{bat} should fulfil the necessary demand of the consumer during off-sunshine hours. Its embodied energy during n_{sys} is:

$$E_{bat} = K_{bat} C_{bat}\left[1 + \text{int}\left(\frac{n_{sys} - 1}{n_{inv}}\right)\right] \tag{7.24}$$

Minor energy is required during the activities of installation of the system and water proofing. Components such as junction boxes, cables, screws, *etc.*, may be reused or changed with new ones. However, a BIPVT system, with a shorter lifespan than a PV system, needs to be dismantled and installed more frequently, which increases the cost of installation, while a system with a longer lifespan PV system has a relatively low cost of installation. In addition to this, energy is also required for periodic maintenance and operation of the system. Therefore:

$$E_{\text{inst+M\&O}} = E_{\text{inst}} + E_{\text{repl}} + E_{\text{M\&O}} \tag{7.25}$$

7.16 A Case Study with a Roof-mounted BIPVT System

7.16.1 Assumptions

1. The system is installed to work for a lifetime equal to that of the building. The building requires only nominal maintenance during its lifetime.
2. The lifetime of a group of components is the same. As soon as the lifetime of the group of components is over, they are replaced by a similar group of new components.
3. The same technology is used to manufacture the items that are being replaced over the time period. So, the embodied energy does not change with the time period.
4. The battery stores 50% of the average electrical energy produced during a day. The excess energy generated during sunshine hours is of no use. Similarly, excess demand during off-sunshine hours is fulfilled from grid.
5. The peak load demand of the user is 5 kW. This is helpful in deciding the size of inverter.
6. The charge controller capacity is according to the peak power generated from the PV array.

7.16.2 Components

The proposed energy-sustained system consists of the following three components.

1. The building. The life of the building is 60 years. Nearly 5300 and 13 500 burnt clay bricks are used for laying the foundations and erecting the walls, respectively. The total mass of bricks used is about 33 183 kg. If the size of each brick is 230 × 110 × 75 mm and the specific heat content is 4.25 MJ kg^{-1}, then the total embodied energy of the bricks is 39 750 MJ. A total of 45 bags of mortar, each of 50 kg, are used. If the embodied energy of the mortar is 4.2 MJ kg^{-1}, then the total embodied energy of the mortar is 9450 MJ. Nearly 135 00 kg of sand are used for mortar. If the embodied energy of sand transportation is 0.1 MJ kg^{-1}, then the total embodied energy of the sand is 1350 MJ. Nearly 10 750 kg of concrete is used below the foundation. If the embodied energy of concrete is 1 MJ kg^{-1}, then the total embodied energy of the concrete is 10 750 MJ. Nearly

4800 kg of mild steel are used for erecting the columns of the building. If the embodied energy of steel is 36 MJ kg^{-1}, then the total embodied energy of the steel is 172 800 MJ. Obviously, use of a lesser quantity of mild steel will reduce the total embodied energy of the building and thereby the energy payback time of the whole system. Stabilized mud block masonry is used for the flooring, which consumes 418 MJ m^{-2}. Thus for 61 m^2 surface the total embodied energy of the stabilized mud block is 25 498 MJ. In addition to above, nearly 1.5 kg of lime is used as whitewash for the walls, 18 kg of plywood is used for the doors, 395 kg of glass is used for windows and 0.5 kg paint is used for painting the windows and doors. Thus the overall embodied energy of the building is 367 366 MJ, summarized in Table 7.13.

2. The PVT system. The PVT system consists of a PV array to obtain electrical energy and plywood for ducting to collect the thermal energy and cool the solar cells. The assigned value for the specific energy content of the PV system with different technologies considered is summarized in Table 7.14.
3. Balance of systems. The balance of systems includes the dc/dc charge controller, the inverter, the lead–acid (PbA) battery with 80% depth of discharge, the frame for the photovoltaic modules, the junction box, the

Table 7.13 Embodied energy of the material used for the structure and frame of the building.

		Specific heat content		
Building construction material	*Quantity used*	*MJ* kg^{-1}	*Assigned value (MJ* kg^{-1}*)*	*Embodied energy (MJ)*
Clay bricks ($23 \times 11 \times 7.5\,cm^3$ for foundation)	9353 kg (for 5300)	2–7	4.25	39 750
Clay bricks ($23 \times 11 \times 7.5\,cm^3$ for walls)	23 830 kg (for 13 500)	2–7	4.25	101 248
Cement (for mortar)	2250 kg (45 bags)	4–8	4.2	9450
Sand (for mortar)	13 500 kg	<0.5	0.1	1350
Concrete (for foundation)	10 750 kg	0.8–1.5	1	10 750
Lime (for whitewash on walls)	1.5 kg	3–5	4	6
Mild steel (for columns)	4800 kg	30–60	36	172 800
Glass (for windows)	395 kg	12–25	15.9	6280
Paint (on doors & windows)	0.5 kg	80–150	93.3	47
Plywood (for two doors)	18 kg	8–12	10.4	187
Stabilized mud block masonry vault floor	61 m^2	–	418 MJ m^{-2}	25 498
Total				367 366

Table 7.14 Module efficiency, service period and specific energy content assigned for silicon and non-silicon PV modules.

Solar cell technology	Assigned value for module effciency, η_{PV} (%)	Expected life, n_{PV} (years)	Assigned value for specific energy content, ε_{PV}	
			$MJ\,m^{-2}$	$MJ\,kW_p^{-1}$
sc-Si	16	30	4284	26 775
mc-Si	14	30	3276	23 400
rc-Si	12	25	2214	18 450
a-Si	6	20	1360.8	22 680
CdTe	8	15	957.6	11 970
CIGS	10	5	87.48	874.8

Table 7.15 Energy content coefficients and service periods assigned for the BOS components, the system installation and M&O stages, and the diesel electric generator.

BIPVT system component	Lifetime (years)	Assigned value for specific heat content
Charge controller (n_{cc}, ξ_{cc})	10	$756\,MJ\,kW^{-1}$
Inverter, safety factor 0.3 (n_{inv}, ζ_{inv})	10	$756\,MJ\,kW^{-1}$
PbA battery, depth of discharge 80% (n_{bat}, k_{bat})	5	$1.03\,MJ\,Wh^{-1}$
PV frame (n_{frame}, ε_{frame})	30	$403.2\,MJ\,m^{-2}$
Installation	–	$25\,MJ\,m^{-2}$
Junction box, cabling, *etc.*	30	$80.8\,MJ\,m^{-2}$
M&O	–	$5.4\,MJ\,m^{-2}\,year^{-1}$
Foundation, array support, *etc.*(n_{sup}, ε_{sup})	30	$756\,MJ\,m^{-2}$

cabling, the installation and the maintenance & operation. The assigned value for the specific energy content of the PV system with different technologies is summarized in Table 7.15.

7.16.3 Overall Embodied Energy and Energy Payback Time

For a BIPVT system, the overall embodied energy is the sum of the embodied energies of the components. Table 7.16 summarizes the embodied energies used by the different components in a lifespan of 60 years, *i.e.* that of the building. The table shows that the overall embodied energies for monocrystalline silicon (c-Si), multi-crystalline silicon (p-Si), ribbon silicon (r-Si), amorphous silicon (a-Si), cadmium telluride (CdTe) and copper indium gallium selenide (CIGS) technological BIPVT systems are 1 215 225, 1 081 254, 983 318, 816 973, 847 935 and 767 226 MJ, respectively. The overall exergy calculations for the climatic conditions of New Delhi show that c-Si, p-Si, r-Si, a-Si, CdTe and CIGS BIPVT systems covering 45 m^2 of roof area generate 16 224, 14 352, 12 512,

Table 7.16 Embodied energy in MJ for a lifetime of 60 years.

PV technology	n_{sys}	E_{PV}	E_{ce}	E_{inv}	E_{bat}	$E_{ins+M\&O}$	E_{frame}	$E_{building} + E_{support} + E_{duct}$	E_{TOT}	EPBT (years)
c-Si	30	385 560	9072	29 484	320 238	24 102	36 288	410 481	1 215 225	20.81
p-Si	30	294 840	7938	29 484	278 121	24 102	36 288	410 481	1 081 254	20.93
r-Si	25	239 112	6804	29 484	236 597	24 552	36 288	410 481	983 318	21.83
a-Si	20	183 708	3402	29 484	128 383	25 227	36 288	410 481	816 973	29.13
CdTe	15	172 368	4536	29 484	168 426	26 352	36 288	410 481	847 935	24.67
CIGS	5	47 239	5670	29 484	202 712	35 352	36 288	410 481	767 226	19.31

7790, 9547 and 11 037 kW of overall exergy output, respectively. Thus, the EPBT for c-Si, p-Si, r-Si, a-Si, CdTe and CIGS are 20.81, 20.93, 21.83, 29.13, 24.67 and 19.31 years, respectively. The higher value of the EPBT is due to (1) using exergy as the divisor and (2) including the embodied energy of the building itself, which had been excluded earlier by the researchers.

Thus, the above calculation concludes that from the energy point of view, the CIGS technology is the most suitable for use in a BIPVT system. The electricity production factors (EPFs) for c-Si, p-Si, r-Si, a-Si, CdTe and CIGS on a life-time basis are 2.8, 2.87, 2.75, 2.06, 2.43 and 3.10, respectively, while the life cycle conversion efficiencies (LCCEs) are 0.349, 0.308, 0.262, 0.132, 0.185 and 0.247, respectively.

References

1. I. Boustead and G. F. Hancock, *Handbook of Industrial Energy Analysis*, Ellis Horwood, Chichester, 1979.
2. R. H. Crawford, G. J. Treloar, R. J. Fuller and M. Bazilian, Life-cycle energy analysis of building integrated photovoltaic systems (BiPVs) with heat recovery unit, *Renewable Sustainable Energy Rev.*, 2006, **10**, 559–575.
3. G. J. Treloar, *Comprehensive Embodied Energy Analysis Framework*, Thesis, Deakin University, 1998.
4. G. J. Treloar, Extracting embodied energy paths from input–output tables: towards an input–output-based hybrid energy analysis method, *Econ. Syst. Res.*, 1997, **9**, 375–391.
5. M. Lenzen, Errors in conventional and input–output-based life-cycle inventories, *J. Ind. Ecol.*, 2001, **4**, 127–148.
6. S. Suh and G. Huppes, Missing inventory estimation tool using input–output analysis, *Int. J. Life Cycle Assess.*, 2002, **7**, 134–140.
7. R. E. Miller and P. D. Blair, *Input–Output Analysis—Foundations and Extensions*, Prentice Hall, Upper Saddle River, NJ, 1985.
8. V. M. Fthenakis and H. C. Kim, Photovoltaics: life-cycle analyses, *Sol. Energy*, 2010, in press.
9. B. Azzopardi and J. Mutale, Life cycle analysis for future photovoltaic systems using hybrid solar cells, *Renewable Sustainable Energy Rev.*, 2010, **14**, 1130–1134.
10. E. Alsema, Energy pay-back time and CO_2 emissions of PV systems, *Prog. Photovoltaics*, 2000, **8**, 17–25.
11. A. Meijer, M. Huijbregts, J. Schermer and L. Reijnders, Life-cycle assessment of photovoltaic modules: comparison of mc-Si, InGaP and InGaP/mc-Si solar modules, *Prog. Photovoltaics*, 2003, **11**, 275–287.
12. N. Jungbluth, Life cycle assessment of crystalline photovoltaics in the Swiss ecoinvent database, *Prog. Photovoltaics*, 2005, **13**, 429–446.
13. W. Palz and H. Zibetta, Energy pay-back time of photovoltaic modules, *Int. J. Sol. Energy*, 1991, **10**, 211–216.

14. G. Hagedorn, *Kumulierter Energieaufwand von Photovoltaik und Windkraftanlagen*, Technischer Verlag Resch, Münich, Germany, 1992.
15. K. M. Hynes, A. E. Baumann and R. Hill, An assessment of the environmental impacts of thin-film cadmium telluride modules based on life cycle analysis, *Proceedings of the IEEE 1st World Conference on Photovoltaic Conversion (WCPEC)*, Hawaii, 1994, pp. 958–961.
16. K. Kato, T. Hibino, K. Komoto, S. Ihara, S. Yamamoto and H. Fujihara, A life-cycle analysis on thin-film CdS/CdTe PV modules, *Sol. Energy Mater. Sol. Cells*, 2001, **67**, 279–287.
17. B. V. Venkatarama Reddy and K. S. Jagadish, Embodied energy of common and alternative building materials and technologies, *Energy Build.*, 2003, **35**, 129–137.
18. G. N. Tiwari and M. K. Ghosal, *Renewable Energy Sources: Basic Principles and Applications*, Narosa, New Delhi, 2005.
19. G. Milne and C. Reardon, Embodied energy material use, http://www.buildwise.org/library/energy/sustainable/green-technical-guide/fs31.pdf.

CHAPTER 8

Carbon Dioxide Mitigation and Credit Earned

8.1 Introduction

The direct and encompassing impact of unchecked human activity on climate change is now clearly documented as a global concern.[1] Among the most dangerous anthropogenic interference with the climate system is the emission of greenhouse gases (GHGs) into the atmosphere, directly causing the rise in global surface temperatures. GHGs in the atmosphere absorb and trap the heat of the Sun reflected from the Earth's surface, causing global warming incrementally as the volume of GHGs increases across time. Carbon dioxide (CO_2), methane (CH_4), nitrous oxide (N_2O) and fluorinated gases are the principal anthropogenic GHGs emitted into the atmosphere.

CO_2 emission is considered to be one of the most significant among these. Since the beginning of the age of industrialization, human activities such as the combustion of fossil fuels and deforestation have caused the atmospheric concentration of CO_2 to increase by about 35%, resulting in an imbalance in the carbon cycle. Fossil fuel burning releases roughly 5.5 billion tonnes of carbon per year into the atmosphere and deforestation contributes roughly 1.6 billion tonnes of carbon per year, resulting in a release of 7.1 billion tonnes of carbon per year by human activities. Nearly 3.2 billion tonnes of carbon remain in the atmosphere, resulting in an increase in atmospheric CO_2 and 2 billion tonnes of carbon diffuses into the world's oceans, leaving 1.9 billion tonnes of carbon unaccounted for.

Scientists believe that the increase in CO_2 is already causing important changes in the global climate.[2] Many attribute the observed 0.6 °C increase in the global average temperature over the past century mainly to increases in

RSC Energy Series No. 4
Building Integrated Photovoltaic Thermal Systems: For Sustainable Developments
By Basant Agrawal and G. N. Tiwari
© B. Agrawal and G. N. Tiwari 2011
Published by the Royal Society of Chemistry, www.rsc.org

atmospheric CO_2. It is estimated that the global mean temperature will increase between 1.4 °C and 5.8 °C over the next century as a result of increases in atmospheric CO_2 and other GHGs. This kind of increase in global temperature could cause a significant rise in average sea-levels (0.09–0.88 m), exposing low-lying coastal cities or cities located by tidal rivers, such as New Orleans, Portland, Washington, and Philadelphia, to increasingly frequent and severe floods. Glacial retreat and species range shifts are also likely to result from global warming, and it remains to be seen whether relatively immobile species such as trees can shift their ranges fast enough to keep pace with warming.

Figure 8.1 shows that the mean concentration of CO_2 in the atmosphere is steadily rising since regular measurements began in 1958 and has reached a new high of 387 parts per million (ppm) by volume or 582 parts per million by mass, in March 2009.[3] The concentration varies seasonally and also considerably on a regional basis, especially near the ground. In urban areas, concentrations are generally higher and indoors they can reach 10 times background levels. Figure 8.2 shows that the share of CO_2 emission on a regional basis. In the figures, the EMU (Economic and Monetary Union) includes Austria, Belgium, Finland, France, Germany, Greece, Ireland, Italy, Luxembourg, Netherlands, Portugal, Slovenia and Spain. The largest emitter of CO_2 is the United States. In the developing world, China and India are the major emitters. Figure 8.3 (*left*) shows that the millions of metric tonnes of CO_2 emissions in industrial countries are mainly linked to the production of electricity and heat. Figure 8.3 (*right*) shows that the million of metric tonnes of CO_2 emissions in developing countries are mostly from deforestation and the production of certain agricultural products. Electricity and heat generation are also an increasing source of emissions.

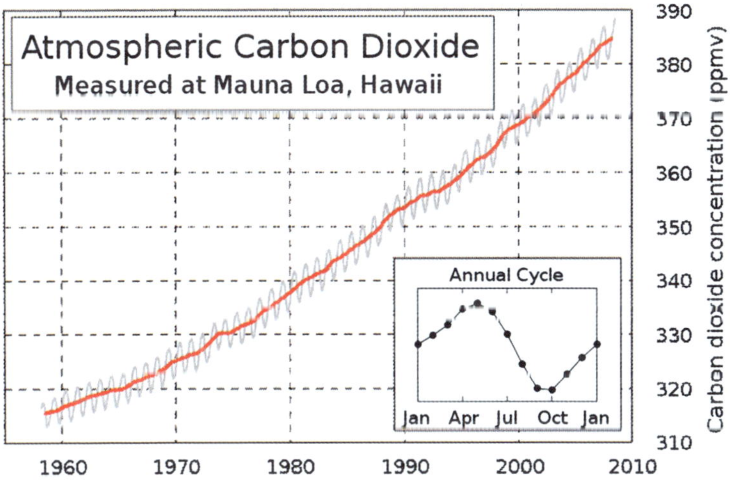

Figure 8.1 Concentration of atmospheric CO_2 measured at Mauna Loa Observatory.

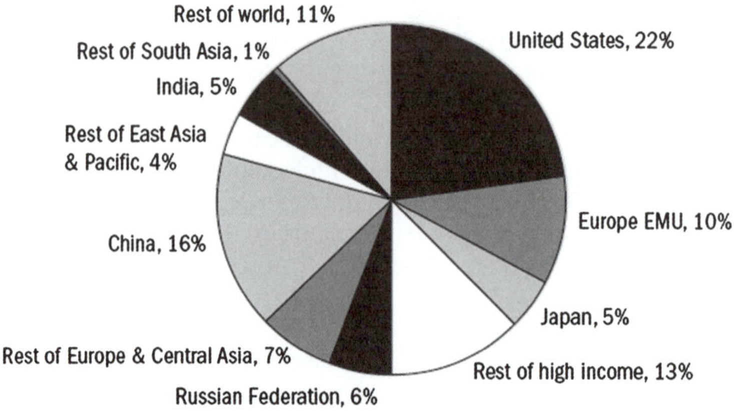

Figure 8.2 Share of CO_2 emission on a regional basis (*source*: World Bank, World Development Indicators 2007).

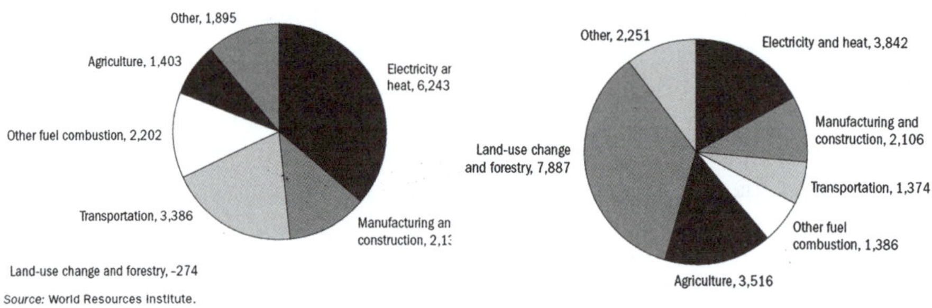

Figure 8.3 Millions of metric tonnes of CO_2 emissions: *left*, industrial countries; *right*, developing countries (*source*: World Resource Institute).

8.2 The Carbon Cycle

The Earth and the air contain only a certain amount of carbon. This carbon moves through nature in a cycle called the "carbon cycle". Figure 8.4 shows the carbon cycle where the limited amount of carbon on the Earth is reused in many forms by all living things. The carbon cycle begins when carbon compounds from the atmosphere are extracted by plants through the process known as photosynthesis. This carbon is combined with other elements in complex ways to form organic molecules important to life. When animals consume or eat plants, this carbon is transferred to them. A part of the carbon is put back into the atmosphere from plants and animals by respiration.

However, much of this carbon is returned to the atmosphere when plants and animals die as the organisms decompose. Every so often, a plant or animal does not decompose right away. Their bodies are trapped, in locations where decomposition simply cannot take place. This is most common at the bottom of

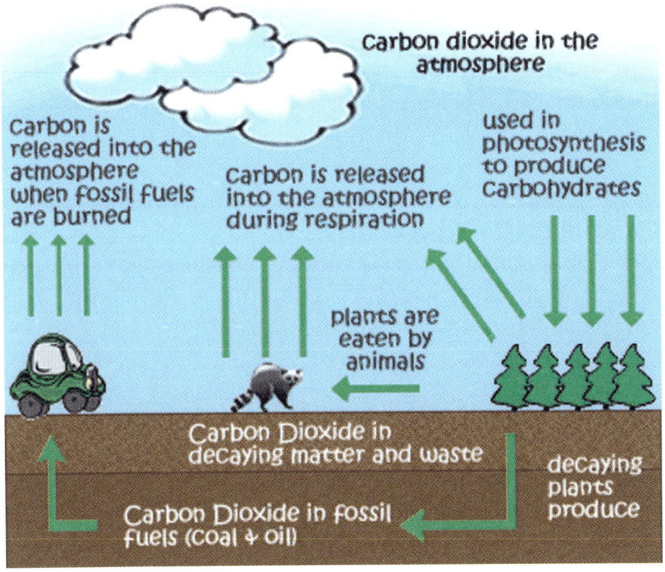

Figure 8.4 The carbon cycle.

oceans and seas, where the life forms become buried by sand. Instead of returning to the atmosphere, the carbon from these life forms is trapped within the Earth. Over millions of years, more and more of the carbon on the Earth has been trapped in this manner. Today, almost 99% of all the carbon on Earth has been locked up deep within the planet.[4] As rocks weather, this carbon is slowly released back into the atmosphere, creating a balance. For the past several hundred million years, the amount of carbon being locked up in the Earth, and the amount being released by weathering rocks, was almost perfectly balanced.

This important balance has been altered significantly in the past century, as humans have begun using fossil fuels to produce energy. By burning the Earth's store of carbon, mankind is able to create the energy needed to operate our communities. However, by releasing more carbon into the atmosphere than is being locked up, there is a risk of damaging the delicate carbon cycle.

8.3 The Kyoto Protocol and the UNFCCC

The Kyoto Protocol is a protocol to the United Nations Framework Convention on Climate Change (UNFCCC or FCCC), aimed at achieving stabilization of greenhouse gas concentrations in the atmosphere at a level that would prevent dangerous anthropogenic interference with the climate system. The Protocol was initially adopted on 11 December 1997 in Kyoto, Japan, and become active on 16 February 2005 when the number of signatories reached 55 (the number required to bring it into force). Meanwhile, the emissions increased to 4 billion metric tonnes. As of November 2009, 187 states have signed and ratified the protocol. The most notable non-member of the Protocol is the United States, which is a signatory of the UNFCCC and was responsible for 36.1% of the 1990 emission levels.[5] The parties agreed upon the following:

1. The largest share of historical and current global emissions of GHGs originated in developed countries.
2. The per capita emissions in developing countries are still relatively low.
3. The share of global emissions originating in developing countries will grow to meet social and development needs.

The United Nations Framework Convention on Climate Change (UNFCCC or FCCC) is an international environmental treaty produced at the United Nations Conference on Environment and Development (UNCED), informally known as the Earth Summit, held in Rio de Janeiro on 3–14 June 1992. Parties to UNFCCC are classified as:

Annex I countries. These include the industrialized countries that were members of the Organisation for Economic Co-operation and Development (OECD) in 1992, plus countries with economies in transition (the EIT Parties), including the Russian Federation, the Baltic States, and several Central and Eastern European States. They are: Australia, Austria, Belarus, Belgium, Bulgaria, Canada, Croatia, Czech Republic, Denmark, Estonia, Finland, France, Germany, Greece, Hungary, Iceland, Ireland, Italy, Japan, Latvia, Liechtenstein, Lithuania, Luxembourg, Monaco, the Netherlands, New Zealand, Norway, Poland, Portugal, Romania, the Russian Federation, Slovakia, Slovenia, Spain, Sweden, Switzerland, Turkey, Ukraine, the United Kingdom and the United States of America (40 countries and separately the European Union).

Annex II countries. These consist of the 24 original members of the OECD and EU, but not the EIT Parties. They are required to provide financial resources to enable developing countries to undertake emission reduction activities under the convention and to help them adapt to adverse effects of climate change. In addition, they have to take all practicable steps to promote the development and transfer of environmentally friendly technologies to EIT Parties and developing countries. They are: Australia, Austria, Belgium, Canada, Denmark, Finland, France, Germany, Greece, Iceland, Ireland, Italy, Japan, Luxembourg, the Netherlands, New Zealand, Norway, Portugal, Spain, Sweden, Switzerland, the United Kingdom and the United States of America (23 countries and separately the European Union; Turkey was removed from the Annex II list in 2001 at its request to recognize its economy as a transition economy).

Non-Annex I countries. They are mostly the developing countries, which reflect their less advanced economic development and their lower GHG emissions to date. These countries' overall emissions are now growing much faster than those of Annex I countries. Non-Annex 1 countries do not have stabilization or reduction commitments unless Annex I countries supply enough funding and technology.

Under the Protocol, Annex I countries commit themselves to reduce their collective emissions by 5.2% of four GHGs (carbon dioxide, methane, nitrous oxide, sulfur hexafluoride) and two groups of long-lived industrial gases (hydrofluorocarbons and perfluorocarbons) produced by them, and all member countries give general commitments. Cuts in carbon dioxide, methane and nitrous oxide will be measured against a base year of 1990, while that of hydrofluorocarbons (HFCs), perfluorocarbons (PFCs) and sulfur hexafluoride can be measured against either a 1990 or 1995 baseline. These reduction targets are in addition to the industrial gases and chlorofluorocarbons (CFCs), which are dealt with under the 1987 Montreal Protocol on substances that deplete the ozone layer. National or joint reduction targets range from 8% for the European Union and others to 7% for the US (non-binding as the US is not a signatory), 6% for Japan and 0% for Russia. The treaty permits emission increases of 8% for Australia and 10% for Iceland. Emission limits do not include emissions by international aviation and shipping. There is no immediate restriction for the Non-Annex I countries for the following purposes:

- To avoid limits on their development, because emissions are strongly linked to industrial capacity.
- They can sell emissions credits to nations whose operators have difficulty meeting their emissions targets.
- They obtain money and technologies for low-carbon investments from Annex II countries.

Non-Annex I countries may volunteer to become Annex I countries when they are sufficiently developed. Some opponents of the convention argue that the split between Annex I and Non-Annex I countries is unfair and both need to reduce their emissions unilaterally. Some countries claim that their costs of following the convention requirements will stress their economy.

8.3.1 The Protocol and Green Growth

The Protocol provides for several flexible mechanisms (also called Kyoto Mechanisms), which enable Annex I countries to meet their GHGs emission targets by acquiring GHG emission reduction credits. The credits are acquired by an Annex I country through financial exchanges, projects that reduce emissions in non-Annex I countries, from other Annex I countries or from Annex I countries with excess credits. Thus, the flexible mechanism calls for emissions trading, the clean development mechanism and joint implementation as follows:

- *Emissions Trading*. The Emissions Trading (ET) mechanism allows Annex I countries to purchase emission permits from other Annex I countries to meet their domestic emission reduction targets.
- *Joint Implementation*. Through the Joint Implementation (JI), Annex I countries can offset their emissions through the purchase of emission reduction units (ERU) from offset projects from any other Annex I countries as an alternative to meeting their domestic emission reduction targets.
- *Clean Development Mechanism*. Through the Clean Development Mechanism (CDM), Annex I countries can meet their domestic emission reduction targets through the purchase of certified emission reductions (CERs) from offset projects from non-Annex I countries that are signatory to the UNFCCC.

Meaningful emission reductions within a trading system can only occur if they can be measured at the level of operator or installation and reported to a regulator. There is an open-source tool for helping operators accurately measure and plan their emissions. For greenhouse gases, all trading countries maintain an inventory of emissions at national and installation level; in addition, the trading groups within North America maintain inventories at the state level through The Climate Registry. For trading between regions these inventories must be consistent, with equivalent units and measurement techniques. In some industrial processes, emissions can be physically measured by inserting sensors and flowmeters in chimneys and stacks, but many types of activity rely on theoretical calculations for measurement. Depending on local legislation, these measurements may require additional checks and verification by government or third party auditors, prior or post submission to the local regulator.

8.3.2 Emissions Trading and the CDM

There is a multitude of emission reduction credits that may fulfil the national targets of a number of Annex I countries. CERs issued through the CDM differ in that they are fully convertible in national emissions trading schemes, such as the European Emissions Trading Scheme (ETS). A single CER is issued to a CDM project for every tonne of CO_2 emission that it offsets. The CDM works by allowing Annex I countries to buy CERs originating from the developing world, as opposed to cutting emissions at home. The additional criterion of the CDM requires that, for a CDM project to be registered, developers must be able to show that the project will provide additional emission reductions, *i.e.* that the project would not have taken place without the CDM.

CDM projects must be validated before they can be registered as delivering CERs. Designated Operational Entities (DOEs) verify that CDM projects have delivered offsets and reduced emissions. Such entities are accredited by the CDM Executive Board and validate standard CDM projects according to the Board's guidelines. Project developers are subsequently issued with CERs according to the volume of GHGs offset, which they can then sell in the carbon market. The carbon market encompasses a number of different participants, but the institutional and legal framework of the global carbon market, of which the CDM plays a crucial role, is set out by the UNFCCC. Countries including Japan, Canada, Italy, the Netherlands, Germany, France, Spain and others are actively promoting government carbon funds, supporting multilateral carbon funds intent on purchasing carbon credits from non-Annex I countries, and are working closely with their major utility, energy, oil and gas and chemicals conglomerates to acquire greenhouse gas certificates as cheaply as possible. Virtually all of the non-Annex I countries have also established a DOE to manage its Kyoto obligations, specifically the "CDM process" that determines which GHG projects they wish to propose for accreditation by the CDM Executive Board.

In practice this means that non-Annex I countries have no GHG emission restrictions but have financial incentives to develop GHG emission reduction projects to receive carbon credits that can then be sold to Annex I countries, encouraging sustainable development. In addition, the flexible mechanisms allow Annex I countries with efficient, low GHG-emitting industries and high prevailing environmental standards to purchase carbon credits on the world market instead of reducing GHGs emission domestically. Annex I countries typically will want to acquire carbon credits as cheaply as possible, while non-Annex I countries want to maximize the value of carbon credits generated from their domestic GHG projects.

8.3.3 Market Value and Volume of Transactions

It is fair to say that the Emissions Trading Scheme (ETS) has been vital for the success of the CDM to date. Approximately 90% of all JI and CDM CERs sold in 2007 went to European buyers. The value of primary CDM transactions was US$ 7.4 billion in 2007, up 28% from 2006.[6] The volume of transactions in the ETS was in the region of 2 billion tonnes of CO_2 in 2007, up 87% on 2006. The volume of trade in the CDM market was 551 million tonnes of CO_2 emission in 2007, up 3% on 2006. In 2007 and early 2008 the average price for a CER was US$ 13.60, compared to US$ 10.60 in 2006 and US$ 7.20 in 2005. In 2007, China accounted for 73% of total CER supply, providing 402 million CERs to the market.

Hydro and biomass energy projects (including biofuels) have been the most common project type to date, each accounting for 22% of the total of 1133 registered projects. However, such projects do not bring big gains in terms of the number of CERs issued. Most CERs issued have been from HFC or N_2O projects (54% and 19%, respectively). Although these industrial gas projects account for just 5% of the total registered CDM projects, they account for 73% of CERs issued to date. However, the share of industrial gas projects (HFC and N_2O) has declined recently, from around 60% of total CER supply in 2005–2006 to under 10% in 2007. The share of clean energy CDM projects increased from a low of just over 10% in 2005–2006 to 64% of the total CER supply in 2007.

Clean energy projects are not as reliant on the existence of heavy industries and thus, in theory, can be implemented in many more countries. This category includes those projects dealing with energy efficiency and/or fuel switching, such as solar, hydro, wind, biomass and other renewable energies. Biofuels are arguably the most controversial renewable energy source. Capoor and Ambrosi concede that the controversy about the energy balance of biofuels derived from maize in the US has spilled over to the CDM.[6] There is no approved CDM methodology for biofuels derived from other crops or plants, such as sugar cane. Without an approved methodology, there is a risk that biofuels may lose out relative to other "clean" energy types. This may have negative implications for some producers in developing countries. As a result of concerns over the GHG pathways of biofuels, the EU has adopted sustainability criteria. The GHG emission saving from the use of biofuels and other bioliquids should be at least 35%, applicable from April 2013. The European Commission will report on the requirements of a "sustainability scheme" for biomass energy uses by 31 December 2010.

8.3.4 CDM Successes to Date

Larger developing countries such as China, India and Brazil may have been able to tap into the CDM to a greater extent than other developing countries because they have a number of existing industries that require "cleaning up", with emissions ready to be offset. The CDM could be considered a success for these countries, as not only are their industries likely to be less polluting and more energy efficient as a result of participation, project developers are also likely to have gained financially by selling CERs issued to the carbon market.

Although larger and middle income developing countries have been the main participants in the CDM to date, a look at the number of CDM projects registered within country and per capita GNP reveals that there is no clear relationship (correlation of –0.05). The same is true for those awaiting validation and registration (0.06) and between the number of CERs supplied by country and per capita GNP (0.00). We know that China, Brazil, Mexico and India have participated in the CDM to a greater extent than other developing countries, but we also know that clean energy projects are, increasingly, outstripping industrial gas projects, which may be more accessible to more developing countries and may account for this result.

8.3.5 The Post-2012 Climate Change Regime

The second Kyoto commitment period is being negotiated and is scheduled to run from 2012 to 2016. The Bali conference at the end of 2007 provided the best indication so far of the possible content of the successor treaty, with the EU pushing for a commitment of 25–40% cuts from the industrialized states by 2020. However, in the face of opposition from other states such as the US, Canada and Japan, the final declaration of the Bali conference omitted any precise figure for emission cuts.

The EU is attempting to use a form of extra-territoriality to exert pressure for a new set of targets. Its ETS Directive of January 2008 (EC 2008b) states that if no satisfactory international agreement is reached, the amount of CERs allowed into the ETS from 2013 will be limited to the 2008–2012 level of 1400 million tonnes. It has made clear that "additional use of CERs and ERUs should be provided for once there is an international agreement on climate change [post-2012], from parties which have concluded that agreement".

The targets set out in Annex I of the Kyoto Protocol were intended to be met by 2012 through emissions trading. A failure to agree emissions reduction targets after 2012 will remove the need for carbon offsetting and, hence, the CDM. Although this worst case scenario seems highly unlikely, there are widespread concerns over the future level of demand for CERs: the CDM is driven by government targets. The uncertainty of the post-2012 climate change regime may constrain further investment in CDM projects, which may stall the continued growth of clean energy projects.

8.3.6 United Nations Climate Change Conference at Copenhagen in 2009

The 2009 United Nations Climate Change Conference, commonly known as the Copenhagen Summit, was held at the Bella Centre in Copenhagen, Denmark, during 7–18 December 2009. The conference included the 15th Conference of the Parties (COP 15) to the UNFCCC and the 5th Meeting of the Parties (COP/MOP 5) to the Kyoto Protocol. According to the Bali Road Map, a framework for climate change mitigation beyond 2012 was to be agreed there.

The Copenhagen Accord was drafted by the US, China, India, Brazil and South Africa on the final day and judged a "meaningful agreement" by the US government. It was "taken note of", but not "adopted", in a debate of all the participating countries the next day, and it was not passed unanimously. The document recognized that climate change is one of the greatest challenges of the present and that actions should be taken to keep any temperature increases to below 2 °C. The document is not legally binding and does not contain any legally binding commitments for reducing CO_2 emissions. Leaders of industrialized countries were pleased with this agreement, but many leaders of other countries and non-governmental organizations were opposed to it. However, the international media concluded that the climate talks were "in disarray".

8.3.7 Prospects

At present there is considerable discussion both in scientific and industrial circles as how best to limit anthropogenic emissions of CO_2 due to its role in global atmospheric greenhouse warming. A wide range of technologies are being investigated that increase the efficiency of electrical production, decrease the energy needed to power electrical components to limit CO_2 and other GHG emissions, *etc.*

Redshaw predicted in June 2007 that the global trading in carbon credits, then already worth US$ 30 billion, would grow to US$ 1 trillion in a decade.[7] Carbon will be the world's biggest commodity market and it could become the world's biggest market overall. The EU is the largest carbon trading market, accounting for US$ 25 billion, and the US will most likely follow. However, the EU is being criticized for failing to create incentives in cutting emissions, causing a decrease in the price of carbon credits. In addition, the market is highly deregulated as of the present. Hence, certain carbon credits are transacted even if carbon emissions are not actually reduced.

8.4 Earned Carbon Credit

The trading of carbon credits was therefore created to curb the effect of greenhouse gases by reducing the carbon footprint. Carbon credits are defined as "a key component of national and international emissions trading schemes that have been implemented to mitigate global warming". They provide a way to reduce greenhouse effect emissions on an industrial scale by capping the total annual emissions and letting the market assign a monetary value to any shortfall through trading. Credits can be exchanged between businesses or bought and sold in international markets at the prevailing market price. Credits can be used to finance carbon reduction schemes between trading partners around the world. There are also many companies that sell carbon credits to commercial and individual customers who are interested in lowering their carbon footprint on a voluntary basis. These carbon offsetters purchase the credits from an investment fund or a carbon development company that has aggregated the credits from individual projects. The quality of the credits is based in part on the validation process and sophistication of the fund or development company that acted as the sponsor to the carbon project.

8.4.1 Formulation

If unit power is used by a consumer and the losses due to poor domestic appliances is L_a, then the transmitted power should be $\frac{1}{1-L_a}$ units. If the transmission and distribution losses are L_{td}, then the power that has to be generated in the power plant is $\frac{1}{1-L_a} \times \frac{1}{1-L_{td}}$ units.

The average CO_2 equivalent intensity for electricity generation from coal is approximately 0.98 kg of CO_2 per kWh at the source. Thus, for unit power consumption by the consumer, the amount of CO_2 emission is $\frac{1}{1-L_a} \times \frac{1}{1-L_{td}} \times 0.98$.

The annual CO_2 emission can be expressed as:

$$\text{CO}_2 \text{ emission per year} = \frac{E_{\text{in}}}{n_{\text{sys}}} \times \frac{1}{1 - L_{\text{a}}} \times \frac{1}{1 - L_{\text{td}}} \times 0.98\,\text{kg} \tag{8.1}$$

where E_{in} is the embodied energy input and n_{sys} is the lifetime of the system. The CO_2 emission over the lifetime of the system is:

$$E_{\text{in}} \times \frac{1}{1 - L_{\text{a}}} \times \frac{1}{1 - L_{\text{td}}} \times 0.98\,\text{kg} \tag{8.2}$$

The net CO_2 mitigation over the lifetime of the system is:

$$\begin{aligned} &\text{Total CO}_2 \text{ mitigation} - \text{total CO}_2 \text{ emission} \\ &= \left[\left(E_{\text{aout}} \times n_{\text{sys}} \right) \times \frac{1}{1 - L_{\text{a}}} \times \frac{1}{1 - L_{\text{td}}} \times 0.98 - E_{\text{in}} \right] \text{kg} \end{aligned} \tag{8.3}$$

where E_{aout} is the overall exergy gain, which is the sum of the annual electrical exergy ($\dot{E}x_{\text{el}}$) and the annual thermal exergy equivalent ($\dot{E}x_{\text{th}}$), which is given by:

$$E_{\text{aout}} = \left[\eta_{\text{ca}} \times I(t) \times bL \times n_{\text{s}} n_{\text{p}} \right] + \left[\dot{Q}_{\text{u}} \times \left(1 - \frac{T_{\text{a}}}{T_{\text{airout}}} \right) \right] \tag{8.4}$$

The net CO_2 mitigation over the lifetime in tonnes of CO_2 is given by:

$$= \left[\left(E_{\text{aout}} \times n_{\text{sys}} \right) \times \frac{1}{1 - L_{\text{a}}} \times \frac{1}{1 - L_{\text{td}}} \times 0.98 - E_{\text{in}} \right] \times 10^{-3} \tag{8.5}$$

If CO_2 emission is being traded at US\$ C per tonnes of CO_2 mitigation, then the carbon credit earned by the system is

$$= \text{US\$C} \times \left[\left(E_{\text{aout}} \times n_{\text{sys}} \right) \times \frac{1}{1 - L_{\text{a}}} \times \frac{1}{1 - L_{\text{td}}} \times 0.98 - E_{\text{in}} \right] \times 10^{-3} \tag{8.6}$$

8.5 A Case Study with the BIPVT System

If unit power is used by a consumer and the losses due to poor domestic appliances are around 20%, then the transmitted power should be $\frac{1}{1-0.2} - 1.25$ units. If the transmission and distribution losses are 40%, common in Indian conditions, then the power that has to be generated in the power plant is $\frac{1.25}{1-0.4} = 2.08$ units. The average CO_2 equivalent intensity for electricity generation from coal is approximately 0.98 kg of CO_2 per kWh at the source. Thus, for unit power consumption by the consumer the amount of CO_2 emission is

$2.08 \times 0.98 = 2.04$ kg. For the BIPVT system, the annual CO_2 mitigation in tonnes of CO_2 is given by:

$$\left(2.04E_{aout} - \frac{E_{in}}{n_{sys}}\right) \times 10^{-3} \tag{8.7}$$

Assume that the overall embodied energy for monocrystalline silicon (c-Si, $n_{sys} = 30$ years), multicrystalline silicon (p-Si, $n_{sys} = 30$ years), ribbon silicon (r-Si, $n_{sys} = 25$ years), amorphous silicon (a-Si, $n_{sys} = 20$ years), cadmium telluride (CdTe, $n_{sys} = 15$ years) and copper indium gallium selenide (CIGS, $n_{sys} = 5$ years) technological BIPVT systems are 607 613, 540 628, 409 716, 272 324, 211 984 and 63 937 MJ, respectively. The overall exergy calculations for the climatic conditions of New Delhi show that c-Si, p-Si, r-Si, a-Si, CdTe and CIGS BIPVT systems covering 45 m^2 of roof area generate 16 224, 14 352, 12 512, 7790, 9547 and 11 037 kW of overall exergy output, respectively. Thus the annual CO_2 mitigation for c-Si, p-Si, r-Si, a-Si, CdTe and CIGS systems are 98.9, 87.4, 75.5, 43.6, 56.0 and 68.3 tonnes, respectively. If CO_2 emissions are being traded at US$ 20 per tonnes of CO_2 mitigation, then the carbon credit earned by the BIPVT system with the c-Si, p-Si, r-Si, a-Si, CdTe and CIGS technologies are US$ 1978, 1748, 1510, 872, 1120 and 1365, respectively. This shows that the monocrystalline silicon BIPVT system gives the highest earnings through carbon credit trading.

References

1. *Carbon Credit and Trading*, Ecoseed, http://www.ecoseed.org/en/general-reference/green-reference/carbon-trading.
2. J. A. Harrison, *The Carbon Cycle: What Goes Around Comes Around*, Visionlearning, http://www.visionlearning.com/library/module_viewer.php?mid=95.
3. P. Tans, *Trends in Atmospheric Carbon Dioxide*, NOAA/ESRL, www.esrl.noaa.gov/gmd/ccgg/trends.
4. *The Carbon Cycle, Geography for Kids*, http://www.kidsgeo.com/geography-for-kids/0159-the-carbon-cycle.php.
5. *Kyoto Protocol: Status of Ratification*, United Nations Framework Convention on Climate Change, http://unfccc.int/files/kyoto_protocol/status_of_ratification/application/pdf/kp_ratification.pdf.
6. K. Capoor and P. Ambrosi, *State and Trends of the Carbon Market 2008*, World Bank, Washington, DC, 2008, http://carbonfinance.org/docs/State___Trends--formatted_06_May_10pm.pdf.
7. J. Kanter, Carbontrading: where greed is green, *The New Tork Times*, 20 June 2007.

CHAPTER 9

Life Cycle Cost Assessments

9.1 Introduction

Economic analysis is a process whereby the strengths and weaknesses of an economy are analyzed. Instead of merely looking at the facility in terms of the cost to design and build, owners can broaden their perspective to include operations, maintenance, repair, replacement and disposal costs. The sum of the initial and future costs associated with the construction and operation of a building over a period of time is called the life cycle cost (LCC) of a facility. Thus, life cycle cost assessment is an economic evaluation technique that is used to determine the total cost of owning and operating a facility over period of time. LCC is associated with an energy delivery system over its lifetime or over a selected period of analysis and takes into account the time value of money. This chapter discusses the basics of LCC analysis in the context of the BIPVT system.

9.2 Estimating the Cost of a Project

Before the LCC of any engineering project can be evaluated, it is necessary to reasonably estimate the various costs and revenues that describe the project.

9.2.1 Capital Costs

Capital costs are the costs incurred to bring a project to the operable status. They are referred to as one-time expenses and are independent of the level of output. Capital costs do not include labour costs except for the labour used for construction. For a BIPVT system the capital costs are the purchase of the land on which the building is to be built, permit and legal costs, the equipment needed to bring the system into operation, the cost of the construction, the cost of financing and the cost of commissioning incurred prior to operation of the system.

RSC Energy Series No. 4
Building Integrated Photovoltaic Thermal Systems: For Sustainable Developments
By Basant Agrawal and G. N. Tiwari
© B. Agrawal and G. N. Tiwari 2011
Published by the Royal Society of Chemistry, www.rsc.org

9.2.2 Variable Costs

Variable costs are costs that vary or change in proportion to the activity of a system. They include the costs associated with production and manufacture. The BIPVT system includes only the labour costs needed for periodical cleaning of the surface of the PV array exposed to the Sun's rays and the maintenance and repair during the operational phase.

9.2.3 Step-variable Costs

Step-variable costs stay the same up to a certain level of activity, but when the level of activity increases beyond this point the cost steps up. The jump in costs after a certain level of activity will form the steps. For a BIPVT system, replacement of the components and batteries calls for step-variable costs. The number and timing of the replacements depend on the lifespan of the components.

9.2.4 Non-product Costs

Non-product costs are the costs that have nothing to do with manufacturing, selling, marketing or administrative purposes.

9.3 Depreciation

Depreciation (C_d) is the reduction in the cost of an asset used for business purposes during a certain amount of time due to usage, passage of time, wear and tear, technological outdating or obsolescence, depletion, inadequacy, rot, rust, decay or other such factors. It is determined using the following relationship:

$$C_{\mathrm{d}} = C_{\mathrm{i}} - C_{\mathrm{sal}} \tag{9.1}$$

where C_i is the initial cost and C_{sal} is the salvage value.

The salvage value is defined as the estimated value of an asset at the end of its useful life. The book value (B_n) represents the cost of the asset after deducting the depreciation from the purchase cost. It can be obtained by subtracting the total amount of annual depreciation charges to date from the initial cost. The book value is usually determined at the end of each year and does not bear any relation to the resale value:

$$B_n = C_{\mathrm{i}} - C_{\mathrm{d}} \tag{9.2}$$

There are several methods for calculating depreciation, generally based on either the passage of time or the level of activity (or use) of the asset.

9.3.1 Straight-line Depreciation

Straight-line depreciation of an asset assumes that the asset loses an equal amount of value each year. The annual depreciation (D_n) is calculated by subtracting the salvage value of the asset from the purchase price and then dividing this number by the estimated useful life (N) of the asset:

$$D_n = \frac{C_d}{N} = \frac{C_i - C_{sal}}{N} \tag{9.3}$$

The book value (B_n) at the end of n years is calculated by:

$$B_n = C_i - nD_n \tag{9.4}$$

or

$$B_n = C_{sal} + \left(1 - \frac{n}{N}\right) C_d \tag{9.5}$$

9.3.2 Accelerated Depreciation

Accelerated depreciation provides for a higher depreciation charge in the first year of an asset's life and gradually decreasing charges in subsequent years. This pattern occurs because the efficiency of an asset trends to decline and maintenance costs tend to increase, with the likelihood that better equipment will become available with time.

One popular accelerated method is the declining-balance method. Under this method the book value is multiplied by a fixed rate. The rate is a multiple of the straight-line rate; typically it is twice the straight-line rate and popularly known as double-declining-balance depreciation. For the nth year the depreciation charge is given by:

$$D_n = \alpha C_i (1 - \alpha)^{n-1} \tag{9.6}$$

where fraction $\alpha = \text{multiplier}/N$.

The total declining balance (TDB) depreciation by the end of n years can be estimated by:

$$\text{TDB} = C_i \times [1 - (1 - \alpha)^n] \tag{9.7}$$

The book value at the end of n years is calculated by:

$$B_n = C_i - \text{TDB} = C_i \times (1 - \alpha)^n \tag{9.8}$$

Another accelerated method for allocating the cost of an asset is called sum-of-years'-digit (SOYD) depreciation. In this method, the numbers 1, 2, 3, . . . , N are summed, where N is the estimated years of useful life, *i.e.*:

$$\text{SOYD} = 1 + 2 + 3 + \ldots + N = \frac{N(N+1)}{2} \tag{9.9}$$

The annual depreciation rate is a fraction in which the denominator is the SOYD and the numerator is N for the first year, $N - 1$ for the second year, $N - 2$ for the third year, and so on. Thus the depreciation charge for each year is estimated by:

$$D_n = \frac{N - n + 1}{\text{SOYD}} \times (C_i - C_{sal}) \tag{9.10}$$

Example 9.1

The present cost of a PV battery is US$ 1000 and whose estimated salvage value is US$ 200 at the end of a useful life of five years. Compute the annual depreciation using the straight-line depreciation method and plot the variation in the book value with time.

Solution

Given that $C_i = \text{US\$ } 1000$, $C_{sal} = \text{US\$ } 200$ and $N = 5$ years, the annual depreciation is given by:

$$D_n = \frac{C_i - C_{sal}}{N} = \frac{1000 - 200}{5} = \text{US\$ } 160$$

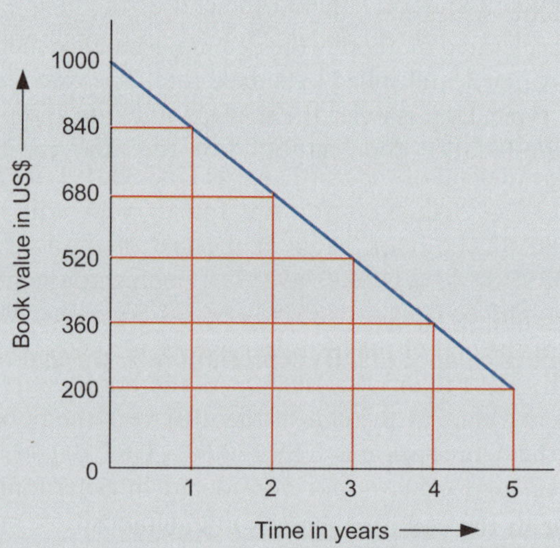

Figure 9.1 Variation in book value with time; straight-line depreciation.

The book value at the end of first year is $B_1 = 1000 - 1 \times 160 = \text{US\$ } 840$. The book value at the end of the second year is $B_2 = 1000 - 2 \times 160 = \text{US\$ } 680$. The book value at the end of the third year is $B_3 = 1000 - 3 \times 160 = \text{US\$ } 520$. The book value at the end of the fourth year is $B_4 = 1000 - 4 \times 160 = \text{US\$ } 360$.

The variation in the book value with time is shown in Figure 9.1.

Example 9.2

The present cost of a PV battery is US$ 1000, whose estimated salvage value is US$ 100 at the end of a useful life of five years. Compute the annual depreciation allowances and the resulting book value using: (a) the double-declining-depreciation method and (b) the sum-of-years'-digits depreciation method.

Solution

Given that $C_i = \text{US\$ } 1000$, $C_{sal} = \text{US\$ } 100$ and $N = 5$ years, then:

(a) For double-declining depreciation the multiplier is 2; therefore the fraction is:

$$\alpha = \frac{2}{5} = 40\%$$

The depreciation deduction for the first year will be 40% × US$ 1000 = US$ 400; the book value at the end of the first year is then US$ 1000 – US$ 400 = US$ 600. The depreciation deduction for the second year will be 40% × US$ 600 = US$ 240; the book value at the end of the second year is therefore US$ 600 – US$ 240 = US$ 360. The depreciation deduction for the third year will be 40% × US$ 360 = US$ 144; thus the book value at the end of the third year is US$ 360 – US$ 144 = US$ 216. The depreciation deduction for the fourth year will be 40% × US$ 216 = US$ 86.40; the book value at the end of the fourth year is then US$ 216 – US$ 86.40 = US$ 129.60. The depreciation deduction for the fifth year would be 40% × US$ 129.60 = US$ 51.84, which gives the book value at the end of the fifth year as US$ 129.60 – US$ 51.84 = US$ 77.76, which is less than US$ 100.00. As the book value cannot be less than the scrap value, at the end of the fifth year the book value is US$ 100.00 and the depreciation is US$ 129.60 – US$ 100.00 = US$ 29.60.

Note: The salvage value is not considered in determining the annual depreciation in the case of the declining-depreciation method, but the book value of the asset is never brought below its salvage value. The process continues until the salvage value, or the end of the asset's useful life, is reached. In the last year of depreciation a subtraction might be needed in order to prevent the book value from falling below the estimated scrap value.

The variation in the book value with time is shown in Figure 9.2.

(b) For the SOYD method:

$$\text{SOYD} = \frac{5(5+1)}{2} = 15$$

The depreciation deduction for the first year will be (5/15) × (1000 − 100) = US$ 300; the book value at the end of the first year is then US$ 1000 − US$ 300 = US$ 700. The depreciation deduction for the second year will be (4/15) × (1000 − 100) = US$ 240; the book value at the end of the second year is therefore US$ 700 − US$ 240 = US$ 460. The depreciation deduction for the third year will be (3/15) × (1000 − 100) = US$ 180; thus the book value at the end of the third year is US$ 460 − US$ 180 = US$ 280. The depreciation deduction for the fourth year will be (2/15) × (1000 − 100) = US$ 120; the book value at the end of the fourth year is then US$ 280 − US$ 120 = US$ 160. The depreciation deduction for the fifth year will be (1/15) × (1000 − 100) = US$ 60; the book value at the end of the fifth year is US$ 160 − US$ 60 = US$ 100.

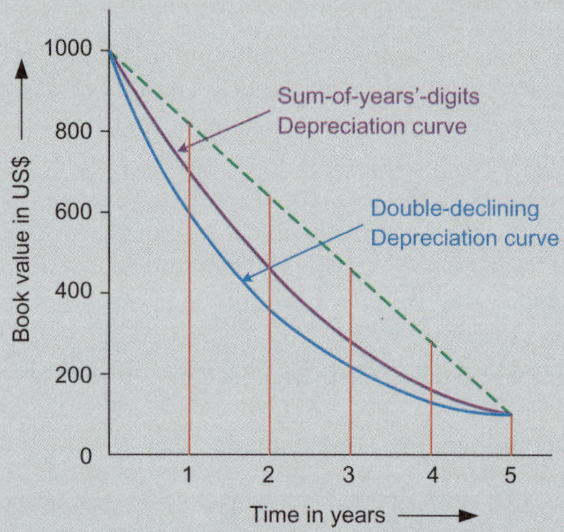

Figure 9.2 Variation in book value with time; accelerated depreciation.

The variation in the book value with time is shown in Figure 9.2.

9.4 Interest

Interest plays a central role in the LCC assessment. It is compensation to the lender and for forgoing other useful investments that could have been made with the loaned asset. These forgone investments are known as the opportunity cost. Instead of the lender using the assets directly, they are advanced to the borrower. The borrower then enjoys the benefit of using the assets ahead of the effort required to obtain them while the lender enjoys the benefit of the fee paid by the borrower for the privilege. The amount lent, or the value of the assets lent, is called the principal. This principal value is held by the borrower on credit. Interest is therefore the price of credit. The two computational schemes for calculating this earned interest yield either simple interest or compound interest.

Simple interest is calculated only on the original principal, or on that portion of the principal amount which remains unpaid. The simple interest (I_s) is calculated using the following expression:

$$I_s = P \times i \times n \tag{9.11}$$

where P is the initially invested or credited amount for n years at i rate of interest.

However, if the borrower does not pay the interest to the lender periodically, as agreed between them, then the lender may charge interest on the previous interest. This is called compounding of interest and the sum paid at the end of the overall duration of credit, over and above, is called compound interest. The effect of compounding depends on the frequency with which the interest is compounded and the periodic interest rate which is applied. Therefore, in order to define accurately the amount to be paid, the frequency of compounding (yearly, half-yearly, quarterly, monthly, daily, *etc.*) and the interest rate must be clearly specified. Assuming that no part of the principal and subsequent interest has been paid, the compound interest (I_c) is calculated using:

$$I_c = P\left\{\left(1 + \frac{i}{p}\right)^{np} - 1\right\} \tag{9.12}$$

where p is the number of times the interest is compounded per year and i is the annual rate of interest.

In the case of compounding done annually, eqn (9.12) reduces to:

$$I_c = P\{(1 + i)^n - 1\} \tag{9.13}$$

Compound interest is standard in finance and economics and simple interest is used infrequently. Economists generally prefer to use effective annual rates to allow for comparability. It is the total accumulated interest that would be payable up to the end of one year, divided by the principal:

$$\text{effective annual rate} = \left(1 + \frac{i}{p}\right)^p - 1 \tag{9.14}$$

The effective annual rate reflects the effective rate as if annual compounding were applied:

$$\text{For } p = 1, \left\{\left(1 + \frac{i}{p}\right)^p - 1\right\} = i \text{ and for } p > 1, \left\{\left(1 + \frac{i}{p}\right)^p - 1\right\} > i \tag{9.15}$$

Example 9.3

A loan of US$ 1200 is taken for a period of 18 months at a simple interest rate of 8% to purchase a PV charger. What is the amount due at the end of the loan period?

Solution

Given that P = US$ 1200, I = 8% and n = 18 months, then:

$$\begin{aligned}\text{Future value} &= \text{Present value} + \text{interest}\\ &= P + I_s = 1200\left(1 + \frac{18}{12} \times 0.08\right)\\ &= \text{US\$ } 1344\end{aligned}$$

Example 9.4

If the annual rate of interest is 10%, what time would an asset take to be doubled (a) at a simple interest rate and (b) when compounded annually?

Solution

Given $I = 0.10$. Let P be the initial amount; after n years this amount is doubled and becomes $2P$. Therefore the interest received is:

$$I_c = \text{Amount} - \text{Principle} = 2P - P = P.$$

(a) On substituting the values in eqn (9.11), we have:

$$\begin{aligned}I_s &= P \times i \times n\\ &\Rightarrow P = P \times 0.10 \times n\\ &\Rightarrow n = 10\,\text{years}\end{aligned}$$

(b) On substituting the values in eqn (9.13), we have:

$$\begin{aligned}I_c &= P\{(1 + i)^n - 1\}\\ &\Rightarrow P = P\{(1 + 0.10)^n - 1\}\\ &\Rightarrow 2 = 1.1^n\\ &\Rightarrow n = \frac{\ln 2}{\ln 1.1} = 7.27\,\text{years}\end{aligned}$$

Example 9.5

Calculate the effective rate of return for a rate of interest of 10%, compounding (a) monthly and (b) half-yearly.

Solution

Given $i = 0.10$.

(a) When compounding is done monthly, the number of times compounding is made in a year is $p = 12$. Therefore, from eqn (9.14) the effective rate of interest is:

$$\left(1+\frac{i}{p}\right)^p - 1 = \left(1+\frac{0.10}{12}\right)^{12} - 1 = 10.47\%$$

(b) When compounding is done half-yearly, the number of times compounding is made in a year is $p = 2$. Therefore, from eqn (9.14) the effective rate of interest is:

$$\left(1+\frac{i}{p}\right)^p - 1 = \left(1+\frac{0.10}{2}\right)^{2} - 1 = 10.25\%$$

9.5 Cash Flow Diagram

A cash flow diagram is a pictorial representation of the transactions that will take place over the course of a given project. These diagrams are helpful in solving problems of economic analysis. Transactions can include initial investments, maintenance costs, projected earnings or savings resulting from the project, as well as salvage and resale value of the equipment at the end of the project.

A cash flow diagram normally has a horizontal axis that is marked off in equal increments, one per period, up to the duration of the project. It also addresses payment and receipt of funds along the horizontal line at the end of the period of occurrence. Funds that are payouts such as deposits or lease payments are called negative cash flows and are represented by arrows which extend downwards from the time line, with their bases at the appropriate positions along the line. Funds that are received such as proceeds from a mortgage or withdrawals are called positive cash flows and are represented by arrows extending upwards from the line. Arrow lengths are approximately proportional to the magnitude of the cash flow. Figure 9.3 (*left*) shows a cash flow diagram for loan transactions where upward and downward arrows show

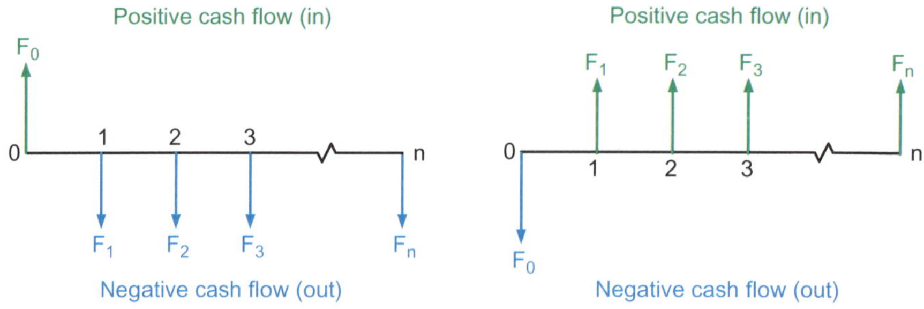

Figure 9.3 Cash flow diagram: *left*, loan transaction; *right*, investment transaction.

the loan received and paid off, respectively. Figure 9.3 (*right*) shows a cash flow diagram for investment transactions where upward and downward arrows show the loan payback and investment, respectively.

In practice, cash flows can occur at the beginning or in the middle of an interest period or at practically any point in time. One of the simplifying assumptions made in economic analysis is the end-of-period convention, which is the practice of placing all cash flow transactions at the end of an interest period. This assumption relieves us of the responsibility of dealing with the effects of interest within an interest period, which would greatly complicate the calculations.

9.6 Present or Future Value Calculations for a Regular Pattern of Cash Flows

When patterns in cash flow transactions can be identified, a concise expression for computing either a present or future value of the series can be derived. This would be useful to determine and compare the LCC of a system with a regular pattern of cash flows.

9.6.1 Single Payment Future Value Factor

Compounding is a process for calculating the future value of a present value invested at a specified interest rate. Thus the compounding translates a BIPVT system whole-life costs into a future value. Let P be the amount invested at present at an annual interest rate i, shown in Figure 9.4; then at the end of n years its future value (S) will be:

$$S_n = P(1+i)^n \tag{9.16}$$

If S_n is written as S, then eqn (9.16) can be written as:

$$S = P(1+i)^n = PF_{PS} \tag{9.17}$$

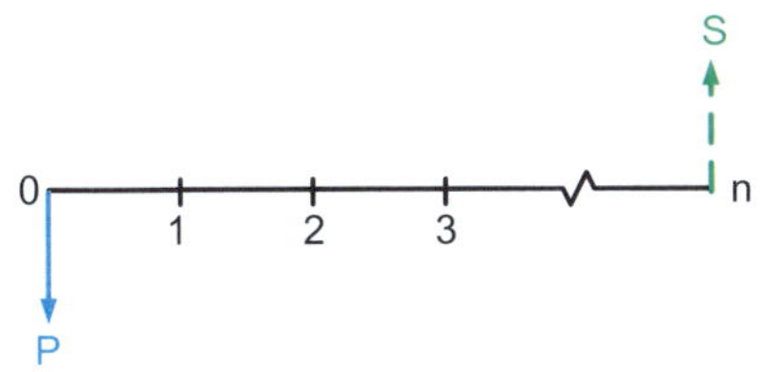

Figure 9.4 Single payment; compounding process.

where $F_{PS} = (1 + i)^n$ and is called the compound interest factor or future value factor, which converts the present value (P) into the future value (S):

$$Future\ value = Present\ value \times Future\ value\ factor$$

F_{PS} is more completely written as $F_{PS,i,n}$. It may be noted that for a positive rate of interest the future value will be greater than the present value invested, *i.e.*

$$\text{for } i > 0,\ S > P \tag{9.18}$$

If the compounding is done periodically, p times in a year, then the expression for the future value is:

$$S = P\left(1 + \frac{i}{p}\right)^{np} \tag{9.19}$$

Using spreadsheet computer software, *e.g.* MS Excel, the future value can be computed by writing the expression as =FV(i,n, – P).

Example 9.6

If US$ 1200 is invested now at 10%, how much would this be worth in eight years?

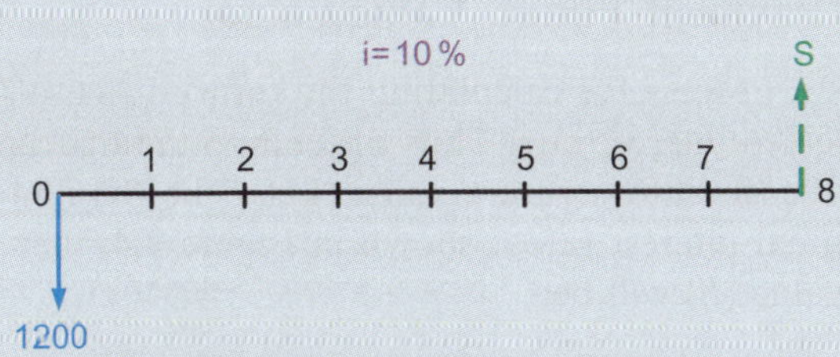

Figure 9.5 Cash flow diagram.

Solution

Given that $P = 1200$, $i = 10\%$ or 0.10 and $n = 8$ years, then Figure 9.5 shows the cash flow diagram.

$$S = P(1+i)^n = 1200(1+0.10)^8 = \text{US\$ } 2572.31.$$

Using MS Excel: =FV(10%,8,, – 1200) = US$ 2572.31.

Example 9.7

If US$ 20 000 compounds to US$ 28 240 in four years, what is the rate of return?

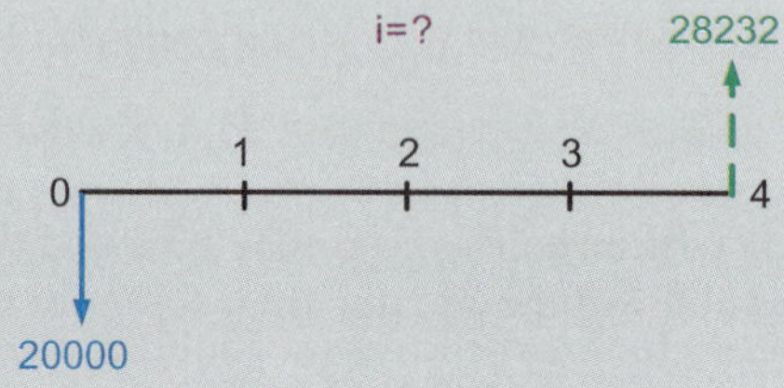

Figure 9.6 Cash flow diagram.

Solution

Given that P = US$ 20 000, S = US$ 28 232 and $n = 4$ years, then Figure 9.6 shows the cash flow diagram.

$$S = P(1+i)^n$$
$$\Rightarrow 28232 = 20000(1+i)^4$$
$$\Rightarrow i = 9\%$$

Using MS Excel: =RATE(4, – 20000,28232) = 9%.

Example 9.8

A person has purchased a solar house for US$ 36 000. If he expects the solar house price to increase by 14% per year, how long need he to wait to obtain a US$ 50 000 market price?

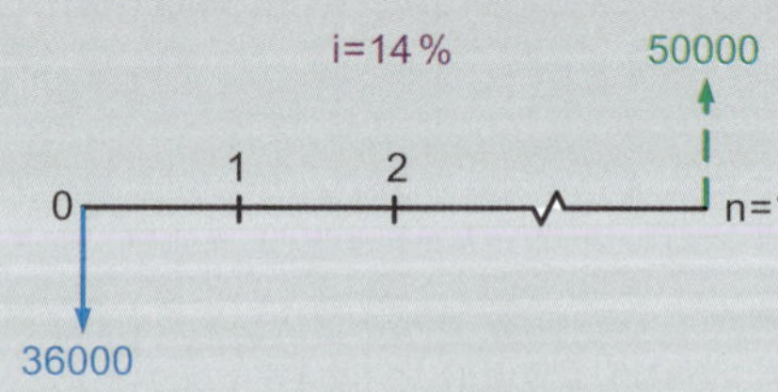

Figure 9.7 Cash flow diagram.

Solution

Given that P = US\$ 36 000, S = 50 000 and i = 14%, then Figure 9.7 shows the cash flow diagram.

$$\begin{aligned} S &= P(1+i)^n \\ &\Rightarrow 50000 = 36000 \times (1+0.14)^n \\ &\Rightarrow n = \frac{\ln(50000/36000)}{\ln(1.14)} = 2.5 \text{ years} \end{aligned}$$

Using MS Excel: =NPER(14%,0, – 36000,50000) = 2.5 years.

Example 9.9

It is targeted to increase the electrification of houses in a certain city with a PV grid from the present 12 million to 59 million houses in five years. Calculate the growth rate required to achieve the target.

Solution

Given that P = 12 million, S = 59 million and n = 5 years, then:

$$\begin{aligned} S &= P(1+i)^n \\ &\Rightarrow 59 \times 10^6 = 12 \times 10^6(1+i)^5 \\ &\Rightarrow i = 37.51\% \end{aligned}$$

Thus, 37.51% growth rate is required to achieve the target.
Using MS Excel: =RATE(5,, – 12E06,59E06) = 37.51%.

Example 9.10

A construction company wants to set aside US\$ 2000 today in an interest bearing account in order to have US\$ 2100 at the end of six months from now. Calculate the interest rate payable to the account.

Solution

Given that P = US\$ 2000, S = US\$ 2100 and n = 6 months, then:

$$\begin{aligned} 2100 &= 2000 \times \left[1 + \frac{6}{12} \times i\right] \\ &\Rightarrow i = 10\% \end{aligned}$$

Example 9.11

A person purchases a solar system for US$ 10 000. In retrospect, if he had invested US$ 10 000 in a savings account that earns 10% annual compound interest, how much would it be worth at the end of four years and four months?

Solution

Given that P = US$ 10 000, n = 4 years and 4 months and i = 0.10, then Figure 9.8 shows the cash flow diagram.

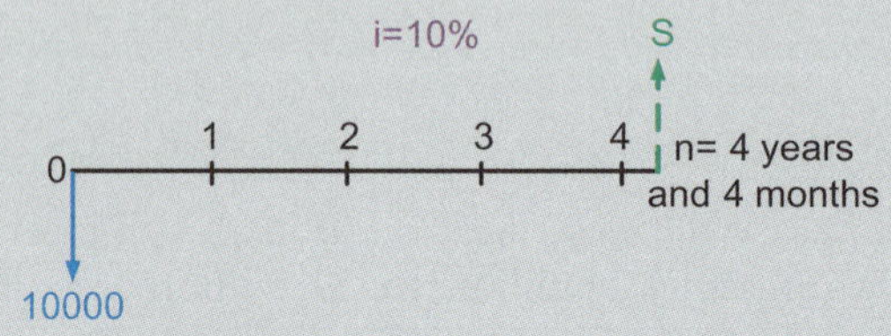

Figure 9.8 Cash flow diagram.

From eqn (9.14), after five years the value would be:

$$S_4 = P(1+i)^n = 10\,000 \times (1+0.10)^4 = \text{US\$ } 14\,641.$$

This amount is charged at a rate of 10% for the last four months. Therefore, at the end of the total duration he has to pay:

$$S = \text{US\$ } 14\,641 \times \left[1 + \frac{4}{12} \times 0.10\right] = \text{US\$ } 15\,129$$

9.6.2 Single Payment Present Value Factor

Discounting is a process for calculating the present value for an amount in the future using a specified discount rate. Thus, the discounting translates a projected BIPVT system whole-life costs into their present value.

Let S be the future value, at the end of n years, shown in Figure 9.9, then the present value (P) at an annual discount (or interest) rate i is given by rewriting eqn (9.17) as:

$$P = \frac{S}{(1+i)^n} = S(1+i)^{-n} = SF_{SP} \tag{9.20}$$

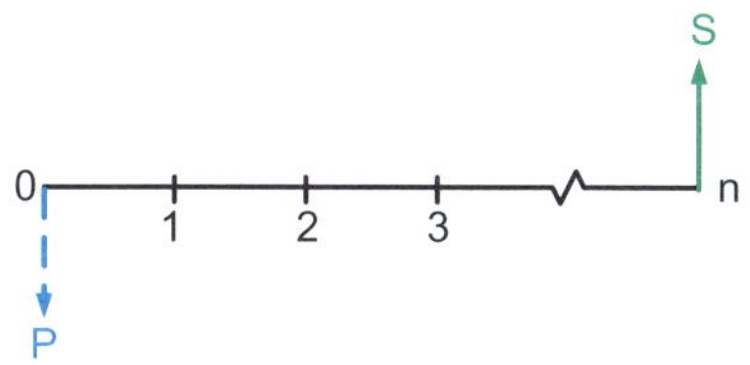

Figure 9.9 Single payment; discounting process.

where $F_{SP} = (1+i)^{-n}$ and is called the present value factor or discounting factor which converts the future value (S) into the present value (P). Therefore:

$$\textit{Present value} = \textit{Future value} \times \textit{Present value factor}$$

F_{SP} is more completely written as $F_{SP,i,n}$. It is the reciprocal of the future value factor. Therefore:

$$F_{PS} \cdot F_{SP} = 1 \tag{9.21}$$

Using MS Excel, the present value can be computed by writing the expression as =PV(i,n, – S).

Example 9.12

Suppose a person is offered the alternative of receiving either US$ 15 000 at the end of 10 years or P dollars today. Because he does not have money requirements today, he deposits the P dollars in an account that pays 8% interest compounded annually. What value of P would make him indifferent to his choice between the P dollars today and the promise of US$ 15 000 at the end of 10 years?

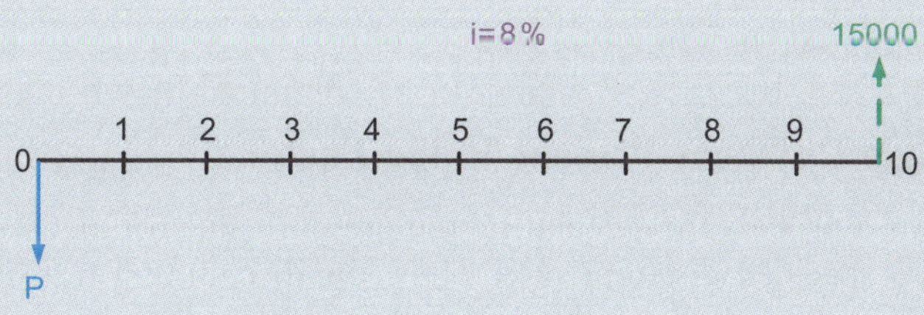

Figure 9.10 Cash flow diagram.

Solution

Given that $S =$ US$ 15 000, $n = 10$ years and $i = 0.08$, then Figure 9.10 shows the cash flow diagram.

Using eqn (9.20), the present value of US\$ 15 000 is:

$$P = \frac{S}{(1+i)^n} = \frac{15000}{(1+0.08)^{10}} = \text{US\$ } 6947.90$$

Using MS Excel: =PV(8%,10,, – 15000) = US\$ 6947.90.

9.6.3 Equal Payment Series Present Value Factor

Let R be the amount to be recovered in equal payments in n years at an annual interest rate i, as shown in Figure 9.11. The total accumulated fund in terms of the present value P is simply the sum of R payments multiplied by the appropriate single payment present value factor for years 1 through n. This summation is a geometric series as follows:

$$\begin{aligned} P &= \sum_{j=1}^{n} \frac{R}{(1+i)^j} \\ &= R\left[\frac{1}{(1+i)^1} + \frac{1}{(1+i)^2} + \frac{1}{(1+i)^3} + \ldots + \frac{1}{(1+i)^{n-1}} + \frac{1}{(1+i)^n}\right] \end{aligned} \quad (9.22)$$

$$\Rightarrow P = RR \cdot F_{RP} + G \cdot F_{PG} = R \cdot F_{RP} \quad (9.23)$$

where $F_{RP} = x\frac{(1+i)^n - 1}{i(1+i)^n}$ and is called the equal payment series present value factor or annuity present value factor. Therefore:

$$\textit{Present value} = \textit{Annual uniform cost} \times \textit{Annuity present value factor}$$

F_{RP} is more completely written as $F_{RP,i,n}$.

Using MS Excel, the present value can be computed by writing the expression as =PV(i,n, – R).

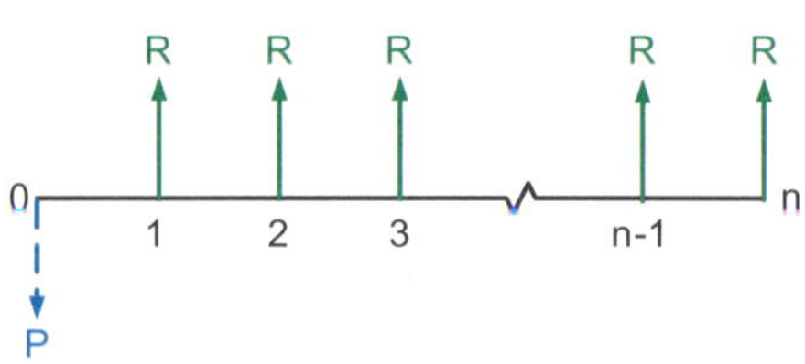

Figure 9.11 Equal payment series present value factor.

Example 9.13

A person wants to make a deposit for his family such that his family can withdraw US$ 2500 at the end of each year, for 12 years. Calculate the amount he needs to deposit now. Take the interest rate for deposit as 10%.

Solution

Given that $R =$ US$ 2500, $n = 12$ years and $i = 10\%$, then Figure 9.12 shows the cash flow diagram.

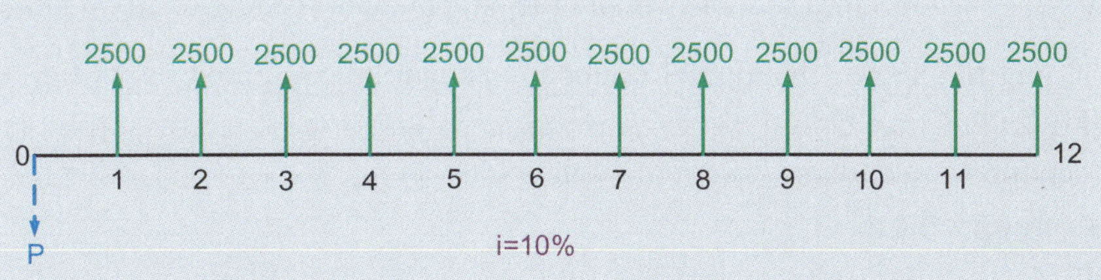

Figure 9.12 Cash flow diagram.

Using eqn (9.23), the present value of the equal payment series is given by:

$$P = R \cdot F_{RP} = R\frac{(1+i)^n - 1}{i(1+i)^n}$$

$$= 2500 \times \frac{(1+0.10)^{12} - 1}{0.10(1+0.10)^{12}} = \text{US\$ } 17\,034.23.$$

Using MS Excel: =PV(10%,12, − 2500) = US$ 17 034.23.

9.6.4 Equal Payment Series Capital Recovery Factor

Let P be the amount invested at present at an annual interest rate i which is to be recovered in n years with equal payments of R, as shown in Figure 9.13. The value of R can be determined by rewriting eqn (9.23) as:

$$R = P\frac{i(1+i)^n}{(1+i)^n - 1} = PF_{PR} \tag{9.24}$$

where $F_{PR} = \frac{i(1+i)^n}{(1+i)^n - 1}$ and is called the capital recovery factor (CRF). Therefore:

Annual uniform cost = Present value × Capital recovery factor

F_{RP} is more completely written as $F_{RP,i,n}$. It is the reciprocal of the annuity

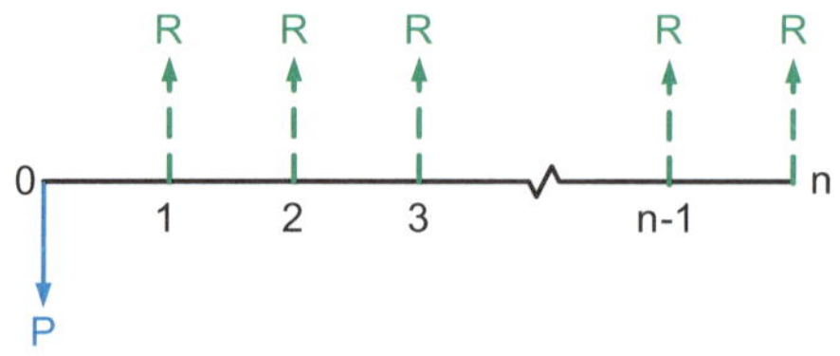

Figure 9.13 Equal payment series capital recovery factor.

present value factor. Therefore:

$$F_{PR} \cdot F_{RP} = 1 \tag{9.25}$$

Using MS Excel, the annual uniform cost can be computed by writing the expression as =PMT(i,n,P).

Example 9.14

A person borrows US$ 150 000 for constructing a room having a roof integrated with a PVT system. The loan carries an interest rate of 10% per year and is to be repaid in equal instalments over the next seven years. Calculate the amount of this annual instalment.

Solution

Given that P = US$ 150 000, $n = 7$ years and $i = 10\%$, Figure 9.14 shows the cash flow diagram.

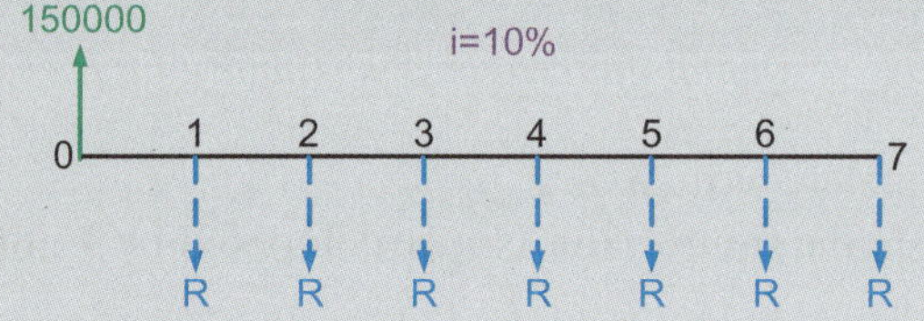

Figure 9.14 Cash flow diagram.

Using eqn (9.24) the annual instalment is given by:

$$R = P \cdot F_{PR} = P \frac{i(1+i)^n}{(1+i)^n - 1}$$

$$= 150\,000 \times \frac{0.10(1+0.10)^7}{(1+0.10)^7 - 1} = \text{US\$ } 30\,810.82.$$

Using MS Excel: =PMT(10%,7, – 150000)= US$ 30 810.82.

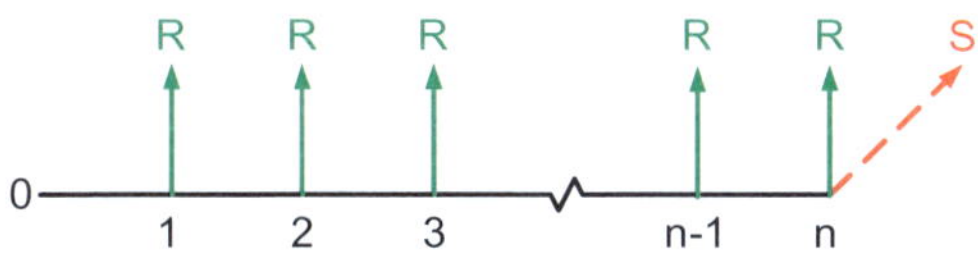

Figure 9.15 Equal payment series future value factor.

9.6.5 Equal Payment Series Future Value Factor

Let R be the amount invested at the end of every year for n years at an annual interest rate i, as shown in Figure 9.15. Then the total accumulated fund (S) at the end of duration is simply the sum of R payments multiplied by the appropriate single payment future value factor for years 1 through n. This summation is a geometric series as follows:

$$
\begin{aligned}
S &= \sum_{j=n-1}^{0} R(1+i)^j \\
&= R[(1+i)^{n-1} + (1+i)^{n-2} + (1+i)^{n-3} + \ldots + (1+i)^2 + (1+i)^1 + 1]
\end{aligned}
\tag{9.26}
$$

or

$$
S = R\frac{(1+i)^n - 1}{i} = R \cdot F_{RS} \tag{9.27}
$$

where $F_{RS} = \frac{(1+i)^n - 1}{i}$ and is called the equal payment series future value factor or annuity future value factor. Therefore:

Future value = Annual uniform cost × Annuity future value factor

F_{RS} is more completely written as $F_{RS,i,n}$.

Using MS Excel, the future value can be computed by writing the expression as =FV(i,n, – R).

Example 9.15

A person wants to purchase a solar system after eight years. For this he contributes annually US$ 1000 to his savings account at the end of each year for the eight years. If the savings account earns 10% interest annually, how much can he withdraw to purchase the solar system?

Solution

Given that R = US$ 1000, n = 8 years and i = 0.10, Figure 9.16 shows the cash flow diagram.

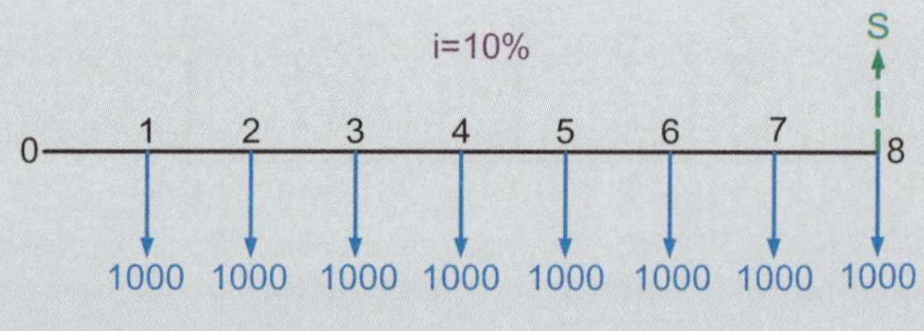

Figure 9.16 Cash flow diagram.

Using eqn (9.27), the future value equal payment series is:

$$S = R\frac{(1+i)^n - 1}{i} = 1000 \times \frac{(1+0.10)^8 - 1}{0.10} = \text{US\$ } 11\,435.89.$$

Using MS Excel: =FV(10%,8,–1000) = US$ 11 435.89.

Example 9.16

Suppose a person wants to purchase a solar system after eight years. For this he contributes annually US$ 1000 to his savings account at the beginning of each year for the eight years. If the savings account earns 10% interest annually, how much he can withdraw to purchase the solar system?

Solution

Given that $n = 8$ years and $i = 0.10$, then Figure 9.17 shows the cash flow diagram.

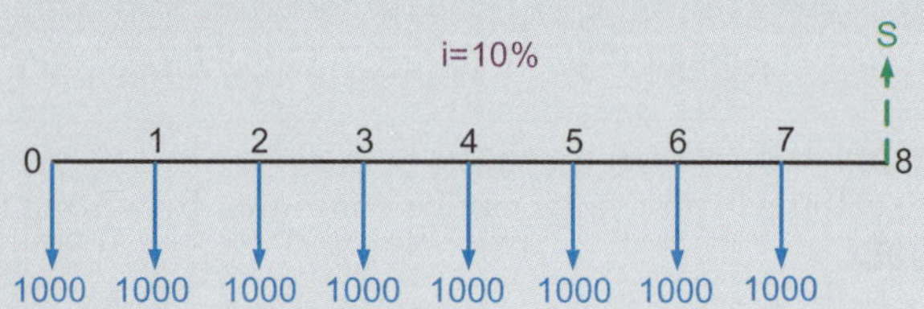

Figure 9.17 Cash flow diagram.

The future value equal payment series is given by:

$$\begin{aligned} S &= P \cdot F_{PS,10\%,8} + R \cdot F_{RS,10\%,8} - 1000 \\ &= 1000 \times (1+0.10)^8 + 1000 \times \frac{(1+0.10)^8 - 1}{0.10} - 1000 \\ &= \text{US\$ } 12\,579.48 \end{aligned}$$

Using MS Excel: =FV(10%,8, – 1000,,1) = US$ 12 579.48.

Example 9.17

A person plans to create a forborne annuity by depositing US$ 1000 at the end of each year for eight years. He wants to withdraw the money at the end of 14 years from now to buy energy-efficient domestic appliances. Find the accumulated value at the end of the 14th year, if the money is worth 10% per year.

Solution

Given that $n = 8$ years and $i = 10\%$, then Figure 9.18 shows the cash flow diagram.

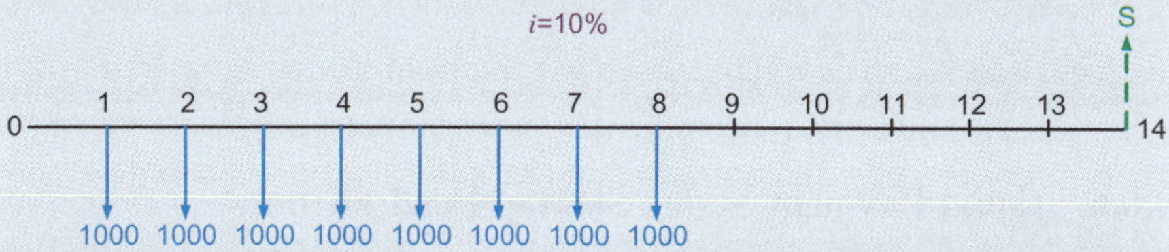

Figure 9.18 Cash flow diagram.

The future value equal payment series is given by

$$
\begin{aligned}
S &= R \times \left(F_{RS,10\%,14} - F_{RS,10\%,6}\right)\\
&= 1000 \times \left[\frac{(1+0.10)^{14} - 1}{0.10} - \frac{(1+0.10)^{6} - 1}{0.10}\right]\\
&= \text{US\$ } 20\,259.37
\end{aligned}
$$

Using MS Excel: = FV(10%,14, – 1000)-FV(10%,6, – 1000) = US$ 20 259.37.

Example 9.18

Find the equivalent value of a series of receipts, shown by the cash flow series in Figure 9.19, at the end of the fourth year. Consider the money is worth 10%.

Solution

For the given series, the present value (at 0 years) is:

$$
\begin{aligned}
P &= 100 \times F_{RP,10\%,10} + 50 \times F_{SP,10\%,7}\\
&= 100 \times \frac{(1+0.10)^{10} - 1}{0.10 \times (1+0.10)^{10}} + 50 \times \frac{1}{(1+0.10)^{7}}\\
&= \text{US\$ } 640.11
\end{aligned}
$$

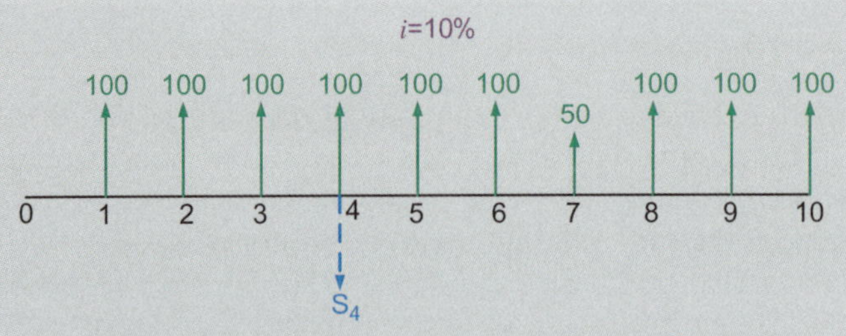

Figure 9.19 Cash flow diagram.

The value after four years is given by:

$$S_4 = P(1+i)^n = 640.11(1+0.10)^4 = \text{US\$}\,899.63$$

9.6.6 Equal Payment Series Sinking Fund Factor

Let S be the amount that has to be received at the end of n years, as shown in Figure 9.20.

For this the amount that has to be invested or deposited at an annual interest rate i at the end of each year from now can be determined by rewriting eqn (9.27) as:

$$R = S\frac{i}{(1+i)^n - 1} = P \cdot F_{SR} \quad (9.28)$$

where $F_{SR} = \frac{i}{(1+i)^n - 1}$ and is called the sinking fund factor (SFF). Therefore:

Annual uniform cost = Future value × Sinking fund factor

F_{SR} is more completely written as $F_{SR,i,n}$. It is the reciprocal of the annuity future value factor. Therefore:

$$F_{RS} \cdot F_{SR} = 1 \quad (9.29)$$

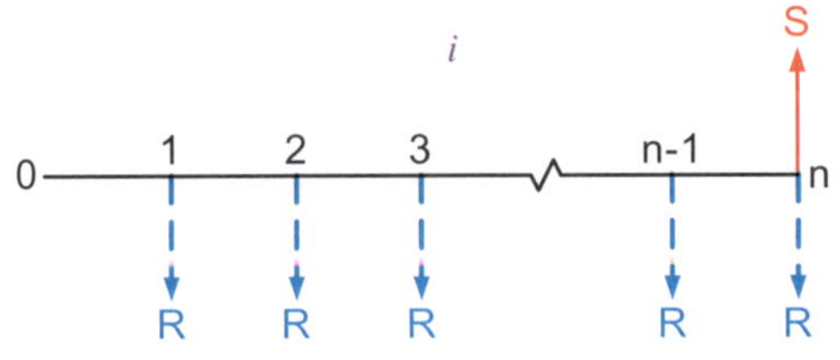

Figure 9.20 Equal payment series sinking fund factor.

Using MS Excel, the annual uniform cost can be computed by writing the expression as =PMT(i,n, – S).

Example 9.19

A person wants to purchase energy-efficient equipment after six years whose expected cost at that time is US$ 30 000. For this he opens a recurring deposit account of six years in which he deposits an amount *R* at the end of each year at an interest rate of 10%, as shown by cash flow diagram in Figure 9.21. Determine the value of *R*.

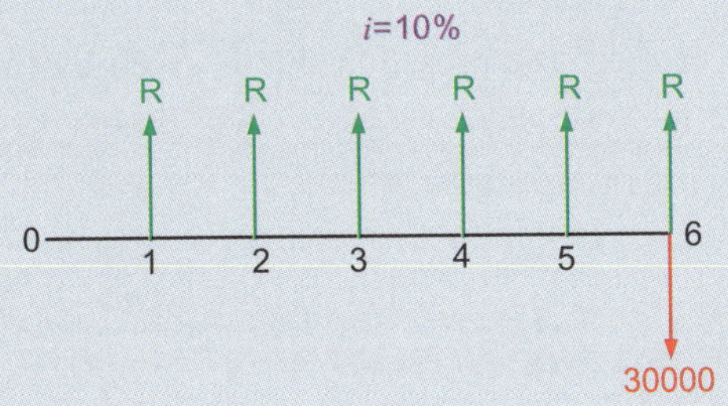

Figure 9.21 Cash flow diagram.

Solution

Given that S = US$ 30 000, i = 10% and n = 6 years, then using eqn (9.28) we have:

$$R = S \cdot F_{SR} = S \frac{i}{(1+i)^n - 1}$$

$$= 30000 \times \frac{0.10}{(1+0.10)^6 - 1} = \text{US\$ } 3888.22$$

Using MS Excel: =PMT(10%,6,,-30000)= US$ 3888.22.

9.6.7 Linear Gradient Series Present Value Factor

Sometimes the situation involves periodic payments that increase or decrease by a constant amount (G) from period to period, as shown in Figure 9.22. The gradient G can be either positive or negative. When G is positive ($G > 0$) it is called an increasing gradient series and when G is negative ($G < 0$) it is called a decreasing gradient series.

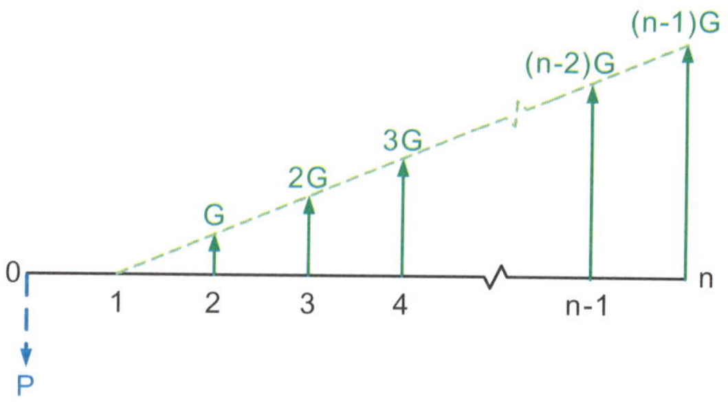

Figure 9.22 Linear gradient series present value factor.

The payment in terms of the present value is given by the following series:

$$P = \sum_{j=1}^{n} \frac{(n-1)G}{(1+i)^j}$$

$$= G\left[0 + \frac{1}{(1+i)^2} + \frac{2}{(1+i)^3} + \frac{3}{(1+i)^4} + \ldots + \frac{n-2}{(1+i)^{n-1}} + \frac{n-1}{(1+i)^n}\right] \quad (9.30)$$

$$\Rightarrow P = G\left[\frac{(1+i)^n - i \cdot n - 1}{i^2(1+i)^n}\right] = G\left[\frac{(1+i)^n - 1}{i^2(1+i)^n} - \frac{n}{i(1+i)^n}\right] \quad (9.31)$$

$$= G \cdot F_{GP}$$

where $F_{GP} = \left[\frac{(1+i)^n - i \cdot n - 1}{i^2(1+i)^n}\right]$ is called the gradient series present value factor and is more completely written as $F_{GP,i,n}$. Hence:

Present value = Linear gradient value × gradient series present value factor

Example 9.20

A firm installs a PV system on the rooftops of buildings that has a useful life of 15 years. It is estimated that the maintenance costs will increase from the first year of US$ 100 at the rate of US$ 25 per year over the lifetime. Assume the maintenance costs occur at the end of each year. The firm sets up a maintenance account that earns 10% annual interest through which all the maintenance expenses will be paid out. How much has the firm to deposit now?

Solution

Given that $n = 15$ years, $R = \text{US\$ } 100$, $G = \text{US\$ } 25$ and $i = 10\%$, Figure 9.23 shows the cash flow diagram.

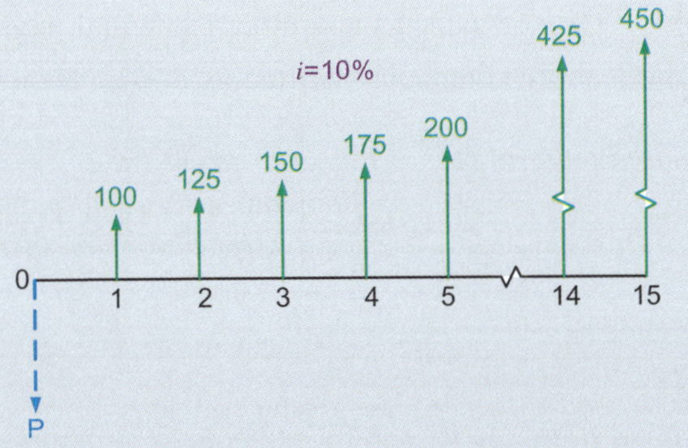

Figure 9.23 Cash flow diagram.

The present value equal payment series is given by:

$$P = R \cdot F_{RP} + G \cdot F_{GP} = R \cdot \frac{(1+i)^n - 1}{i(1+i)^n} + G \cdot \frac{(1+i)^n - i \cdot n - 1}{i^2(1+i)^n}$$
$$= 100 \times \frac{(1+0.10)^{15} - 1}{0.10(1+0.10)^{15}} + 25 \times \frac{(1+0.10)^{15} - 0.10 \times 15 - 1}{(0.10)^2(1+0.10)^{15}}$$
$$= \text{US\$ } 760.61 + 1003.80 = \text{US\$ } 1764.41$$

9.6.8 Gradient to Equal Payment Series Conversion Factor

As depicted in Figure 9.24, the gradient series may be converted into an equivalent equal payment series using the empirical relationship obtained by substituting the value of P from eqn (9.31) into eqn (9.28):

$$R = G\left[\frac{(1+i)^n - i \cdot n - 1}{i^2(1+i)^n} \cdot \frac{i(1+i)^n}{(1+i)^n - 1}\right] \quad (9.32)$$

or

$$R = G\left[\frac{(1+i)^n - i \cdot n - 1}{i\{(1+i)^n - 1\}}\right] = G\left[\frac{1}{i} - \frac{n}{(1+i)^n - 1}\right] = G \cdot F_{GR} \quad (9.33)$$

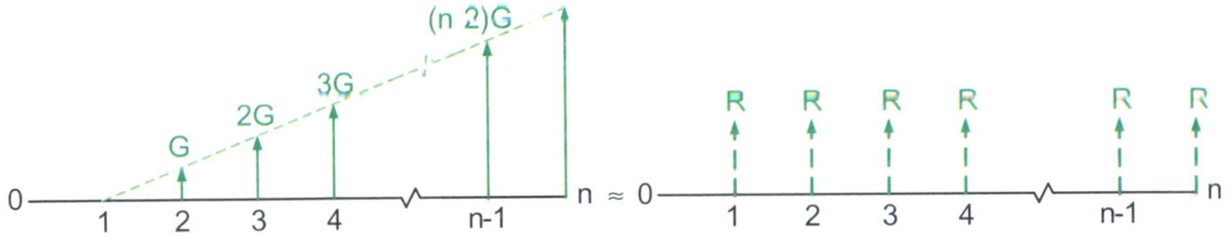

Figure 9.24 Converting a gradient series into an equivalent uniform series.

where $F_{GR} = \left[\frac{1}{i} - \frac{n}{(1+i)^n - 1}\right]$ is called the gradient to equal payment series conversion factor and is more completely written as $F_{GR,i,n}$. Hence:

$$\textit{Annualised uniform cost} = \textit{Linear gradient value} \times \left(\begin{array}{c}\textit{Gradient to equal payment}\\ \textit{series conversion factor}\end{array}\right)$$

Example 9.21

The two cash flows in Figure 9.25 are equivalent at an interest rate of 10%, compounded annually. Determine the value of R.

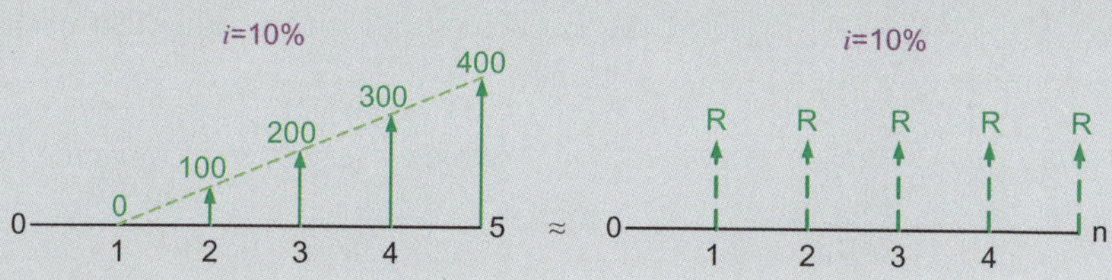

Figure 9.25 Cash flow diagram.

Solution

Given that $G =$ US\$ 100, $i = 10\%$ and $n = 5$ years, then using eqn (9.33) we have:

$$\begin{aligned} R &= G \cdot F_{GR} = G\left[\frac{(1+i)^n - i \cdot n - 1}{i\{(1+i)^n - 1\}}\right] \\ &= 100 \times \left[\frac{(1+0.10)^5 - 0.10 \times 5 - 1}{0.10 \times \{(1+0.10)^5 - 1\}}\right] \\ &= \text{US\$ } 181.01 \end{aligned}$$

9.6.9 Linear Gradient Series Future Value Factor

Figure 9.26 shows a gradient series whose equivalent future value S can be obtained by substituting the value of R from eqn (9.33) into eqn (9.27):

$$S = G\left[\frac{(1+i)^n - i \cdot n - 1}{i\{(1+i)^n - 1\}} \cdot \frac{(1+i)^n - 1}{i}\right] \tag{9.34}$$

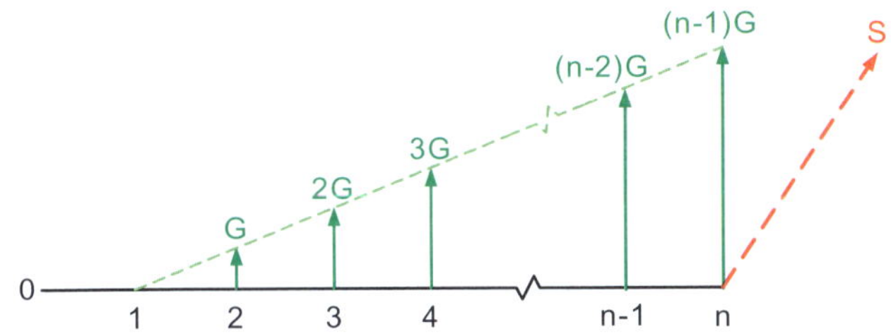

Figure 9.26 Linear gradient series future value factor.

$$\Rightarrow S = G\left[\frac{(1+i)^n - i \cdot n - 1}{i^2}\right] = G \cdot F_{GS} \quad (9.35)$$

where $F_{GS} = \frac{(1+i)^n - i \cdot n - 1}{i^2}$ is called the gradient series future value factor and is more completely written as $F_{GS,i,n}$. Hence:

Future value = Linear gradient value × Gradient series future value factor

9.7 Cost Comparison with Equal Duration

If two energy-efficient systems have the same service life but different associated costs, they can be compared by submitting all the costs in terms of either the present value or the future value. Consider the following solved example.

Example 9.22

A customer has to choose between two available PVT systems. The associated costs and the relevant details of the systems are as follows:

Detail	*System A*	*System B*
Initial costs (US$)	30000	20000
Annual maintenance costs (US$)	2000	4000
Overhaul costs at the end of the third year (US$)	–	2000
Salvage value (US$)	4000	6000
Life of the system (years)	5	5
Benefit from quality control as a uniform end-of-year amount per year (US$)	1000	–

Determine the system which is more economical to the customer. Consider the money is worth 10% per year.

Solution

The cash flow diagrams for both systems are shown in Figure 9.27.

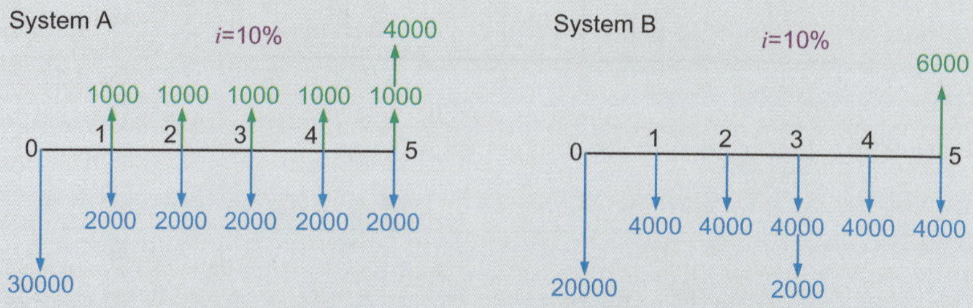

Figure 9.27 Cash flow diagrams: *left*, System A; right, System B.

(a) Combining all the associated costs for System A, its present value is:

$$\begin{aligned} P_{\rm A} &= 30000 + (2000 - 1000)F_{RP,10\%,5} - 4000F_{SP,10\%,5} \\ &= 30000 + (2000 - 1000) \times \frac{(1+0.10)^5 - 1}{0.10 \times (1+0.10)^5} - 4000 \times \frac{1}{(1+0.10)^5} \\ &= 30000 + 3790.79 - 2483.69 = \text{US\$}\,31\,307.10 \end{aligned}$$

(b) Combining all the associated costs for System B, its present value is

$$\begin{aligned} P_{\rm B} &= 20000 + 4000F_{RP,10\%,5} + 2000F_{SP,10\%,3} - 6000F_{SP,10\%,5} \\ &= 20000 + 4000 \times \frac{(1+0.10)^5 - 1}{0.10 \times (1+0.10)^5} + 2000 \times \frac{1}{(1+0.10)^3} - 6000 \times \frac{1}{(1+0.10)^5} \\ &= 20000 + 15163.15 + 1502.63 - 3725.53 = \text{US\$}\,32\,940.25 \end{aligned}$$

The above calculation shows that the customer should purchase System A, because it is more economical than System B.

9.8 Cost Comparison with Unequal Duration

If two energy-efficient systems have different service lives and different associated costs, they can be compared by making the common denominator of the service time and submitting all the costs in terms of some common base. Consider the following solved example.

Example 9.23

A customer has to choose between two available PVT systems. The associated costs and the relevant details of the systems are as follows:

Detail	*System A*	*System B*
Initial costs (US$)	20000	30000
Annual maintenance costs (US$)	4000	3000
Salvage value (US$)	500	1500
Life of the system (years)	2	3

Determine the system which is more economical to the customer. Consider the money is worth 10% per year.

Solution

(a) Single present value method. As the systems have different service lives, consider the systems are needed for six years and that, during this period, whenever the system service life is over it is replaced by a new one. Cash flow diagrams for both the systems under these circumstances are shown in Figure 9.28.

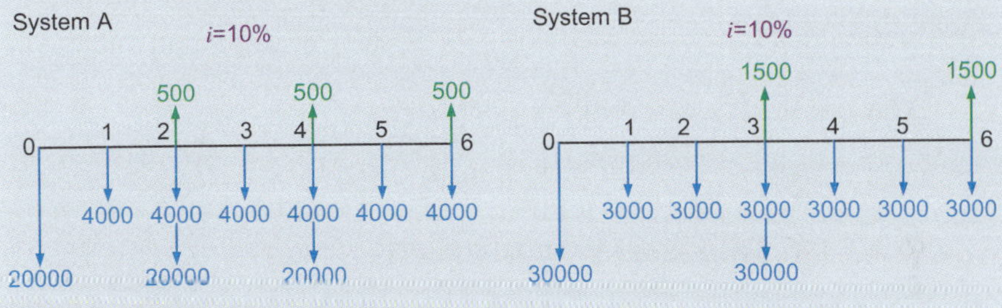

Figure 9.28 Cash flow diagram: *left*, System A; *right*, System B.

Combining all the associated costs for System A, its present value is:

$$P_A = 20000 + 4000\, F_{RP,10\%,6} + (20000 - 500)F_{SP,10\%,2} + (20000 - 500)F_{SP,10\%,4} - 500F_{SP,10\%,6}$$

$$= 20000 + 4000 \times \frac{(1+0.10)^6 - 1}{0.10 \times (1+0.10)^6} + \frac{19500}{(1+0.10)^2} + \frac{19500}{(1+0.10)^4} - \frac{500}{(1+0.10)^6}$$

$$= 20000 + 17421.04 + 16115.70 + 13318.76 - 282.24 = \text{US\$ } 66573.27$$

Combining all the associated costs for System B, its present value is:

$$\begin{aligned} P_{\mathrm{B}} &= 30000 + 3000F_{RP,10\%,6} + (30000 - 1500)F_{SP,10\%,3} - 1500F_{SP,10\%,6} \\ &= 30000 + 3000 \times \frac{(1+0.10)^6 - 1}{0.10 \times (1+0.10)^6} + \frac{28500}{(1+0.10)^3} - \frac{1500}{(1+0.10)^6} \\ &= 30000 + 13065.78 + 21412.47 - 846.71 = \text{US\$}\,63631.54 \end{aligned}$$

The above calculation shows that the customer should purchase System B.

(b) Equal payment annual cost method. As the systems have different service lives, the systems may be compared on the basis of the annual cost method. Cash flow diagrams for both systems under these circumstances are shown in Figure 9.29.

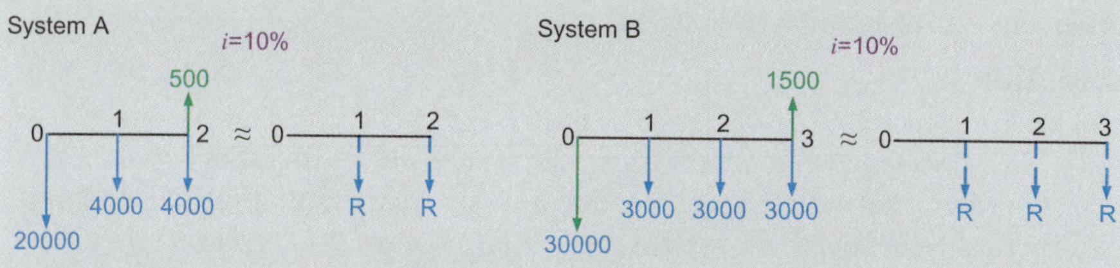

Figure 9.29 Cash flow diagrams: *left*, System A; *right*, System B.

Combining all the associated costs for System A in the equal payment annual cost we have:

$$\begin{aligned} R_{\mathrm{A}} &= 20000F_{PR,10\%,2} + 4000 - 500F_{SR,10\%,2} \\ &= 20000 \times \frac{0.10(1+0.10)^2}{(1+0.10)^2 - 1} + 4000 - 500 \times \frac{0.10}{(1+0.10)^2 - 1} \\ &= 11523.81 - 238.1 = \text{US\$}\,15\,285.71 \end{aligned}$$

Combining all the associated costs for System B in the equal payment annual cost we have:

$$\begin{aligned} R_{\mathrm{B}} &= 30000F_{PR,10\%,3} + 3000 - 1500F_{SR,10\%,3} \\ &= 30000 \times \frac{0.10(1+0.10)^3}{(1+0.10)^3 - 1} + 3000 - 1500 \times \frac{0.10}{(1+0.10)^3 - 1} \\ &= 12063.44 + 3000 - 453.17 = \text{US\$}\,14\,610.27 \end{aligned}$$

The above calculation shows that the customer should purchase System B.

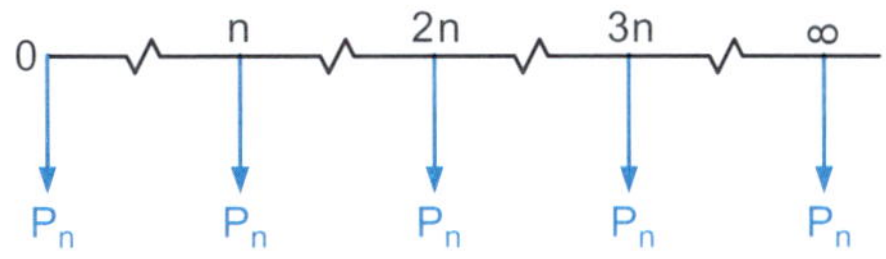

Figure 9.30 Infinite cash flow series.

9.9 Cost Comparison using Capitalized Cost

Capitalized cost is the present value on an infinite time basis. It is used for the analysis of systems whose service life is perpetual or the horizon is extremely long (say, 30 years or more), such as bridges, hydroelectric dams, waterway constructions, *etc*. Consider the cash flow series shown in Figure 9.30.

For a system costing P_n with a service life of n years, the present value replacing out to infinity is given by:

$$K = P_n \sum_{x=0}^{\infty} \frac{1}{(1+i)^{xn}} = P_n \left[1 + \frac{1}{(1+i)^n} + \frac{1}{(1+i)^{2n}} + \frac{1}{(1+i)^{3n}} + \ldots \right] \quad (9.36)$$

$$= P_n \cdot \frac{1}{1-(1+i)^{-n}} = P_n \cdot \frac{(1+i)^n}{(1+i)^n - 1} = P_n \cdot F_{PK,i,n} \quad (9.37)$$

where K is called the capitalized cost and $F_{PK,i,n} = \frac{(1+i)^n}{(1+i)^n-1}$ is called the capitalized cost factor and is used to convert the present value to the capitalized cost.

Capitalized cost = Present value basis for n years duration × Capitalized cost factor

From eqn (9.37) the capitalized cost factor is:

$$F_{PK,i,n} = \frac{(1+i)^n}{(1+i)^n - 1} \quad (9.38)$$

and from eqn (9.24) the capital recovery factor (CRF) is:

$$F_{PR,i,n} = \frac{i(1+i)^n}{(1+i)^n - 1} \quad (9.39)$$

Combining eqns (9.38) and (9.39) we have:

$$F_{PR,i,n} = i \cdot F_{PK,i,n} \quad (9.40)$$

Capital recovery factor = Worth of money × Capitalized cost factor

$$\text{Similarly, } R = i \cdot K : \quad (9.41)$$

Annualized uniform cost = Worth of money × Capitalized cost

As $K = i/R$ (where i is a constant), Example 9.23 can also be solved using the capitalized cost method.

9.10 Payback Period

The payback period refers to the period of time required for the return on an investment to repay the sum of the original investment. If C_i is the initial cost on the investment and F is the annual cash flow, then the payback period is given by:

$$\text{Payback period} = \frac{\text{Initial investment}}{\text{Annual cash flow}} = \frac{C_i}{F} \tag{9.42}$$

A short payback period indicates that the investment provides revenues early in its life sufficient to cover the initial outlay. Thus, an investment with a short payback period can be viewed as having a higher degree of liquidity than one with a longer payback period.

Example 9.24

A PVT system initially costs US$ 450 000 and the expected uniform annual benefit is US$ 56 250. Determine the payback period of the system.

Solution

Given that C_i = US$ 450000 and R = US$ 56250, then:

$$\text{Payback period} = \frac{C_i}{R} = \frac{450000}{56250} = 8 \text{ years}$$

Payout time without interest. This is the length of time required to recover the initial cost of the investment from the net cash flow produced by that investment for an interest rate of zero. If C_i is the initial cost on the investment and F_t is the net cash flow in period t, then the payback period is defined as the smallest value of n that satisfies the equation:

$$\sum_{t=0}^{n} F_t \geq 0 \tag{9.43}$$

Payout time with interest. This is the length of time required until the investment's equivalent receipts exceed the equivalent capital outlays. If C_i is the initial cost of the investment and F_t is the net cash flow in period t, then the payback period with an annual interest rate i is the smallest value of n' that satisfies the equation:

$$\sum_{t=0}^{n'} F_t(1+t)^{-t} \geq 0 \tag{9.44}$$

Example 9.25

A PVT system initially costs US$ 1000 and the expected annual benefits by the end of each year are US$ 400, 300, 200, 200, 165 and 165 for six consecutive years. Determine the payback period (a) without interest and (b) with an interest rate of 10%.

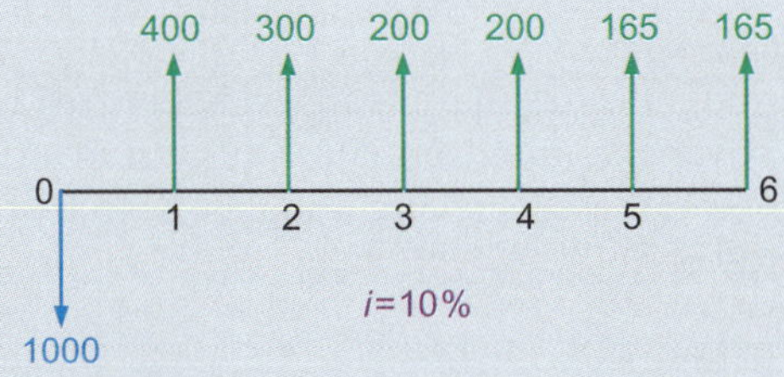

Figure 9.31 Cash flow diagram.

Solution

Figure 9.31 shows the cash flow diagram.

(a) The payback period without interest is:

$$-1000 + 400 + 300 + 200 + 200 + 165 + 165 \geq 0$$

In this example the shortest payback period is between 3 and 4 years, *i.e.* 3.5 years.

(b) The payback period with interest is:

$$-1000 + 400F_{SP,10\%,1} + 300F_{SP,10\%,2} + 200F_{SP,10\%,3} + 200F_{SP,10\%,4} + 165F_{SP,10\%,5} + 165F_{SP,10\%,6} \geq 0$$

In this example the shortest payback period is 5 years.

Example 9.26

A solar water heating system purchased for US$ 18 000 is expected to generate annual revenues of US$ 3000 and have a salvage value of US$ 3000 at any time during 10 years of anticipated ownership. If a 15% per year required return is imposed on the purchase, compute the payback period.

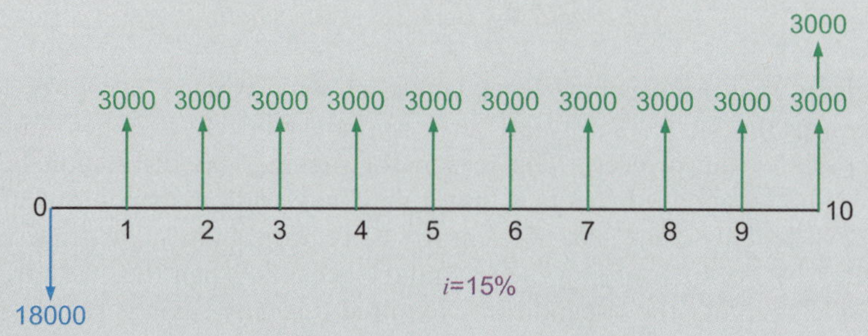

Figure 9.32 Cash flow diagram.

Solution

Figure 9.32 shows the cash flow diagram.

(a) The payback period with interest is:

$$-18000 + \sum_{n=1}^{10} 3000F_{SP,15\%,n} + 3000F_{SP,15\%,10} \geq 0$$

Solving the above expression we have $n = 14.21$ years, which is beyond the life of the system; hence the system is not suitable from an economical point of view.

However, if the interest rate is 12%, the invested amount would have been recovered in 10 years and the system would be neither at profit nor at loss. Any interest rate below 12% will lead to a benefit.

9.11 Cost–Benefit Analysis

Cost–benefit analysis (CBA) is a method of comparing the investment costs of a project with its expected benefits. The analysis is based on the ratio of benefits-to-costs with a particular project, indicating the total return expected per unit of money spent. Benefits and costs are often expressed in money terms, and are adjusted for the time value of the money, so that all flows of benefits and flows of project costs over time are expressed on a common basis in terms of their present value. This analysis generally excludes consideration of factors that are not measured ultimately in economic terms. The CBA is frequently used by government agencies for projects whose benefits are reaped by the common public and the costs are incurred by the government, to evaluate the desirability of a given intervention.

To begin a benefit–cost analysis, all the benefits and disbenefits are identified, and their difference gives the net benefit of the project as follows:

$$Net\ benefits = Benefits - Disbenefits$$

In identifying the benefits, we classify each as a primary benefit (a benefit directly attributable to the project) or a secondary benefit (a benefit indirectly attributable to the project). The reason for making this distinction between primary and secondary benefits is that it may make our analysis more efficient. If primary benefits alone are sufficient to justify project costs, we can save time and effort by not quantifying the secondary benefits. Then the costs are identified by classifying the expenditures required and any savings (or revenues), including both capital investments and annual operating costs. Therefore, the costs are given by:

$$Costs = Capital\ costs + Operating\ and\ maintenance\ costs - Revenues \quad (9.45)$$

For a given benefit–cost profile, let B and C be the present worth of the benefits and costs, defined by:

$$B = \sum_{j=0}^{n} \frac{B_j}{(1+i)^j} \quad (9.46)$$

$$C = \sum_{j=0}^{n} \frac{C_j}{(1+i)^j} \quad (9.47)$$

where B_j and C_j are the benefits and costs at the end of period of j, respectively; n is the project life and i is the worth of the money.

The present value cost C may be split into two components: (i) the initial capital expenditure and (ii) the annual costs accrued in each successive period. Let us assume that a series of initial investments is required during the first m periods, while annual operating and maintenance costs accrue in each following period. Then the equivalent present worth for each component is:

$$C = \sum_{j=0}^{m} \frac{C_j}{(1+i)^j} + \sum_{j=m+1}^{n} \frac{C_j}{(1+i)^j} \quad (9.48)$$

The conventional benefit-to-cost ratio is calculated as:

$$B/C = \frac{\text{Benefits} - \text{Disbenefits}}{\text{Capital costs} + \text{Operating and maintenance costs} - \text{Revenues}} \quad (9.49)$$

$$\Rightarrow B/C = \frac{\sum_{j=0}^{n} \frac{B_j}{(1+i)^j}}{\sum_{j=0}^{m} \frac{C_j}{(1+i)^j} + \sum_{j=m+1}^{n} \frac{C_j}{(1+i)^j}} \tag{9.50}$$

If B/C is greater than unity the project may be accepted, otherwise it should be rejected.

The advantage of CBA is that it gives planners a method to try and "put all relevant costs and benefits on a common temporal footing" in order to help people make informed decisions. It provides people with an understanding as to the economic costs of decisions, and allows arguments to be made for or against a change based upon economic considerations. The downside of CBA is that it is used to measure effects that may be difficult, or improper, to measure in financial terms.

The modified B/C ratio, which is gaining support, includes operation and maintenance costs in the numerator and treats them in a manner similar to disbenefits:

$$\text{Modified } B/C = \frac{\text{Benefits} - \text{Disbenefits} - \text{Operating and maintenance costs}}{\text{Capital costs} - \text{Revenues}} \tag{9.51}$$

$$\Rightarrow \text{Modified } B/C = \frac{\sum_{j=0}^{n} \frac{B_j}{(1+i)^j} - \sum_{j=m+1}^{n} \frac{C_j}{(1+i)^j}}{\sum_{j=0}^{m} \frac{C_j}{(1+i)^j}} \tag{9.52}$$

Example 9.27

A non-profit organization is contemplating an investment of US$ 100 000 to install a solar water-heating system. The grant would extend over a 10-year period and would create an estimated saving of US$ 20 000 per year. The organization uses a rate of return of 6% per year on all grant investments. An estimated US$ 4000 a year would have to be released from other sources for expenses. In order to make this programme successful a US$ 2000 per year operating expense will be incurred by the organization from its regular O&M budget. Use the following analysis methods to determine whether the programme is justified over a 10-year period: (a) conventional B/C analysis; (b) modified B/C analysis.

Solution

Benefits = US$ 20 000 per year, investment cost = $100\,000 F_{PR,6\%,10}$ = US$ 13587 per year, O&M cost = US$ 2000 per year, disbenefits = US$ 4000 per year.

(a) Conventional method:

$$B/C = \frac{20000 - 4000}{13587 + 2000} = 1.026$$

(b) Modified method:

$$\text{Modified } B/C = \frac{20000 - 4000 - 2000}{13587} = 1.03$$

The project is justified as B/C is greater than unity.

9.12 Internal Rate of Return

The internal rate of return (IRR) is a widely accepted discounted measure of investment worth and is used as an index of profitability for the appraisal of projects. The IRR is defined as the rate of interest that equates the present value of a series of cash flows to zero. Mathematically, the internal rate of return is the interest rate i_{IRR} that satisfies the equation:

$$\text{NAV}(i_{\text{IRR}}) = \sum_{j=0}^{n} \frac{B_j - C_j}{(1 + i_{\text{IRR}})^j} = 0 \tag{9.53}$$

Alternatively, the internal rate of return is the interest rate that causes the discounted present value of the benefits in a cash flow to be equal to the present value of the costs, *i.e.*:

$$\sum_{j=0}^{n} \frac{B_j}{(1 + i_{\text{IRR}})^j} = \sum_{j=0}^{n} \frac{C_j}{(1 + i_{\text{IRR}})^j} \tag{9.54}$$

Multiplying both sides of eqn (9.54) by $(1 + i_{\text{IRR}})^j$ we have:

$$(1 + i_{\text{IRR}})^n \times \text{NAV}(i_{\text{IRR}}) = (1 + i_{\text{IRR}})^n \times \sum_{j=0}^{n} \left\{ \frac{B_j - C_j}{(1 + i_{\text{IRR}})^j} \right\}$$

$$\text{NAV}(i_{\text{IRR}})(1 + i_{\text{IRR}})^n - \sum_{j=0}^{n} (B_j - C_j)(1 + i_{\text{IRR}})^{n-j} = 0 \tag{9.55}$$

The IRR is widely used in the appraisal of projects because (i) the IRR on a project is its expected rate of return; (ii) it employs a percentage rate of return as the decision variable which suits the banking community; and (iii) for situations in which the IRR exceeds the cost of the funds used to finance the project, a surplus would remain after paying for the capital.

The IRR can be calculated using the iterative procedure at a trial. Using a guess rate of interest, the NPV of all disbursements and receipts is calculated. If the calculated value of the NPV is positive, then the receipts from the investments are worth more than the disbursements of the investments and the actual value of the IRR would be more than the trial rate. On the other hand, if the NPV is negative the actual value of the IRR would be less than the trial rate of interest. Adjust the estimate of the trial rate of return accordingly and recalculate the NAV, until at a predicted value of $i\ (=i_1)$ it is found such that the NAV is positive and at the next higher value of $i\ (=i_2)$ the NAV is negative. Solve for the value of the IRR by interpolation using the following relationship:

$$\mathrm{IRR} = \left(\frac{i_2 - i_1}{\mathrm{NPV}_1 - \mathrm{NPV}_2}\right) i_1 \times \mathrm{NPV}_1 \tag{9.56}$$

An important aspect of the iterative method of computing the IRR is making the initial estimate. If the initial estimate is too far from the actual value of the IRR, a large number of trials will have to be made to obtain the two consecutive values of interest rate (i_1 and i_2) to permit accurate interpolation. It should be noted that the initial estimate of the IRR will always be somewhat in error and several iterations will normally be required to determine i_1 and i_2. A simple approach for making a guess of the first trial rate of return is given below.

The NPV of a capital investment C_i resulting in uniform net annual cash flows of amount A for an infinite time horizon can be expressed as:

$$\mathrm{NPV} = -C_i + \left[\frac{A}{(1+i)} + \frac{A}{(1+i)^2} + \frac{A}{(1+i)^3} + \ldots\right]$$

$$\Rightarrow \mathrm{NPV} = -C_i + \frac{A}{(1+i)}\left[\frac{1}{1-(1+i)^{-1}}\right] = -C_i + \frac{A}{i} \tag{9.57}$$

Since NPV = 0 at i = IRR, we have:

$$-C_i + \frac{A}{i} = 0$$

$$\Rightarrow i = \frac{A}{C_i} \tag{9.58}$$

In actual practice, for investment projects with a finite life the IRR will be less than A/C_i. However, to begin with, for cases with uniform periodic cash

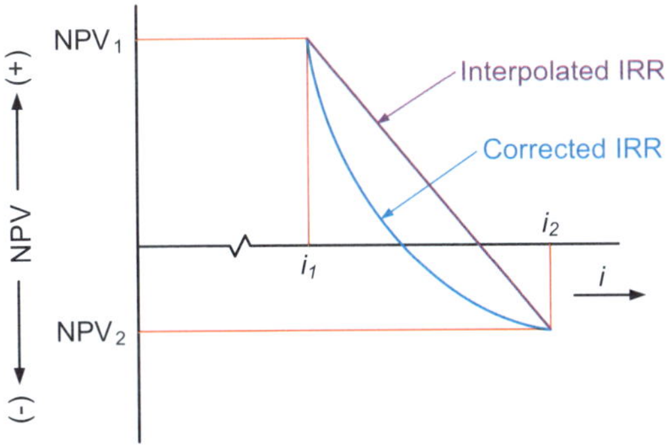

Figure 9.33 Interpolation of the IRR.

flows the figure A/C_i or a value close to it may be used as the trial rate of return in the iterative procedure used for determining the IRR.

The above interpolation between two consecutive values of interest rates that bracket the IRR always overestimates its true value. This is because the linear interpolation technique makes an implicit assumption that between the two interest rates i_1 and i_2 the IRR changes, following a straight line, whereas the true value of the IRR follows a concave curvilinear function between the two values. However, the error introduced by interpolation is usually very small. Referring to Figure 9.33, the true value of the IRR is that value of i for which the NPV(i) function intersects the horizontal axis, whereas the interpolated value of the IRR is somewhat higher than the true value. Obviously, the interpolation error would become less and less as the incremental change in the trial values of i used in iteration is made smaller and smaller.

Example 9.28

Calculate the internal rate of return for the investment in a heat exchanger which will cost US$ 500 000 to purchase and install, will last 10 years and will result in fuel savings of US$ 145 000 per year. Assume that the salvage value of the heat exchanger at the end of 10 years is negligible.

Solution

Let the first guess at the value of the IRR be 25%, then:

$$\text{NPV at } 25\% = 145000\left[\frac{(1+0.25)^{10}-1}{0.25(1+0.25)^{10}}\right] - 500000 = \text{US\$ } 17722$$

Since the NPV at 25% is positive, the IRR will be greater than 25%. If the next trial value is chosen at 30%, then:

$$\text{NPV at } 30\% = 145000\left[\frac{(1+0.3)^{10}-1}{0.3(1+0.3)^{10}}\right] - 500000 = \text{US\$ } 51\,724$$

Obviously, the true IRR lies between 25% and 30%. By interpolating between the two, the IRR can be estimated as:

$$\text{IRR} = \left(\frac{0.3-0.25}{17722+51724}\right)0.25 \times 17722 = 0.2627 = 26.27\%$$

A better estimate of the true IRR may be obtained by using smaller incremental changes in the interest rate.

Multiple values of the IRR. The NPV of a set of cash receipts and disbursements can be expressed as an *n*th degree polynomial of the form:

$$\text{NPV}(i_{\text{IRR}}) = 0 = F_0 + F_1x + F_2x^2 + F_3x^3 + \ldots + F_nx^n \qquad (9.59)$$

where $x = \frac{1}{1+i}$ and the F_i symbols are coefficients of the n terms in the polynomial.

For the above polynomial, in principle there may be n different roots or values of x which satisfy eqn (9.59). Thus, it is possible that the NPV(i) function crosses the i axis several times, as shown in Figure 9.34. It may be noted that a unique value of the IRR of special interest in applying the IRR method and consequently multiple values of the IRR essentially hinder the application of the IRR criterion. In fact, in the case with multiple IRR values, use of the IRR criterion is normally not recommended.

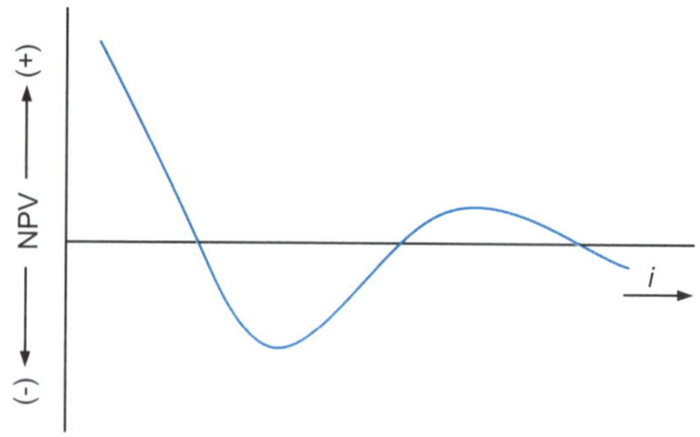

Figure 9.34 Multiple values of the IRR.

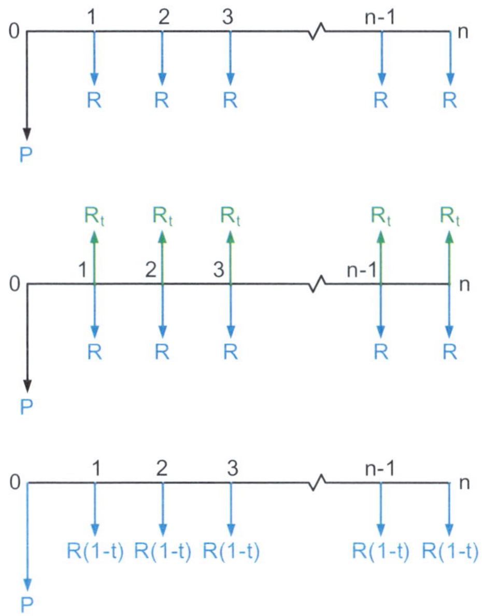

Figure 9.35 Cash flow chart: *top*, before taxes; *middle*, after taxes as given; *bottom*, after taxes, simplified.

9.13 Cost Comparison after Taxes

(a) *Without depreciation*. If i is the rate of return before taxes and t is the tax rate, then the rate of return (r) after taxes will be expressed as:

$$r = i(1 - t)$$

Before taxes, the flow chart is shown in Figure 9.35 (*top*). In this case, an investment that compounds at a rate i and first cost P can be expressed in terms of the annualized uniform costs, R, as:

$$P = R\,F_{RP,i,n} = R\frac{(1+i)^n - 1}{i(1+i)^n}$$

After taxes, the flow chart is shown by Figure 9.35 (*middle*) and simplified in Figure 9.35 (*bottom*). In this case, it will compound at a rate r and an expression for first cost P can be written as:

$$P = \frac{R(1-t)}{(1+r)} + \frac{R(1-t)}{(1+r)^2} + \frac{R(1-t)}{(1+r)^3} + \cdots\cdots\cdots + \frac{R(1-t)}{(1+r)^n}$$
$$= R(1-t)\frac{(1+r)^n - 1}{r(1+r)^n} = R(1-t)\,F_{RP,r,n}$$

From the above equation, an expression for the annualized uniform cost after taxes can be expressed as:

$$R = \frac{P}{(1-t)} \cdot \frac{r(1+r)^n}{(1+r)^n - 1} = \frac{P}{(1-t)} F_{PR,r,n}$$

Also, $R = \frac{S}{(1-t)} \cdot \frac{r}{(1+r)^n - 1} = \frac{S}{(1-t)} F_{SR,r,n}$
Here, $S = P(1+r)^n$.

(b) *With depreciation*. Consider C_i is the initial cost of an article that lasts n years with a salvage value C_{sal}. The depreciable cost will be:

$$C_d = C_i - C_{sal}$$

There is no tax consideration at the time of purchase of an article.
Let $D_{f1}, D_{f2}, D_{f3}, \ldots, D_{f(n-1)}, D_{fn}$ be the fractional depreciation for each year; then the time–cost diagram without tax can be shown by Figure 9.36 (*top*). Here:

$$\begin{aligned} C_d &= D_{f1}C_d + D_{f2}C_d + \ldots + D_{fn}C_d \\ &= D_f C_d + D_f C_d + \ldots + D_f C_d \text{ (for same fractional depreciation)} \end{aligned}$$

Also $D_{f1} = D_{f2} = D_{f3} = \ldots = D_{fn} = \frac{1}{n}$
Now, the taxable base is reduced to $D_{f1}C_d$ and a saving or reduction in taxes amounting to $D_{f1}C_d t$ is realized. In this case the time–cost diagram can be shown by Figure 9.36 (*bottom*). By using the above time – cost diagram and assuming $D_{f1} = D_{f2} = D_{f3} = \ldots = D_{fn} = \frac{1}{n}$, the present value

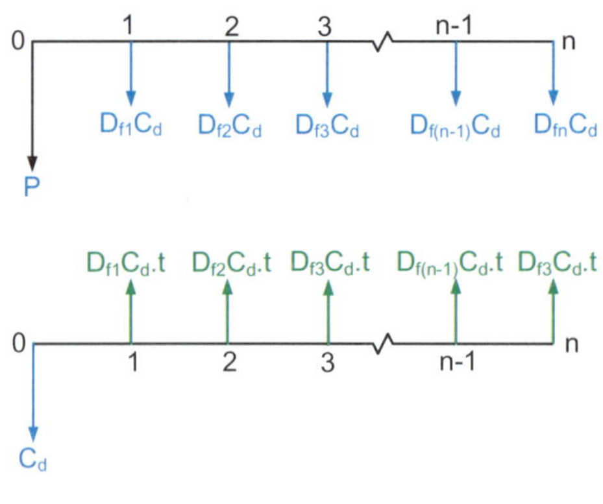

Figure 9.36 *Top*, Cash flow chart at the time of purchase; *bottom*, cash flow chart after time t from the date of purchase.

is:

$$\begin{aligned}P &= C_d - D_f C_d t\left[\frac{1}{1+r} + \frac{1}{(1+r)^2} + \ldots + \frac{1}{(1+r)^n}\right]\\ &= C_d - C_d t\frac{1}{n}\left[\frac{1}{1+r} + \frac{1}{(1+r)^2} + \ldots + \frac{1}{(1+r)^n}\right]\\ &= C_d - C_d t F_{SLP,r,n}\\ &\Rightarrow P = C_d(1 - t F_{SLP,r,n})\end{aligned}$$

After knowing an expression for the present value, the expressions for the uniform annualized cost (R) and capitalized cost (K) as follows:

$$R = P F_{PR,r,n} = C_d(1 - t F_{SLP,r,n})F_{PR,r,n}$$

and

$$K = P F_{PK,r,n} = C_d(1 - t F_{SLP,r,n})F_{PK,r,n}$$

Here, it is important to note that an expression for the conversion factor from straight-line depreciation to the present value with tax is given by:

$$F_{SLP,r,n} = \frac{1}{n}\cdot\frac{(1+r)^n - 1}{r(1+r)^n}$$

Example 9.29

Calculate the conversion factor from straight-line depreciation to the present value with and without tax for a 10% rate of return and a 30% tax for a period of five years.

Solution

The conversion factor from straight-line depreciation to the present value is given by:

(a) Without tax:

$$F_{SLP,i,n} = \frac{1}{5}\frac{(1+0.10)^5 - 1}{0.10(1+0.10)^5} = 0.7581$$

(b) With tax, here replacing $r = i(1 - t) = 0.10(1 - 0.30) = 0.07$ we have:

$$F_{SLP,r,n} = \frac{1}{5}\frac{(1+0.07)^5 - 1}{0.07(1+0.07)^5} = 0.82$$

Example 9.30

Derive an expression for the present value, P, for a uniform end-of-year cost, R, occurring simultaneously with the tax instant t.

Solution

The uniform end-of-year cost after taxes at the end of each year = $R(1 - t)$. The $R(1 - t)$ will be same up to n years at end of each year. In order to obtain the expression for P, convert the uniform annualized cost $R(1 - t)$ into the present value as:

$$P = R(1 - t)F_{RP,r,n}$$

$$\Rightarrow P = R(1 - t)\frac{(1 + r)^n - 1}{r(1 + r)^n}$$

Example 9.31

Derive an expression for the present value P for a given salvage C_{sal} value at end of the nth year by treating an expense as a nondepreciable first cost.

Solution

The C_{sal}, a nondepreciable expense, is invested now and fully recovered at the end of n years with no tax consideration. The present value is:

$$P = C_{sal} - \frac{C_{sal}}{(1 + r)^n} = C_{sal}\frac{(1 + r)^n - 1}{(1 + r)^n}$$

9.14 Case Studies with BIPV and BIPVT Systems

There are numerous costs associated with acquiring, operating, maintaining and disposing of a system. Figure 9.37 shows the line diagram of the cash flows at different intervals of time for a BIPVT system. Here for the life cycle cost (LCC) assessment, all relevant present and future costs associated with the system are combined in the present value. The purpose is to estimate the overall cost of project alternatives and to select the design that ensures the facility will provide the lowest overall cost of ownership.

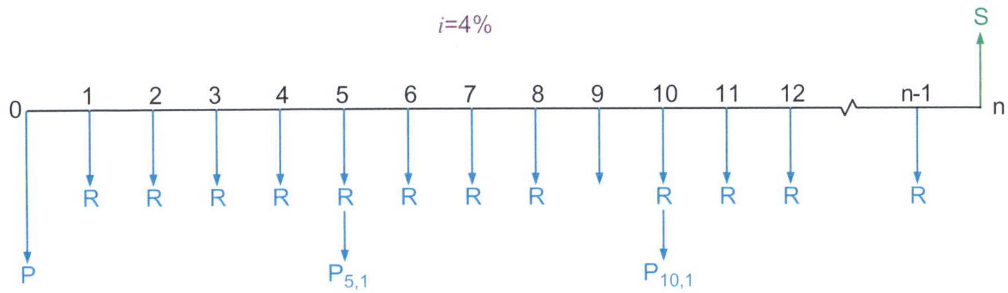

Figure 9.37 Cash flow diagram for BIPV and BIPVT systems.

9.14.1 Estimation of Cost

The LCC assessment of a BIPVT system takes into account the following.

(i) Initial costs (P_I). This is the sum of the costs involved in the system technologies (such as the BIPVT system, charge controller, battery bank, inverter, *etc.*), utility interconnection costs and associated costs for building permits.

(ii) Maintenance and repair costs. The supplier quoted and published estimating guides sometimes provide information on maintenance and repair costs. However, there is great variation in these costs for BIPVT systems of the same type and life. For the present analysis the annual maintenance and repair cost of a BIPVT system is considered as M = US$ 150. Then the maintenance and repair costs in terms of the present value are given by

$$P_{MR} = M \times \left[\frac{(i+1)^n - 1}{i(i+1)^n}\right] \tag{9.60}$$

(iii) Replacement costs. This is the cost incurred during the replacement of batteries and components. The number and timing of capital replacements of components of a BIPVT system depend on the life of the components and system. We assume that the service life of a battery is five years and that all other replacements are made at the end of five years. If R_5, R_{10}, R_{15}, … R_n is the replacement cost incurred for batteries and other components made every five years, then the net replacement cost in terms of present values is:

$$P_R = R_5 \times \left[\frac{1}{(i+1)^5}\right] + R_{10} \times \left[\frac{1}{(i+1)^{10}}\right] + R_{15} \times \left[\frac{1}{(i+1)^{15}}\right] + \ldots + R_n \times \left[\frac{1}{(i+1)^n}\right] \tag{9.61}$$

(iv) Salvage value. This is defined as the costs incurred in demolition and disposal of the system. If S is the salvage value at the end of the system,

Table 9.1 Capital costs of various items of a BIPVT system in US$.

PV technology	*Initial cost of the system*	*Maintenance and repair cost in terms of present value*	*Replacement cost in terms of present value*	*Salvage value in terms of present value*	*Net present value*
c-Si	28989.49	2593.80	1913.01	104.86	33391.45
p-Si	24208.84	2593.80	1661.44	87.72	28376.37
r-Si	19796.79	2343.31	1229.60	134.16	23235.53
a-Si	8251.36	2038.55	545.85	158.39	10677.37
CdTe	10640.37	1667.76	522.41	279.72	12550.82
CIGS	13015.86	667.77	0	644.52	13039.12

then the net salvage value in terms of the present value is:

$$P_S = S \times \left[\frac{1}{(i+1)^n}\right] \tag{9.62}$$

Table 9.1 gives the initial capital costs, maintenance and repair costs, replacement costs and salvage value in terms of the present value for different technologies of a BIPVT system.

9.14.2 Modelling for Annualized Uniform Cost

Thus, the overall LCC of a BIPVT system in terms of the present value is given by:

$$\begin{aligned}\text{LCC} &= P_I + P_{MR} + P_R - P_S \\ &= P_I + R\left[\frac{(i+1)^n - 1}{i(i+1)^n}\right] + \frac{R_5}{(i+1)^5} + \frac{R_{10}}{(i+1)^{10}} + \ldots + \frac{R_n}{(i+1)^n} - \frac{S}{(i+1)^n}\end{aligned} \tag{9.63}$$

The capital recovery factor over the lifetime is:

$$\text{CRF} = \frac{i(i+1)^n}{(i+1)^n - 1} \tag{9.64}$$

Therefore the annualized uniform cost (unacost) is given by:

$$\text{Unacost} = \text{LCC} \times \text{CRF} \tag{9.65}$$

The cost per unit of electricity generated by a BIPVT system is determined as the ratio of the annualized uniform cost to the electrical energy consumed by the load in a year.

9.14.3 Methodology

The approach uses an individual BIPVT system to compute the energy, exergy and outlet air temperatures using basic thermal modelling and heat transfer relations. Figure 9.38 shows the annual exergy output from BIPV and BIPVT systems using different solar cells. The net mass flow-rate of air inside the duct is considered as $1\,kg\,s^{-1}$ and the velocity of air inside the duct is $3.2\,m\,s^{-1}$. The annual interest rate usually offered by government sectors in India to promote the use of renewable energy applications is 4%.

The annualized uniform cost (unacost) is obtained from eqns (9.46) and (9.47). Figure 9.39 shows the annualized uniform cost of BIPVT systems with different technologies. It also shows the annualized uniform cost of BIPV systems for comparison. Dividing the annualized uniform cost by the net exergy output gives the cost per unit power generation. Figure 9.40 shows the cost of unit power generation from BIPVT systems with different technologies. It also shows the cost of unit power generation from BIPV systems for comparison.

Table 9.2 summarizes the annual electrical output, thermal outputs, overall thermal and exergy efficiencies, annualized uniform cost and cost per unit power generation for BIPVT systems with different solar cell technologies.

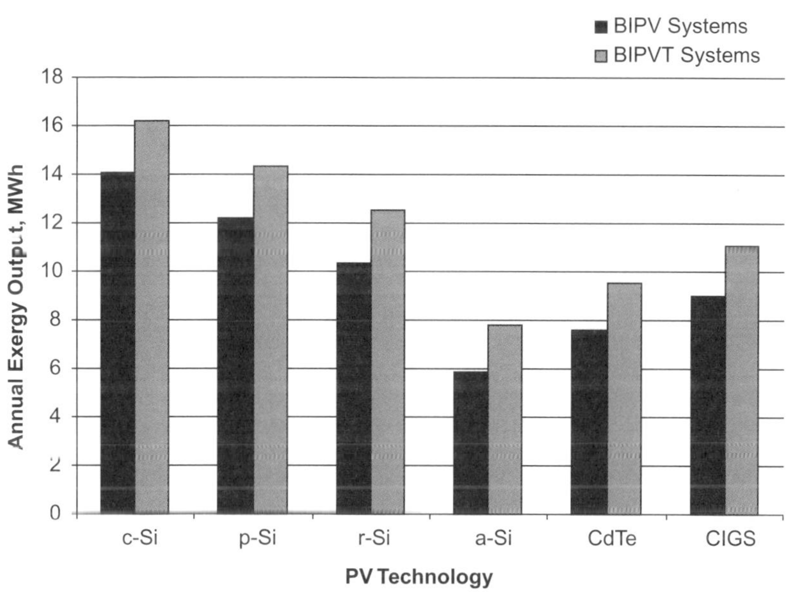

Figure 9.38 Annual exergy output from BIPV and BIPVT systems with different solar cells.

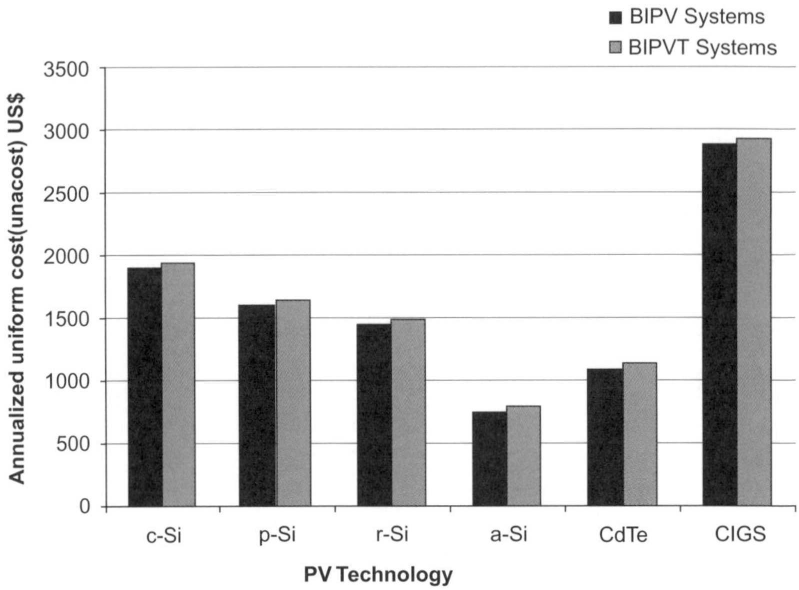

Figure 9.39 Annualized uniform cost of BIPV and BIPVT systems with different solar cells.

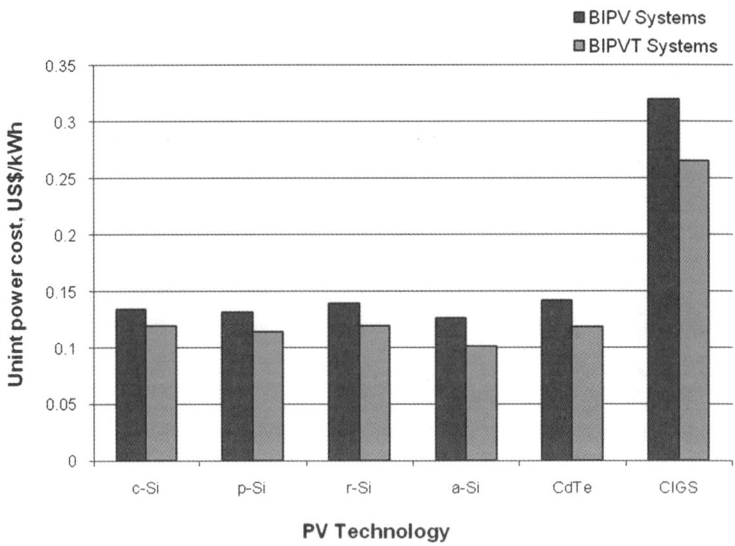

Figure 9.40 Unit power generation cost of BIPV and BIPVT systems using different solar cells.

Table 9.2 The annual outputs, overall efficiencies and annualized cost of power generation for BIPVT systems.

PV technology	*Electrical output from BIPVT system (kWh)*	*Thermal output from BIPVT system (kWh)*	*Overall thermal efficiency of BIPVT system (%)*	*Overall exergy efficiency of BIPVT system (%)*	*Annualized cost of BIPVT system (US$)*	*Cost per unit power generation for BIPVT system (US$ kWh^{-1})*	*Cost per unit power generation for BIPV system (US$ kWh^{-1})*	*Reduction in cost (%)*
c-Si	1[illegible] 131	16 764	51.99	14.91	1931.03	0.1190	0.1338	12.44
p-Si	1[illegible] 141	17 535	47.88	13.19	1641.01	0.1143	0.1309	14.51
r-Si	11 179	18 306	43.84	11.50	1487.35	0.1189	0.1395	17.35
a-Si	6066	20 615	33.54	7.13	785.66	0.1009	0.1260	24.97
CdTe	7958	19 845	37.41	8.75	1128.83	0.1182	0.1416	19.77
CIGS	9578	19 074	40.65	10.13	2928.94	0.2654	0.3195	20.41

9.14.4 Results and Discussion

Figure 9.38 shows that the annual exergy output is a maximum (16 225 kWh) in the case of a c-Si BIPVT system and a minimum (7790 kWh) in the case of an a-Si BIPVT system. Moreover, the annual exergy output of BIPVT systems are 15–30% higher than that of similar BIPV systems. This is because the solar cells of a BIPVT system are cooled, which helps in producing higher electrical energy than a BIPV system. Also, the thermal gain from a BIPVT system helps in space heating.

Figure 9.39 shows that the copper indium gallium selenide (CIGS) BIPVT system has a relatively higher annualized cost (US$ 2928.94) owing to the higher initial investment made for a short lifespan, whereas the a-Si BIPVT system has a relatively lower annualized cost (US$ 785.66). The figure also shows that the annualized costs of such systems are 2–7% higher than similar BIPV systems. Figure 9.40 shows that the use of BIPVT systems reduces the unit power generation cost by 12–25% compared with similar BIPV systems. The figure also shows that the unit power generation cost of the CIGS BIPVT system is highest (US$ 0.2654 per kW) while that of the a-Si system is lowest (US$ 0.1009 per kW). Thus, from an economic point of view the a-Si BIPVT systems are more suitable for a rooftop. Also, the cost of unit power generation from the a-Si BIPVT system is quite close to the cost of unit power generation through the conventional grid. Therefore the application of such systems in residential and commercial buildings will help in reducing greenhouse gas emissions, which is necessary for sustainable development.

Further Reading

1. G. N. Tiwari, *Solar Energy Fundamentals, Design, Modelling and Applications*, Narosa, New Delhi, 2002.
2. G. N. Tiwari and S. Dubey, *Fundamentals of Photovoltaic Modules and their Applications*, RSC, Cambridge, 2010.
3. B. M. Ayyub, *Risk Analysis in Engineering and Economics*, Chapman & Hall/CRC, Boca Raton, FL, 2003.
4. G. J. Thuesen and W. J. Fabrycky, *Engineering Economy*, Prentice Hall of India, New Delhi, 2002.
5. C. S. Park, *Fundamentals of Engineering Economics*, Pearson, Upper Saddle River, NJ, 2004.
6. B. Agrawal and G. N. Tiwari, Life cycle cost assessment of building integrated photovoltaic thermal (BIPVT) systems, *Energy Build.*, 2010, in press.

Appendix

A1 Conversion of Units

A1.1 Length, SI Unit: Metre (m)

1 centimetre (cgs unit)	cm	$= 10^{-2}$ m
1 Ångström	Å	$= 10^{-10}$ m
1 micron	μ	$= 1\ \mu m = 10^{-6}$ m
1 millimicron	mμ	$= 1$ nm $= 10^{-9}$ m
1 inch	in	$= 2.54 \times 10^{-2}$ m
1 foot	ft	$= 12$ in $= 0.3048$ m
1 yard	yd	$= 3$ ft $= 0.9144$ m
1 mile	mi	$= 1760$ yd $= 1609.344$ m
1 astronomical unit	AU	$= 1.496 \times 10^{11}$ m

A1.2 Area, SI Unit: Square Metre (m^2)

1 acre		$= 4046.856\ m^2$
1 are	a	$= 100\ m^2$
1 hectare	ha	$= 10^4\ m^2$
1 square foot	sq ft	$= 9.290 \times 10^{-2}\ m^2$
1 square inch	sq in	$= 6.453 \times 10^{-4}\ m^2$
1 square yard	sq yd	$= 0.836\ m^2$

A1.3 Volume, SI Unit: Cubic Metre (m^3)

1 litre	L, l	$= 10^{-3}\ m^3$
1 lambda	λ	$= 1\ \mu L = 10^{-9}\ m^3$

RSC Energy Series No. 4
Building Integrated Photovoltaic Thermal Systems: For Sustainable Developments
By Basant Agrawal and G. N. Tiwari
© B. Agrawal and G. N. Tiwari 2011
Published by the Royal Society of Chemistry, www.rsc.org

(*Continued*).

1 barrel (US)	bl, bbl	$= 42\,\text{gal} = 158.987 \times 10^{-3}\,\text{m}^3$
1 gallon (US fluid)	gal (US)	$= 3.785 \times 10^{-3}\,\text{m}^3$
1 gallon (UK)	gal (UK)	$= 4.546 \times 10^{-3}\,\text{m}^3$
1 fluid ounce	fl oz	$= 28.41 \times 10^{-6}\,\text{m}^3$
1 cubic foot	cu ft	$= 0.028\ 317\,\text{m}^3$
1 cubic inch	cu in	$= 16.387 \times 10^{-6}\,\text{m}^3$
1 cubic yard	cu yd	$= 0.764\,55\,\text{m}^3$

A1.4 Mass, SI Unit: Kilogram (kg)

1 gram (cgs unit)	g	$= 10^{-3}\,\text{kg}$
1 electron mass (au)	m_e	$= 9.109\ 39 \times 10^{-31}\,\text{kg}$
1 tonne (mts unit)	t	$= 1\,\text{Mg} = 10^3\,\text{kg}$
1 pound (avoirdupois)	lb	$= 0.4536\,\text{kg}$
1 pound (metric)		$= 0.5\,\text{kg}$
1 ounce (avoirdupois)	oz	$= 0.028\ 35\,\text{kg}$
1 ounce (troy)	oz (trou)	$= 0.031\ 14\,\text{kg}$
1 grain	gr	$= 64.8 \times 10^{-6}\,\text{kg}$
1 slug		$= 14.5939\,\text{kg}$
1 quintal		$= 100\,\text{kg}$
1 microgram	µg	$= 10^{-6}\,\text{g} = 10^{-3}\,\text{kg}$
1 carat (metric)	ct	$= 200\,\text{mg} = 0.2 \times 10^{-3}\,\text{kg}$

A1.5 Density, SI Unit: Kilogram ($\text{kg}\,\text{m}^{-3}$)

1 gram per cubic centimetre	$\text{g}\,\text{cm}^{-3}$	$= 1000\,\text{kg}\,\text{m}^{-3}$
1 gram per millilitre	$\text{g}\,\text{mL}^{-1}$	$= 1000\,\text{kg}\,\text{m}^{-3}$
1 pound per cubic feet	$\text{lb}\,\text{ft}^{-3}$	$= 16.0185\,\text{kg}\,\text{m}^{-3}$
1 pound per cubic inch	$\text{lb}\,\text{in}^{-3}$	$= 2.768\,\text{kg}\,\text{m}^{-3}$

A1.6 Pressure, SI Unit: Pascal (Pa)

1 pascal	Pa	$= 1\,\text{N}\,\text{m}^{-2} = 1\,\text{kg}\,\text{m}^{-1}\,\text{s}^{-2}$
1 atmosphere	atm	$= 1.01325 \times 10^5\,\text{Pa}$
1 bar	bar	$= 10^5\,\text{Pa}$
1 pound per square inch	psi	$= 6.895 \times 10^3\,\text{Pa}$
1 pound per square foot	psf	$= 47.88\,\text{Pa}$
1 inch of mercury	inHg	$= 3.386 \times 10^3\,\text{Pa}$

A1.7 Velocity, SI Unit: Metre Per Second ($m\,s^{-1}$)

1 kilometre per hour	$km\,h^{-1}$	$= 0.2778\,m\,s^{-1}$
1 mile per hour	mph	$= 0.447\,m\,s^{-1}$
1 mile per second	mps	$= 1609.344\,m\,s^{-1}$
1 foot per hour	fph	$= 8.4667 \times 10^{-5}\,m\,s^{-1}$
1 foot per second	fps	$= 0.3048\,m\,s^{-1}$
speed of light in vacuum	c	$= 299\,792\,458\,m\,s^{-1}$
speed of sound in air	s	$= 340\,m\,s^{-1}$ to $295\,m\,s^{-1}$ at aircraft altitudes

A1.8 Acceleration, SI Unit: Metre Per Square Second ($m\,s^{-2}$)

1 galileo	Gal	$= 10^{-2}\,m\,s^{-2}$
foot per hour per second	$fph\,s^{-1}$	$= 8.4667 \times 10^{-5}\,m\,s^{-2}$
foot per minute per second	$fpm\,s^{-1}$	$= 5.08 \times 10^{-3}\,m\,s^{-2}$
foot per second squared	fps^2	$= 0.3048\,m\,s^{-2}$
knot per second	$kn\,s^{-1}$	$= 0.5144\,m\,s^{-2}$
inch per second squared	ips^2	$= 2.54 \times 10^{-2}\,m\,s^{-2}$
mile per hour per second	$mph\,s^{-1}$	$= 0.4470\,m\,s^{-2}$
mile per second squared	mps^2	$= 1.609 \times 10^3\,m\,s^{-2}$
standard gravity	g	$= 9.806\,65\,m\,s^{-2}$

A1.9 Force, SI Unit: Newton (N)

1 newton	N	$= 1\,kg\,m\,s^{-2}$
1 dyne (CGS)	dyn	$= 10^{-5}\,N$
1 kilogram force	kgf	$= 9.8066\,N$
1 atomic unit of force		$= 8.2387 \times 10^{-8}\,N$
1 pound (or pound force)	lb	$= 4.448\,N$
1 ounce force	ozf	$= 0.278\,N$
1 ton force	tnf	$= 8.896 \times 10^3\,N$

A1.10 Discharge, SI Unit: Cubic Metre Per Kilogram ($m^3\,kg^{-1}$)

1 cubic foot per minute	CFM	$= 1\,ft^3\,min^{-1} = 4.719 \times 10^{-4}\,m^3\,s^{-1}$
1 cubic foot per second	$ft^3\,s^{-1}$	$= 0.028\,317\,m^3\,s^{-1}$
1 cubic inch per minute	$in^3\,min^{-1}$	$= 2.731 \times 10^{-7}\,m^3\,s^{-1}$
1 cubic inch per second	$in^3\,s^{-1}$	$= 1.6387 \times 10^{-5}\,m^3\,s^{-1}$
1 litre per minute	LPM	$= 1.6 \times 10^{-5}\,m^3\,s^{-1}$
1 litre per second	LPS	$= 10^{-3}\,m^3\,s^{-1}$

A1.11 Energy, SI Unit: Joule (J)

1 joule	J	$= 1\,\text{kg}\,\text{m}^2\,\text{s}^{-2}$
1 erg (cgs unit)	Erg	$= 1\,\text{g}\,\text{cm}^2\,\text{s}^{-2} = 10^{-7}\,\text{J}$
1 rydberg	Ry	$= 2.179\,87 \times 10^{-18}\,\text{J}$
1 electronvolt	EV	$= 1.602 \times 10^{-19}\,\text{J}$
1 calorie	Cal	$= 4.18\,\text{J}$
1 British thermal unit	Btu	$= 1055.06\,\text{J}$
1 megajoule	MJ	$= 10^6\,\text{J}$
1 gegajoule	GJ	$= 10^9\,\text{J}$
1 terajoule	TJ	$= 10^{12}\,\text{J}$
1 exajoule	EJ	$= 10^{18}\,\text{J}$

A1.12 Power, SI Unit: Watt (W)

1 watt	W	$= 1\,\text{kg}\,\text{m}^2\,\text{s}^{-3} = 1\,\text{J}\,\text{s}^{-1} = 1\,\text{N}\,\text{m}\,\text{s}^{-1}$
1 horsepower	Hp	$= 745.7\,\text{W}$
1 BTU per minute	$\text{Btu}_{\text{IT}}\,\text{min}^{-1}$	$= 17.584\,\text{W}$
1 BTU per second	$\text{Btu}_{\text{IT}}\,\text{s}^{-1}$	$= 1.055 \times 10^3\,\text{W}$
1 lusec	Lusec	$= 1.333 \times 10^{-4}\,\text{W}$
1 poncelet	P	$= 980.665\,\text{W}$
1 ton of air conditioning		$= 3504\,\text{W}$
1 kilowatt	KW	$= 1000\,\text{W}$
1gegawatt	GW	$= 10^9\,\text{W}$

A1.13 Specific heat, SI Unit: Joules Per Kilogram Per Kelvin ($\text{J}\,\text{kg}^{-1}\,\text{K}^{-1}$)

1 joule per kilogram per kelvin	$\text{J}\,\text{kg}^{-1}\,\text{K}^{-1}$	$= \text{J}\,\text{kg}^{-1}\,{}^\circ\text{C}^{-1}$
1 calorie per gram per °C	$\text{cal}\,\text{g}^{-1}\,{}^\circ\text{C}^{-1}$	$= 4186.8\,\text{J}\,\text{kg}^{-1}\,\text{K}^{-1}$
1 kilocalorie per kilogram per °C	$\text{kcal}\,\text{kg}^{-1}\,{}^\circ\text{C}^{-1}$	$= 4186.8\,\text{J}\,\text{kg}^{-1}\,\text{K}^{-1}$
1 BTU per pound per °F	$\text{Btu}\,\text{lb}^{-1}\,\text{F}^{-1}$	$= 4186.8\,\text{J}\,\text{kg}^{-1}\,\text{K}^{-1}$

A1.14 Temperature, SI Unit: kelvin (K)

1 degree Celsius	°C	$[\text{K}] = [{}^\circ\text{C}] + 273.15$
1 degree Fahrenheit	°F	$[\text{K}] = ([{}^\circ\text{F}] + 459.67) \times \frac{5}{9}$
degree Rankine	°R, °Ra	$[\text{K}] = [{}^\circ\text{R}] \times \frac{5}{9}$
degree Delisle	°De	$[\text{K}] = (373.15 - [{}^\circ\text{De}]) \times \frac{2}{3}$
degree Newton	°N	$[\text{K}] = [{}^\circ\text{N}] \times \frac{100}{33} + 273.15$

A.1.15 Dynamic viscosity, SI Unit: Pascal Second (Pa s)

1 pascal second	Pa s	$= 1\ \mathrm{kg\ m^{-1}\ s^{-1}}$
1 poise	P	$= 0.1$ Pa s
1 centipoise	cP	$= 1 \times 10^{-3}$ Pa s
1 pound per foot hour	$\mathrm{lb\ ft^{-1}\ h^{-1}}$	$= 4.134 \times 10^{-4}$ Pa s
1 pound per foot second	$\mathrm{lb\ ft^{-1}\ s^{-1}}$	$= 1.488$ Pa s

A1.16 Kinematic viscosity, SI Unit: Square Metre Per Second ($m^2\,s^{-1}$)

1 stokes	St	$= 10^{-4}\ \mathrm{m^2\ s^{-1}}$
1 square foot per second	$\mathrm{ft^2\ s^{-1}}$	$= 0.0929\ \mathrm{m^2\ s^{-1}}$

A1.17 Electric current, SI Unit: Ampere (A)

1 esu per second, statampere	$\mathrm{esu\ s^{-1}}$	$= 3.3356 \times 10^{-10}$ A
1 electromagnetic unit, abampere	abamp	$= 10$ A

A1.18 Electric charge, SI Unit: Coulomb (C)

1 coulomb	C	$= 1$ A s
1 franklin	Fr, statC	$= 3.3356 \times 10^{-10}$ C
1 electrostatic unit (statcoulomb)	esu, statC	$= 3.3356 \times 10^{-10}$ C
1 electromagnetic unit, abcoulomb	abC	$= 10$ C
1 proton charge (au)	e	$= 1.6022 \times 10^{-19}$ C
		$= 4.8032 \times 10^{-10}$ Fr
1 faraday	F	$= 96\,485.34$ C

A1.19 Radioactivity, SI Unit: Becquerel (Bq)

1 becquerel	Bq	= per second (s^{-1})
1 curie	Ci	$= 3.7 \times 10^{10}$ Bq

A1.20 Angle, SI Unit: Radian (rad)

1 degree	°	$= \mathrm{rad} \times \frac{2\pi}{360} = 17.453 \times 10^{-3}$ rad
1 arc minute	′	$=$ degree/60 $- 0.290\,888 \times 10^{-3}$ rad
1 arc second	″	$=$ minute/60 $= 4.84814 \times 10^{-6}$ rad

(*Continued*).

1 hour angle		= 4 minutes (time)
1 gradian	grad	$= \text{rad} \times \frac{2\pi}{400} = 15.708 \times 10^{-3}$, rad
1 octant		= 0.7854 rad
1 quadrant		= 1.5708 rad
1 sextant		= 1.0472 rad

A1.21 Time, SI Unit: Second (s)

1 atomic unit of time	au	$= 2.418 \times 10^{-17}$ s
1 minute	min	= 60 s
1 hour	h	= 3600 s
1 day (solar day)	d	= 86 400 s
1 week	wk	= 604 800
1 year (calendar year)	a	$= 365.25$ days $= 31.557 \times 10^{6}$ s
1 svedberg	Sv	$= 10^{-13}$ s
1 milliday	md	= 1/1000 d = 86.6 s
1 fortnight		= 2 weeks
1 month (30 days)	mo	$= 2592 \times 10^{3}$ s
1 decade		= 10 years
1 century		= 100 years
1 millennium		= 1000 years

A1.22 Heat flux, SI Unit: Watt Per Square Metre (W/m^2)

1 BTU hour per square foot	Btu h ft^{-2}	= 3.1552 W m^{-2}
1 BTU minute square foot	Btu min ft^{-2}	= 189.275 W m^{-2}
1 horsepower (metric) per square foot	hp ft^{-2}	= 7916.843 W m^{-2}
1 kilocalorie per hour square foot	kcal h^{-1} ft^{-2}	= 12.518 W m^{-2}
1 kilocalorie per hour square metre	kcal h^{-1} m^{-2}	= 1.163 W m^{-2}
1 kilowatt per square metre	kW m^{-2}	= 1000 W m^{-2}
1 watt per square inch	W in^{-2}	= 1550 W m^{-2}

A1.23 Heat transfer coefficient, SI Unit: Watt Per Square Metre Per Kelvin (W m^{-2} K^{-1})

1 watt per square metre per °C	W m^{-2} °C^{-1}	= 1 W m^{-2} K^{-1}
1 joule per second per square metre per K	J s^{-1} m^{-2} °C^{-1}	= 1 W m^{-2} K^{-1}
1 kilocalorie per hour per square metre per °C	kcal h^{-1} m^{-2} °C^{-1}	= 1.163 W m^{-2} K^{-1}

(*Continued*).

1 kilocalorie per hour square foot per °C	kcal h^{-1} ft^{-2} $°C^{-1}$	= 12.518 W m^{-2} K^{-1}
1 calorie per second per square centimetre per °C	cal s^{-1} cm^{-2} $°C^{-1}$	= 41 868 W m^{-2} K^{-1}
1 BTU per hour per square foot per °F	Btu h^{-1} ft^{-2} $°F^{-1}$	= 5.678 W m^{-2} K^{-1}

A1.24 Thermal conductivity, SI Unit: Watt Per Metre Per Kelvin (W m^{-1} K^{-1})

1 watt per centimetre per °C	W cm^{-1} $°C^{-1}$	= 100 W m^{-1} K^{-1}
1 kilowatt per metre per kelvin	kW m^{-1} K^{-1}	= 1000 W m^{-1} K^{-1}
1 calorie per second per centimetre per °C	cal s^{-1} cm^{-1} $°C^{-1}$	= 418.4 W m^{-1} K^{-1}
1 BTU per hour per foot per °F	Btu h^{-1} ft^{-1} $°F^{-1}$	= 1.7296 W m^{-1} K^{-1}

A1.25 Illumination, SI Unit: Lux (lx)

1 lumen per square metre	lm m^{-2}	= 1 lx
1 foot candle		= 10.7639 lx
1 flame		= 43.0556 lx
1 lumen per square foot	lm ft^{-2}	= 1 foot candle = 10.7639 lx
1 watt per square metre	W m^{-2}	= 100 lx
1 nox		= 0.001 lx
1 phot		= 10 000 lx

A1.26 Concentration, SI Unit: Kilogram Per Cubic Metre (kg m^{-3})

1 gram per litre	g L^{-1}	= 1 kg m^{-3}
1 pound per cubic foot	lb ft^{-3}	= 6.236 kg m^{-3}

A1.27 Diffusivity, SI Unit: Square Metre Per Second (m^2 s^{-1})

1 square foot per hour	ft^2 h^{-1}	= 25.81×10^{-6} m^2 s^{-1}
1 square foot per second	ft^2 s^{-1}	= 0.0929 m^2 s^{-1}

A1.28 Frequency, SI Unit: hertz (Hz) = 1 cycle s^{-1}

A2 Average Annual Daily Ground Solar Energy Available During Clear Days on the Horizontal Surface

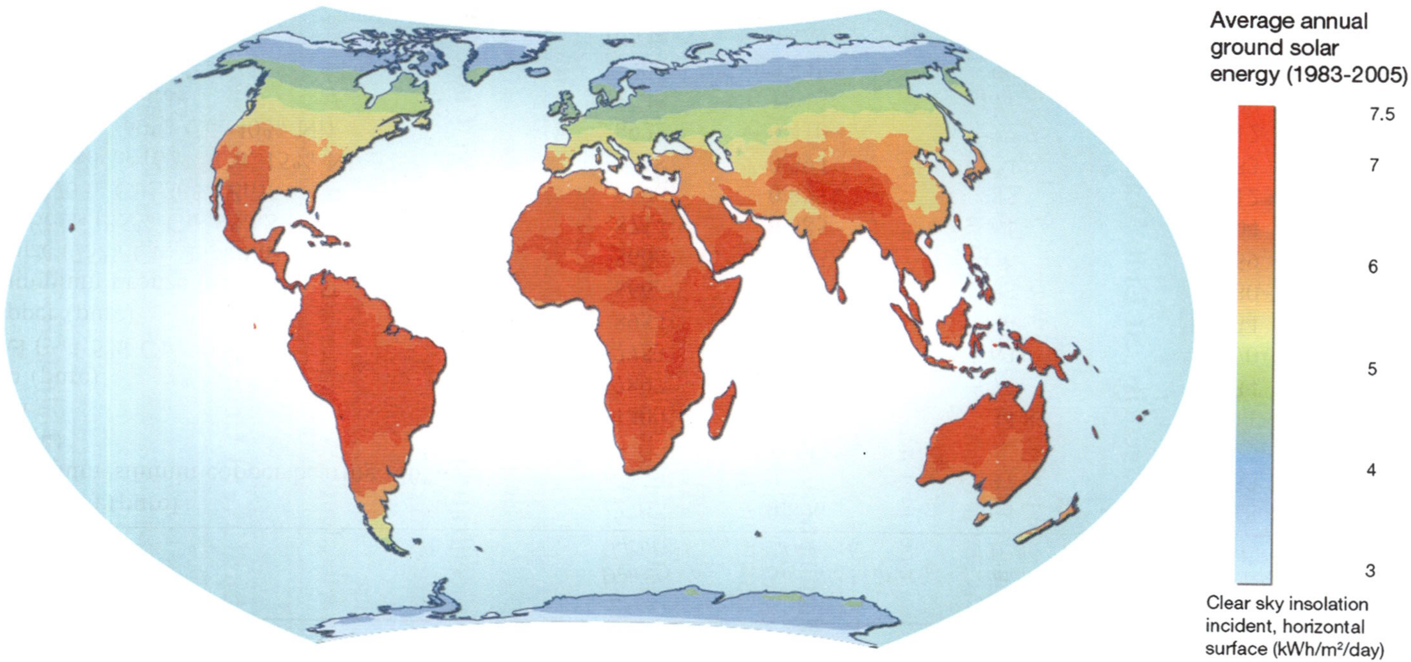

A3 Physical Properties of Metals and Non-metals

A3.1 Physical Properties of Metals at 20 °C

Metal	*Density, ρ ($kg\,m^{-3}$)*	*Specific heat, C_p ($J\,kg^{-1}\,K^{-1}$)*	*Thermal conductivity, K ($W\,m^{-1}\,K^{-1}$)*	*Expansion coefficient, $\alpha \times 10^{-5}$ ($m^2\,s^{-1}$)*
Aluminium (pure)	2707	0.896	204	8.418
Aluminium–silumin copper bearing (86% Al, 1% Cu)	2659	0.867	137	5.933
Lead	11400	0.1298	34.87	7.311
Iron (pure)	7897	0.452	73	2.034
Steel (~1.5% C)	7753	0.486	63	0.970
Copper (pure)	8954	0.3831	386	11.234
Aluminium bronze (95% Cu, 5% Al)	8666	0.410	383	2.330
Bronze (75% Cu, 25% Sn)	8666	0.343	326	0.859
Red brass (85% Cu, 9% Sn, 6% Zn)	8714	0.385	61	1.804
Brass (70% Cu, 30% Zn)	8600	0.877	85	3.412
German silver (62% Cu, 15% Ni, 22% Zn)	8618	0.394	24.9	0.733
Constantan (60% Cu, 40% Ni)	8922	0.410	22.7	0.612
Magnesium (pure)	1746	1.013	171	9.708
Nickel (pure)	8906	0.4459	90	2.266
Silver (pure)	10524	0.2340	419	17.004
Tin (pure)	7304	0.2265	64	3.884
Tungsten	19350	0.1344	163	6.271
Zinc (pure)	7144	0.3843	112.2	4.106

A3.2 Physical Properties of Non-metals

Non-metal	*Temperature, T (°C)*	*Density, ρ ($kg\,m^{-3}$)*	*Specific heat, C_p ($J\,kg^{-1}\,K^{-1}$)*	*Thermal conductivity, K ($W\,m^{-1}\,K^{-1}$)*	*Expansion coefficient, $\alpha \times 10^{-5}$ ($m^2\,s^{-1}$)*
Asbestos	50	470	–	0.08	–
Building brick	20	1600	0.84	0.69	5.2
Common face	–	2000	–	1.32	–
Concrete, cinder	23	–	–	0.76	–
Stone, 1-2-4 mix	20	1900–2300	0.88	1.37	8.2–6.8
Cork board	30	160	1.88	0.043	2–5.3
Cork, regranulated	32	45–120	1.88	0.045	2–5.3
Cotton wool	–	1500	–	0.018	–
Glass, window	20	2700	0.84	0.78 (av)	3.4
Borosilicate	30–75	2200	–	1.09	–
Ground	32	150	–	0.043	–
Glass wool	23	24	0.7	0.38	22.6
Paraffin wax	–	900	2900	0.25	–
Perspex	–	1190	1500	0.20	–
Plaster, gypsum	20	1440	0.84	0.48	4.0

Stone					
Granite	–	2640	0.82	1.73–3.98	8–18
Limestone	100–300	2500	0.90	1.26–1.33	5.6–5.9
Marble	–	2500–2700	0.80	2.07–2.94	10–13.6
Sandstone	40	2160–2300	0.71	1.83	11.2–11.9
Sawdust	23	–	–	0.059	–
Sand, dry	–	–	–	0.039	–
Soil, dry	–	–	–	0.014	–
Soil, wet	–	–	–	0.042	–
Tiles, clay	–	1900	–	0.85	–
Tiles concrete	–	2100	–	1.10	–
Wood (across the grain)					
Fir	23	420	2.72	0.11	0.96
Maple or oak	30	540	2.4	0.166	1.28
Yellow pine	23	640	2.8	0.147	0.82
Wood shavings	23	–	–	0.059	–
Wood chipboard	–	800	–	0.15	–

A3.3 Physical Properties of Some Other Materials

Material	*Density, ρ* *(kg m^{-3})*	*Specific heat, C_p* *(J kg^{-1} K^{-1})*	*Thermal conductivity, K* *(W m^{-1} K^{-1})*
Air	11.177	1006	40.026
Alumina	3800	800	29.0
Aluminium	41–45	0.946	211
Asphalt	1700	1000	0.50
Brick	1700	800	0.84
Carbon dioxide	1.979	871	0.145
Cement	1700	670	0.80
Clay	1458	879	11.28
Concrete	2400	1130	1.279
Copper	8795	–	385
Cork	240	2050	0.04
Cotton wool	1522	1335	–
Fibre board	300	1000	0.057
Glass, crown	2600	670	1.0
Glass, window	2350	712	0.816
Glass wool	50	670	0.042
Ice	920	1930	2.21
Iron	7870	106	80
Limestone	2180	–	1.5
Oxygen	1.301	920	0.027
Plasterboard	950	840	0.16
Polystyrene, expanded	25	1380	0.033
PVC, rigid foam	25–80	–	0.035–0.041
PVC, rigid sheet	1350	–	0.16
Sawdust	188	–	0.57
Thermocole	22	–	0.03
Timber	600	1210	0.14
Turpentine	870	1760	0.136
Water	998	4190	0.591
Water, sea	1025	3900	–
Water vapour	0.586	2060	0.025
Wood wool	500	1000	0.10

A4 Thermophysical Properties of Air and Saturated Water

A4.1 Thermophysical Properties of Air at Atmospheric Pressure

Temp., T (K)	Temp., T (°C)	Density, ρ (kg m^{-3})	Specific heat, C_p (J kg^{-1} K^{-1})	Kinematic viscosity, $\mu \times 10^{-5}$ (kg m^{-1} s^{-1})	Dynamic viscosity, í $\times 10^{-5}$ (m^2 s^{-1})	Thermal conductivity, K (W m^{-1} K^{-1})	Expansion coefficient, $\alpha \times 10^{-5}$ (m^2 s^{-1})	Prandtl number, Pr
100	−173	3.605	1039	0.711	0.197	0.009 41	0.251	0.784
150	−123	2.368	1012	1.035	0.437	0.014 06	0.587	0.745
200	−73	1.769	1007	1.333	0.754	0.018 36	1.031	0.731
250	−23	1.412	1006	1.606	1.137	0.022 41	1.578	0.721
260	−13	1.358	1006	1.649	1.214	0.023 29	1.705	0.712
270	−3	1.308	1006	1.699	1.299	0.024 00	1.824	0.712
280	7	1.261	1006	1.747	1.385	0.02473	1.879	0.711
290	17	1.217	1006	1.795	1.475	0.025 44	2.078	0.710
300	27	1.177	1007	1.857	1.578	0.026 23	2.213	0.713
310	37	1.139	1007	1.889	1.659	0.026 84	2.340	0.709
320	47	1.103	1008	1.935	1.754	0.027 53	2.476	0.708
330	57	1.070	1008	1.981	1.851	0.028 21	2.616	0.708
340	67	1.038	1009	2.025	1.951	0.028 88	2.821	0.707
350	77	1.008	1009	2.090	2.073	0.029 84	2.931	0.707
400	127	0.8821	1014	2.310	2.619	0.033 28	3.721	0.704
450	177	0.7840	1021	2.517	3.210	0.036 56	4.567	0.703
500	227	0.7056	1030	2.713	3.845	0.039 71	5.464	0.704
550	277	0.6414	1040	2.902	4.524	0.042 77	6.412	0.706
600	327	0.5880	1051	3.082	5.242	0.045 73	7.400	0.708
650	377	0.5427	1063	3.257	6.001	0.048 63	8.430	0.712
700	427	0.5040	1075	3.425	6.796	0.051 46	9.498	0.715
750	477	0.4704	1087	3.588	7.623	0.054 25	10.61	0.719

A4.1 10 (*Continued*).

Temp., T (K)	Temp., T (°C)	Density, ρ (kg m^{-3})	Specific heat, C_p (J kg^{-1} K^{-1})	Kinematic viscosity, $\mu \times 10^{-5}$ (kg m^{-1} s^{-1})	Dynamic viscosity, $í \times 10^{-5}$ (m^2 s^{-1})	Thermal conductivity, K (W m^{-1} K^{-1})	Expansion coefficient, $\alpha \times 10^{-5}$ (m^2 s^{-1})	Prandtl number, Pr
800	527	0.4410	1099	3.747	8.497	0.056 99	11.76	0.723
850	577	0.4150	1110	3.901	9.400	0.059 69	12.96	0.725
900	627	0.3920	1121	4.052	10.34	0.062 37	14.19	0.728
950	677	0.3716	1131	4.199	11.30	0.065 01	15.47	0.731
1000	727	0.3528	1142	4.343	12.31	0.067 63	16.79	0.733
1100	827	0.3207	1159	4.622	14.41	0.072 81	19.59	0.736
1200	927	0.2940	1175	4.891	16.64	0.077 92	22.56	0.738
1300	1027	0.2714	1189	5.151	18.98	0.082 97	25.71	0.738
1400	1127	0.2520	1201	5.403	21.44	0.087 98	29.05	0.738
1500	1227	0.2352	1211	5.648	24.01	0.092 96	32.64	0.735

A4.2 Thermophysical Properties of Saturated Water

Temp., T (K)	Temp., T (°C)	Density, ρ (kg m^{-3})	Specific heat, C_p (J kg^{-1} K^{-1})	Kinematic viscosity, $\mu \times 10^{-5}$ (kg m^{-1} s^{-1})	Dynamic viscosity, $í \times 10^{-7}$ (m^2 s^{-1})	Thermal conductivity, K (W m^{-1} K^{-1})	Expansion coefficient, $\alpha \times 10^{-7}$ (m^2 s^{-1})	Prandtl number, Pr
273.16	0.01	999.8	4220	179.064	17.91	0.5610	1.33	13.47
275	2	999.9	4214	168.183	16.82	0.5645	1.340	12.55
280	7	999.9	4201	143.386	14.34	0.5740	1.366	10.63
285	12	999.5	4193	123.938	12.40	0.5835	1.392	8.91
290	17	998.8	4187	108.370	10.85	0.5927	1.417	7.66

295	22	997.8	4183	95.789	9.600	0.6017	1.442	6.66
300	27	996.5	4181	85.380	8.568	0.6103	1.465	5.85
305	32	995.0	4180	76.695	7.708	0.6184	1.487	5.18
310	37	993.3	4179	69.352	6.982	0.6260	1.508	4.63
320	47	989.3	4181	57.696	5.832	0.6396	1.546	3.77
340	67	979.5	4189	42.197	4.308	0.6605	1.610	2.68
360	87	967.4	4202	32.611	3.371	0.6737	1.657	2.03
373.15	100	958.3	4216	28.174	2.940	0.6791	1.681	1.75
400	127	937.5	4256	21.863	2.332	0.6836	1.713	1.36
420	147	919.9	4299	18.674	2.030	0.6825	1.726	1.18
440	167	900.5	4357	16.281	1.808	0.6780	1.728	1.05
460	187	879.5	4433	14.433	1.641	0.6702	1.719	0.955
480	207	856.5	4533	12.967	1.514	0.6590	1.697	0.892
500	227	831.3	4664	11.771	1.416	0.6439	1.660	0.853
520	247	803.6	4838	10.760	1.339	0.6246	1.607	0.833
540	267	772.8	5077	9.876	1.278	0.6001	1.530	0.835
560	287	738.0	5423	9.085	1.231	0.5701	1.425	0.864
580	307	697.6	5969	8.336	1.195	0.5346	1.284	0.931
600	327	649.4	6953	7.572	1.166	0.4953	1.097	1.06
620	347	586.9	9354	6.726	1.146	0.4541	0.8272	1.39
640	367	481.5	25,940	5.528	1.148	0.4149	0.3322	3.46

A5 Absorptivity of Various Surfaces for the Sun's Rays

Surface	*Absorptivity*
Metals	
Polished aluminium/copper	0.26
New galvanized iron	0.66
Old galvanized iron	0.89
Polished iron	0.45
Oxidized rusty iron	0.38
Roofs	
Asphalt	0.89
White asbestos cement	0.59
Cooper sheeting	0.64
Uncoloured roofing tile	0.67
Red roofing tiles	0.72
Galvanized iron, clean	0.77
Brown roofing tile	0.87
Galvanized iron, dirty	0.89
Black roofing tile	0.92
Walls	
White/yellow brick tiles	0.30
White stone	0.40
Cream brick tile	0.50
Burl brick tile	0.60
Concrete/red brick tile	0.70
Red sand line brick	0.72
White sand stone	0.76
Stone rubble	0.80
Blue brick tile	0.88
Surfaces	
White paint	0.12–0.26
Whitewash/glossy white	0.21
Bright aluminium	0.30
Flat white	0.25
Yellow	0.48
Bronze	0.50
Silver	0.52
Dark aluminium	0.63
Bright red	0.65
Brown	0.70
Light green	0.73
Medium red	0.74
Medium green	0.85
Dark green	0.95
Blue/black	0.97

A5 (*Continued*).

Surface	*Absorptivity*
Surroundings	
Sea/lake water	0.29
Snow	0.30
Grass	0.80
Light coloured grass	0.55
Light green shiny leaves	0.75
Sand, grey	0.82
Rock	0.84
Green leaf	0.85
Earth (black ploughed field)	0.92
White leaves	0.20
Yellow leaves	0.58
Aluminium foil	0.39
Unpainted wood	0.60

A6 Measured Radiation and Ambient Air Temperature at Srinagar (India) for Sample Calculations

A6.1 Measured Average Hourly Global, Diffuse and Beam Radiation for "Type a" Weather Conditions (W m^{-2})

		Month											
Solar radiation	*Time*	*Jan*	*Feb*	*Mar*	*Apr*	*May*	*Jun*	*Jul*	*Aug*	*Sep*	*Oct*	*Nov*	*Dec*
Global	8 am	73.61	69.79	221.53	373.06	408.52	434.92	408.89	289.58	291.25	179.42	66.67	30.55
	9 am	252.78	284.72	461.57	594.31	633.52	634.52	626.11	542.71	492.78	379.32	272.22	213.89
	10 am	455.56	481.60	665.74	776.53	793.52	796.43	788.06	711.11	668.47	544.75	452.78	360.19
	11 am	545.84	647.22	800.70	880.56	915.74	904.76	888.89	831.60	784.30	661.83	558.33	485.19
	12 pm	577.78	732.99	891.43	932.50	981.67	951.19	950.28	885.42	842.71	719.34	652.78	543.52
	1 pm	572.22	747.57	905.79	931.81	975.74	963.89	958.89	903.47	832.57	702.06	622.22	513.89
	2 pm	511.11	688.20	841.67	870.70	913.70	919.84	850.55	843.75	760.69	622.53	563.89	470.37
	3 pm	418.06	562.50	703.71	735.69	783.89	795.24	727.78	745.49	612.71	479.84	380.56	362.04
	4 pm	227.78	356.60	498.15	531.11	560.56	626.19	550.55	534.38	400.49	290.84	216.67	201.85
	5 pm	43.06	127.08	245.37	283.06	344.81	417.46	317.78	293.40	196.11	94.86	25.00	52.78
Diffuse	8 am	19.45	45.48	74.31	99.03	104.26	102.78	119.72	106.60	89.31	61.63	33.33	2.78
	9 am	70.83	111.46	105.32	122.92	129.63	125.79	151.11	128.82	121.87	95.68	108.33	29.63
	10 am	101.39	112.50	123.38	138.47	145.56	139.29	164.17	125.35	138.47	113.89	158.33	61.11
	11 am	115.28	127.78	143.52	152.08	152.59	130.56	176.39	133.68	147.15	131.28	172.22	85.18
	12 pm	119.44	136.46	153.70	156.25	151.30	135.32	166.94	127.08	149.86	137.14	136.11	94.44
	1 pm	118.06	134.72	144.91	150.97	147.22	132.14	152.78	127.08	142.29	125.21	127.78	112.96
	2 pm	118.06	113.19	132.64	143.75	147.41	121.43	144.72	116.67	127.36	104.42	77.78	105.55
	3 pm	84.72	95.14	120.37	142.50	134.44	121.03	140.56	107.64	108.54	84.67	58.33	95.37
	4 pm	48.61	77.08	90.74	130.83	134.45	111.51	130.28	100.69	90.00	58.44	33.33	83.33
	5 pm	18.06	44.79	62.04	95.00	109.45	88.49	100.28	78.13	60.00	26.13	22.22	20.37

Beam	8 am	54.16	24.31	147.22	274.03	304.26	332.14	289.17	182.98	201.94	117.79	33.34	27.77
	9 am	181.95	173.26	356.25	471.39	503.89	508.73	475.00	413.89	370.91	283.64	163.89	184.26
	10 am	354.17	369.[illegible]	542.36	638.06	647.96	657.14	623.89	585.76	530.00	430.86	294.45	299.08
	11 am	430.56	519.44	657.18	728.48	763.15	774.20	712.50	697.92	637.15	530.55	386.11	400.01
	12 pm	458.34	596.53	737.73	776.25	830.37	815.87	783.34	758.34	692.85	582.20	516.67	449.08
	1 pm	454.16	612.35	760.88	780.84	828.52	831.75	806.11	776.39	690.28	576.85	494.44	400.93
	2 pm	393.05	575.01	709.03	726.95	766.29	798.41	705.83	727.08	633.33	518.11	486.11	364.82
	3 pm	333.34	467.36	583.34	593.19	649.45	674.21	587.22	637.85	504.17	395.17	322.23	266.67
	4 pm	179.17	279.52	407.41	400.28	426.11	514.68	420.27	433.69	310.49	232.40	183.34	118.52
	5 pm	25.00	82.29	183.33	188.06	235.36	328.97	217.50	215.27	136.11	68.73	2.78	32.41

A6.2 Measured Average Hourly Global, Diffuse and Beam Radiation for "Type b" Weather Conditions (W m^{-2})

		Month											
Solar Radiation	*Time*	*Jan*	*Feb*	*Mar*	*Apr*	*May*	*Jun*	*Jul*	*Aug*	*Sep*	*Oct*	*Nov*	*Dec*
Global	8 am	25.70	86.11	146.43	266.88	359.83	366.03	316.58	267.13	250.82	168.98	46.20	33.33
	9 am	165.51	250.00	369.05	464.96	547.76	572.49	555.69	538.89	463.40	364.82	184.35	202.78
	10 am	315.05	474.31	556.75	676.28	733.98	755.88	713.58	671.29	636.77	525.00	339.72	277.78
	11 am	421.53	719.44	661.51	826.92	857.91	919.23	892.03	864.82	742.16	639.58	462.50	341.67
	12 pm	463.66	812.50	769.05	858.55	893.16	878.53	888.80	899.08	761.93	693.06	518.52	525.00
	1 pm	473.38	811.81	816.67	811.32	880.66	888.94	844.88	800.82	761.44	677.78	521.20	544.44
	2 pm	457.18	750.00	665.87	673.29	711.65	808.60	769.77	730.93	694.77	593.29	473.89	491.67
	3 pm	361.81	585.41	535.72	446.37	355.13	542.84	590.17	637.50	527.29	448.61	357.59	327.78
	4 pm	216.43	359.03	351.98	390.17	318.70	419.07	438.24	457.41	334.80	266.43	203.52	155.56
	5 pm	52.55	70.83	184.53	184.83	206.31	222.60	258.29	293.98	157.35	81.25	58.80	27.78

A6.2 10 (*Continued*).

		Month											
Solar Radiation	*Time*	*Jan*	*Feb*	*Mar*	*Apr*	*May*	*Jun*	*Jul*	*Aug*	*Sep*	*Oct*	*Nov*	*Dec*
Diffuse	8 am	17.13	56.95	54.36	108.76	125.94	128.11	143.91	159.72	124.02	86.34	27.13	8.33
	9 am	87.27	129.86	121.82	166.88	191.72	200.37	203.20	206.02	167.98	139.82	97.59	91.67
	10 am	124.77	231.25	196.43	204.91	256.89	264.56	253.81	243.06	196.73	168.98	143.61	133.33
	11 am	148.84	235.41	226.59	213.25	300.27	321.73	278.92	236.11	222.06	194.68	166.67	138.89
	12 pm	162.73	226.39	267.46	238.67	312.61	307.48	265.32	223.15	213.24	203.93	182.78	197.22
	1 pm	160.88	206.25	223.41	284.62	308.23	311.13	267.14	223.15	215.36	203.24	183.89	175.00
	2 pm	151.62	192.36	223.41	261.54	249.08	283.01	232.25	181.48	199.84	185.19	161.67	158.33
	3 pm	110.19	150.00	212.70	186.54	124.30	190.00	195.69	201.39	173.69	146.53	127.78	133.33
	4 pm	78.24	111.11	170.63	170.51	111.54	146.67	161.95	177.22	131.05	102.78	77.31	61.11
	5 pm	23.61	48.61	96.43	91.88	72.21	77.91	87.10	96.30	79.57	40.28	17.96	11.11
Beam	8 am	8.57	29.16	92.07	158.12	233.89	237.92	172.67	107.41	126.80	82.64	19.07	25.00
	9 am	78.24	120.14	247.23	298.08	356.04	372.12	352.49	332.87	295.42	225.00	86.76	111.11
	10 am	190.28	243.06	360.32	471.37	477.09	491.32	459.77	428.23	440.04	356.02	196.11	144.45
	11 am	272.69	484.03	434.92	613.67	557.64	597.50	613.11	628.71	520.10	444.90	295.83	202.78
	12 pm	300.93	586.11	501.59	619.88	580.55	571.05	623.48	675.93	548.69	489.13	335.74	327.78
	1 pm	312.50	605.56	593.26	526.70	572.43	577.81	577.74	577.67	546.08	474.54	337.31	369.44
	2 pm	305.56	557.64	442.46	411.75	462.57	525.59	537.52	549.45	494.93	408.10	312.22	333.34
	3 pm	251.62	435.41	323.02	259.83	230.83	352.84	394.48	436.11	353.60	302.08	229.81	194.45
	4 pm	138.19	247.92	181.35	219.66	207.16	272.40	276.29	280.19	203.75	163.65	126.21	94.45
	5 pm	28.94	22.22	88.10	92.95	134.10	144.69	171.19	197.68	77.78	40.97	40.84	16.67

A6.3 Measured Average Hourly Global, Diffuse and Beam Radiation for "Type c" Weather Conditions ($W\,m^{-2}$)

Solar radiation	Time	Month Jan	Feb	Mar	Apr	May	Jun	Jul	Aug	Sep	Oct	Nov	Dec
Global	8 am	23.89	35.48	51.39	208.33	204.34	204.85	188.89	177.43	165.97	133.33	33.33	39.75
	9 am	90.00	123.89	94.44	352.08	360.42	217.52	344.44	320.14	295.83	211.11	155.56	158.48
	10 am	191.67	230.67	151.39	592.36	474.48	323.26	372.22	387.85	403.48	258.33	289.81	238.50
	11 am	271.67	338.09	281.94	622.22	550.35	380.51	505.56	532.64	559.72	427.78	399.38	295.91
	12 pm	306.67	399.70	434.72	783.34	668.76	447.54	541.67	529.17	516.66	702.78	428.70	304.76
	1 pm	377.78	396.53	408.33	603.47	651.74	574.04	711.11	622.22	533.33	636.11	400.00	242.92
	2 pm	359.44	392.25	475.00	604.86	600.35	600.71	577.78	559.73	541.67	433.33	367.28	260.59
	3 pm	383.33	321.33	390.28	576.39	574.99	500.37	508.33	443.06	377.78	202.78	255.86	169.61
	4 pm	155.56	157.83	226.38	331.25	284.38	293.83	263.89	290.28	316.67	94.44	124.69	88.34
	5 pm	30.00	38.83	72.22	188.89	229.34	268.86	225.00	180.21	135.42	50.00	32.10	23.85
Diffuse	8 am	19.89	23.26	25.00	153.47	141.97	130.48	55.56	97.64	139.72	100.00	24.07	25.00
	9 am	73.34	79.75	44.44	228.47	183.51	138.55	91.67	144.79	197.92	147.22	100.62	99.67
	10 am	148.33	124.97	63.89	325.00	265.45	205.90	238.89	271.88	304.86	258.33	143.83	150.00
	11 am	201.66	164.23	80.56	375.00	308.68	242.36	219.44	296.52	373.61	350.00	187.35	186.11
	12 pm	257.22	227.73	166.67	355.56	320.31	285.06	308.33	331.60	354.86	372.22	243.52	191.67
	1 pm	281.11	261.95	288.89	423.61	394.62	365.63	427.78	389.93	352.08	386.11	238.89	152.78
	2 pm	212.78	258.06	363.89	313.19	347.91	382.62	385.56	352.50	319.44	316.67	227.78	163.89
	3 pm	160.00	217.73	372.22	277.78	298.24	318.71	346.11	309.51	272.92	191.67	169.45	106.67
	4 pm	76.67	96.25	113.89	226.39	206.77	187.15	211.11	235.07	259.03	94.44	97.22	55.56
	5 pm	18.33	35.29	61.11	115.28	143.26	171.25	202.78	154.31	105.83	41.67	25.31	15.00
Beam	8 am	4.00	12.22	26.39	54.86	62.37	74.37	133.33	79.79	26.25	33.33	9.26	14.75
	9 am	16.66	44.14	50.00	123.61	176.91	78.97	252.77	175.35	97.91	63.89	54.94	58.81
	10 am	43.34	105.73	87.50	267.36	209.03	117.36	133.33	115.97	98.62	0.00	145.98	88.50
	11 am	70.01	173.85	201.38	247.22	241.67	138.15	286.12	236.12	186.11	77.78	212.03	109.80

A6.3 10 (*Continued*).

Solar radiation	Time	Month Jan	Feb	Mar	Apr	May	Jun	Jul	Aug	Sep	Oct	Nov	Dec
	12 pm	49.45	171.97	268.05	427.78	348.45	162.48	233.34	197.57	161.80	330.56	185.18	113.09
	1 pm	96.67	134.58	119.44	179.86	257.12	208.41	283.33	232.29	181.25	250.00	161.11	90.14
	2 pm	146.66	134.19	111.11	291.67	252.44	218.09	192.22	207.23	222.23	116.66	139.50	96.70
	3 pm	223.33	103.55	18.06	298.61	276.75	181.66	162.22	133.55	104.86	11.11	86.41	62.94
	4 pm	78.89	61.58	112.49	104.86	77.61	106.68	52.78	55.21	57.64	0.00	27.47	32.78
	5 pm	11.67	3.54	11.11	73.61	86.08	97.61	22.22	25.90	29.59	8.33	6.79	8.85

A6.4 Measured Average Hourly Global, Diffuse and Beam Radiation for "Type d" Weather Conditions ($W\,m^{-2}$)

Solar radiation	Time	Month Jan	Feb	Mar	Apr	May	Jun	Jul	Aug	Sep	Oct	Nov	Dec
Global	8 am	29.63	56.95	77.78	127.78	177.78	186.11	194.44	163.89	41.67	39.45	37.22	25.00
	9 am	148.15	168.06	205.56	241.67	277.78	263.89	250.00	191.67	100.00	129.16	158.33	112.22
	10 am	258.89	279.17	369.44	394.44	419.44	369.44	319.44	449.44	219.44	294.44	369.44	225.00
	11 am	351.85	286.81	391.67	188.89	489.45	511.39	533.33	544.44	438.89	406.94	375.00	438.89
	12 pm	361.11	364.58	366.67	550.00	733.33	611.11	488.89	688.33	641.67	527.78	413.89	531.11
	1 pm	300.00	323.61	283.33	933.33	933.33	708.33	483.33	593.89	575.00	483.33	391.67	491.67
	2 pm	286.11	287.50	319.00	583.33	883.33	655.55	427.78	569.44	491.67	405.56	319.44	388.89
	3 pm	171.30	221.53	246.50	427.78	655.56	590.28	325.00	375.00	341.67	312.50	283.33	304.44
	4 pm	80.56	100.00	144.44	119.44	94.44	184.72	275.00	197.22	277.78	188.89	100.00	105.56
	5 pm	20.37	42.36	47.22	77.78	108.34	141.67	175.00	116.67	88.89	60.56	32.22	16.67
Diffuse	8 am	25.48	46.53	67.67	98.39	129.11	158.19	180.56	137.67	38.89	36.11	33.33	21.00
	9 am	127.41	147.92	178.84	186.09	193.34	224.31	238.89	161.00	97.22	123.33	129.44	107.22
	10 am	222.65	243.75	321.41	303.72	286.03	314.02	313.89	377.53	216.67	256.94	297.22	213.89

	11 am	302.59	258.33	340.75	145.45	439.25	434.68	363.89	457.33	406.11	352.22	298.33	336.89
	12 pm	310.55	260.42	246.50	423.50	600.50	519.44	441.67	578.20	507.78	415.83	323.89	350.00
	1 pm	258.00	293.75	190.91	718.66	779.41	602.08	462.78	498.87	501.67	380.00	258.33	350.00
	2 pm	246.05	258.33	283.33	449.16	579.32	557.22	416.67	478.33	488.89	358.61	228.33	230.00
	3 pm	147.32	184.03	200.00	329.39	412.28	501.74	322.22	315.00	338.89	256.06	173.22	258.89
	4 pm	69.28	88.19	125.66	91.97	58.28	157.01	272.22	165.66	230.56	163.89	97.22	100.56
	5 pm	17.52	38.20	41.08	59.89	78.70	120.42	172.22	98.00	86.11	57.77	29.44	13.89
Beam	8 am	4.15	10.42	10.11	29.39	48.67	27.92	13.88	26.22	2.78	3.34	3.89	4.00
	9 am	20.74	20.14	26.72	55.58	84.44	39.58	11.11	30.67	2.78	5.83	28.89	5.00
	10 am	36.24	35.42	48.03	90.72	133.41	55.42	5.55	71.91	2.77	37.50	72.22	11.11
	11 am	49.26	28.48	50.92	43.44	50.20	76.71	169.44	87.11	32.78	54.72	76.67	102.00
	12 pm	50.56	104.16	120.17	126.50	132.83	91.67	47.22	110.13	133.89	111.95	90.00	181.11
	1 pm	42.00	29.86	92.42	214.67	153.92	106.25	20.55	95.02	73.33	103.33	133.34	141.67
	2 pm	40.06	29.17	35.67	134.17	304.01	98.33	11.11	91.11	2.78	46.95	91.11	158.89
	3 pm	23.98	37.50	46.50	98.39	243.28	88.54	2.78	60.00	2.78	56.44	110.11	45.55
	4 pm	11.28	11.81	18.78	27.47	36.16	27.71	2.78	31.56	47.22	25.00	2.78	5.00
	5 pm	2.85	4.16	6.14	17.89	29.64	21.25	2.78	18.67	2.78	2.79	2.78	2.78

A6.5 Number of Days Fall in Different Weather Conditions

Type of day	*Jan*	*Feb*	*Mar*	*Apr*	*May*	*Jun*	*Jul*	*Aug*	*Sep*	*Oct*	*Nov*	*Dec*
a	5	7	8	10	13	11	7	6	14	12	5	3
b	17	14	17	17	15	12	17	18	12	12	14	8
c	7	4	3	2	2	4	4	3	3	5	8	19
d	2	3	3	1	1	3	3	4	2	2	3	1

A6.6 Average Ambient Temperature for Different Months (°C)

	Month											
Time	*Jan*	*Feb*	*Mar*	*Apr*	*May*	*Jun*	*Jul*	*Aug*	*Sep*	*Oct*	*Nov*	*Dec*
12 am	−4.00	2.50	3.50	11.00	15.10	20.00	18.30	18.60	14.10	9.50	−1.00	−5.00
1 am	−3.80	2.30	2.50	11.50	15.00	20.10	17.00	18.00	13.00	8.30	−2.80	−4.80
2 am	−3.80	2.00	2.20	10.50	14.50	20.00	16.00	17.00	12.00	7.00	−3.80	−4.80
3 am	−3.80	2.00	2.20	9.00	14.00	20.00	15.00	16.00	11.00	6.00	−3.80	−4.80
4 am	−3.20	2.00	1.00	9.00	13.00	19.10	14.00	16.00	10.00	4.00	−3.00	−3.00
5 am	−3.00	2.00	2.00	8.10	14.00	19.00	15.10	17.10	10.00	4.00	−3.00	−3.00
6 am	−2.00	3.00	2.00	9.00	15.00	20.10	17.10	17.10	11.00	5.00	−2.00	−2.00
7 am	−1.10	3.20	3.10	10.00	16.20	20.60	18.00	18.00	12.10	5.20	−0.10	−0.10
8 am	0.20	4.50	3.80	11.90	17.30	21.20	19.20	19.40	13.00	6.30	2.10	1.40
9 am	−0.50	4.30	3.10	11.60	16.80	20.70	19.20	19.40	12.80	5.80	1.60	1.40
10 am	−0.80	3.90	2.70	11.40	16.00	20.20	18.60	19.40	12.40	5.40	1.50	1.30
11 am	−0.80	4.00	2.50	11.40	16.10	20.30	18.80	19.30	12.30	5.20	1.20	0.90
12 pm	−0.90	4.00	3.30	11.70	17.60	22.50	19.60	19.30	12.40	6.40	1.20	0.80
1 pm	−0.10	3.90	6.50	13.70	18.30	24.40	20.40	20.30	13.80	9.00	3.50	2.30
2 pm	1.60	4.90	9.50	14.80	19.50	26.60	20.40	20.80	14.40	11.50	6.00	4.20
3 pm	2.70	6.30	12.10	15.90	20.70	28.40	21.40	22.80	16.10	14.50	10.00	5.30
4 pm	4.50	7.30	13.00	17.10	23.90	29.20	21.60	24.80	18.50	16.90	12.30	6.70
5 pm	6.40	7.80	14.30	18.80	24.60	30.30	21.50	25.30	19.60	18.60	13.60	7.70
6 pm	4.30	6.50	11.00	17.00	23.10	28.20	20.00	25.00	19.00	16.50	11.30	4.30
7 pm	2.00	5.50	10.00	16.00	22.00	26.20	20.30	24.30	18.60	15.50	8.00	2.00
8 pm	1.10	4.30	8.50	15.00	20.00	25.00	19.80	23.00	17.60	14.30	7.10	1.10
9 pm	−0.50	4.00	6.50	14.10	19.00	24.20	19.70	22.00	17.60	13.00	5.70	−0.50
10 pm	−2.00	3.00	4.50	14.00	18.10	23.50	19.80	21.00	16.00	13.00	3.10	−2.00
11 pm	−3.80	2.50	3.50	12.50	16.20	23.20	19.30	19.80	15.40	10.50	1.80	−4.80

A7 Embodied Energy Coefficients of Materials, PV Cells and Balance of Systems

A7.1 Embodied Energy Coefficients of Materials

Material	$MJ\,kg^{-1}$	$MJ\,m^{-3}$
Aggregate, general	0.1	150
Virgin rock	0.04	63
River	0.02	36
Aluminium, virgin	191	515–700
Extruded	201	542–700
Extruded, anodized	227	612–900
Extruded, factory painted	218	588–600
Foil	204	550–800
Sheet	199	537–300
Aluminium, recycled	8.1	21–870
Extruded	17.3	46–710
Extruded, anodized	42.9	115–830
Extruded, factory painted	34.3	92–610
Foil	20.1	54–270
Sheet	14.8	39–960
Asphalt (paving)	3.4	7–140
Bitumen	44.1	45–420
Brass	62	519–560
Carpet	72.4	–
Felt underlay	18.6	–
Nylon	148	–
Polyester	53.7	–
Poly(ethyleneterephthalate) (PET)	107	–
Polypropylene	95.4	–
Wool	106	–
Cement	7.8	15–210
Cement mortar	2	3–200
Fibre cement board	9.5	13–550
Soil–cement	0.42	819
Ceramic		
Brick	2.5	5–170
Brick, glazed	7.2	14–760
Pipe	6.3	–
Tile	2.5	5–250
Concrete		
Block	0.94	–
Brick	0.97	
GRC	7.6	14–820
Paver	1.2	–
Pre-cast	2.0	–
Ready mix, 17.5 Mpa	1.0	2–350
30 Mpa	1.3	3–180
40 Mpa	1.6	3–890

A7.1 (*Continued*).

Material	$MJ\,kg^{-1}$	$MJ\,m^{-3}$
Roofing tile	0.81	–
Copper	70.6	631–160
Earth, raw		
Adobe block, straw stabilized	0.47	750
Adobe, bitumen stabilized	0.29	–
Adobe, cement stabilized	0.42	–
Rammed soil cement	0.80	–
Pressed block	0.42	–
Fabric		
Cotton	143	–
Polyester	53.7	–
Glass		
Float	15.9	40–60
Toughened	26.2	20–66
Laminated	16.3	41–80
Tinted	14.9	375–450
Insulation		
Cellulose	3.3	112
Fibreglass	30.3	970
Polyester	53.7	430
Polystyrene	117	2–340
Wool (recycled)	14.6	139
Lead	35.1	30–398
Linoleum	116	150–930
Paint	90.4	118 per L
Solvent based	98.1	128 per L
Water based	88.5	115per L
Paper	36.4	33–670
Building	25.5	–
Kraft	12.6	–
Recycled	23.4	–
Wall	36.4	–
Plaster, gypsum	4.5	6–460
Plaster board	6.1	5–890
Plastics		
ABS	111	–
High-density polyethylene (HDPE)	103	97–340
Low-density polyethylene (LDPE)	103	91–800
Polyester	53.7	7–710
Polypropylene	64	57–600
Polystyrene, expanded	117	2–340
Polyurethane	74	44–400
PVC	70	93–620
Rubber		
Natural latex	67.5	62–100
Synthetic	110	–
Sand	0.1	232

A7.1 (*Continued*).

Material	*MJ kg⁻¹*	*MJ m⁻³*
Sealants and adhesives		
Phenol formaldehyde	87	–
Urea formaldehyde	78.2	–
Steel, recycled	10.1	37–210
Reinforcing, sections	8.9	–
Wire rod	12.5	–
Steel, virgin, general	32	200–251
Galvanized	34.8	180–273
Imported, structural	35	274–570
Stone, dimension		
Local	0.79	1–890
Imported	6.8	1–890
Straw, baled	0.24	30.5
Timber, softwood		
Air dried, rough sawn	0.3	165
Kiln dried, rough sawn	1.6	880
Air dried, dressed	1.16	638
Kiln dried, dressed	2.5	1–380
Mouldings, *etc.*	3.1	1–710
Hardboard	24.2	13–310
MDF	11.9	8–330
Glulam	4.6	2–530
Particle board	8	–
Plywood	10.4	–
Shingle	9	–
Timber, hardwood		
Air dried, rough sawn	0.5	388
Kiln dried, rough sawn	2	1–550
Vinyl flooring	79.1	105–990
Zinc	51	364–140
Galvanizing, per kg steel	2.8	–

A7.2 Range of Embodied Energy for Different PV Solar Cells

Solar cell technology	*Module efficiency, η_{PV} (%)*	*Expected life, η_{pv} (years)*	*Specific energy content, ε_{PV}*	
			kWh m⁻²	*MJ m⁻²*
sc-Si	14–18	30	1120–1260	4032–4556
mc-Si	12–16	30	840–980	3024–3528
rc-Si	10–14	25	570–650	2052–2340
a-Si	5–7	20	308–448	1110–1615
CdTe	7–9	15	196–336	705–1210
CIS	9–11	5	20–29	72–105

A7.3 Range of Embodied Energy for the Balance of Systems

BIPVT system components	*Life (years)*	*Specific heat content*	
Charge controller	10	130–290 kWh kW^{-1}	468–1044 MJ kW^{-1}
Inverter, safety factor 0.3	10	140–280 kWh kW^{-1}	504–1008 MJ kW^{-1}
Lead–acid battery, depth of discharge 80%	5	0.24–0.33 kWh Wh^{-1}	0.87–1.19 MJ Wh^{-1}
PV frame	30	84–140 kWh m^{-2}	302–504 MJ m^{-2}
Installation	–	23.6–35 kWh m^{-2}	85–126 MJ m^{-2}
Junction box, cabling, *etc.*	30		
M&O	–	1–2 kWh m^{-2} a^{-1}	3.6–7.2 MJ m^{-2} a^{-1}
Foundation, array support, *etc.*	30	168–252 kWh m^{-2}	604–907 MJ m^{-2}
Diesel incl. installation	5	400–800 kWh kW^{-1}	1440–2880 MJ kW^{-1}

A8 Lower and Higher Heating Values of Fuels

Fuel	*Lower heating value (LHV) (MJ kg^{-1})*	*Higher heating value (HHV) (MJ kg^{-1})*	*Density (kg m^{-3})*
Gaseous fuels @ 0 °C and 1 atm			
Natural gas	47.141	52.225	0.777
Hydrogen	120.21	142.18	0.090
Still gas (in refineries)	46.898	50.951	1.158
Liquid fuels			
Crude oil	42.686	45.543	846.5
Conventional gasoline	43.448	46.536	744.6
Reformulated or low-sulfur gasoline	42.358	45.433	747.5
CA reformulated gasoline	42.500	45.577	747.0
US conventional diesel	42.791	45.766	836.5
Low-sulfur diesel	42.612	45.575	846.8
Petroleum naphtha	44.938	48.075	725.0
Natural gas-based FT naphtha	44.383	47.654	700.2
Residual oil	39.466	42.210	991.0
Methanol	20.094	22.884	794.0
Ethanol	26.952	29.847	789.2
Butanol	34.366	37.334	809.6
Acetone	29.589	31.862	782.9
E-Diesel additives	43.448	46.536	744.6
Liquefied petroleum gas (LPG)	46.607	50.152	507.9
Liquefied natural gas (LNG)	48.632	55.206	428.2
Dimethyl ether (DME)	28.882	31.681	665.1
Dimethoxymethane (DMM)	23.402	25.670	859.8
Methyl ester (biodiesel, BD)	37.528	40.168	887.7

A8 (*Continued*).

Fuel	*Lower heating value (LHV) (MJ kg^{-1})*	*Higher heating value (HHV) (MJ kg^{-1})*	*Density (kg m^{-3})*
Fischer–Tropsch diesel (FTD)	43.247	45.471	796.9
Renewable diesel I (SuperCetane)	43.563	46.628	748.8
Renewable diesel II (UOP-HDO)	43.979	46.817	778.7
Renewable gasoline	43.239	46.314	747.5
Liquid hydrogen	120.07	141.80	70.8
Methyl *tert*-butyl ether (MTBE)	35.108	37.957	742.5
Ethyl *tert*-butyl ether (ETBE)	36.315	39.247	742.2
tert-Amyl methyl ether (TAME)	36.392	39.322	769.4
Butane	45.277	49.210	584.5
Isobutane	44.862	49.096	559.4
Isobutene	44.824	48.238	595.1
Propane	46.296	50.235	507.1
Solid fuels			
Coal (wet basis)	22.732	23.968	
Bituminous coal (wet basis)	26.122	27.267	
Coking coal (wet basis)	28.610	29.865	
Farmed trees (dry basis)	19.551	20.589	
Herbaceous biomass (dry basis)	17.209	18.123	
Corn stover (dry basis)	16.370	17.415	
Forest residue (dry basis)	15.402	16.473	
Sugar cane bagasse	15.058	16.355	
Petroleum coke	29.505	31.308	

A9 Values of Conversion Factors at Different Worth Rates for Economic Analysis

A9.1 Values of Conversion Factors at $i = 2\%$

n	F_{ps}	F_{sp}	F_{rp}	F_{pr}	F_{rs}	F_{sr}	F_{gp}	F_{gr}	F_{gs}
1	1.020	0.980	0.980	1.020	1.000	1.000	0.000	0.000	0.000
2	1.040	0.961	1.942	0.515	2.020	0.495	0.961	0.495	1.000
3	1.061	0.942	2.884	0.347	3.060	0.327	2.846	0.987	3.020
4	1.082	0.924	3.808	0.263	4.122	0.243	5.617	1.475	6.080
5	1.104	0.906	4.713	0.212	5.204	0.192	9.240	1.960	10.202
6	1.126	0.888	5.601	0.179	6.308	0.159	13.680	2.442	15.406
7	1.149	0.871	6.472	0.155	7.434	0.135	18.903	2.921	21.714

A9.1 (*Continued*).

n	F_{ps}	F_{sp}	F_{rp}	F_{pr}	F_{rs}	F_{sr}	F_{gp}	F_{gr}	F_{gs}
8	1.172	0.853	7.325	0.137	8.583	0.117	24.878	3.396	29.148
9	1.195	0.837	8.162	0.123	9.755	0.103	31.572	3.868	37.731
10	1.219	0.820	8.983	0.111	10.950	0.091	38.955	4.337	47.486
11	1.243	0.804	9.787	0.102	12.169	0.082	46.998	4.802	58.436
12	1.268	0.788	10.575	0.095	13.412	0.075	55.671	5.264	70.604
13	1.294	0.773	11.348	0.088	14.680	0.068	64.948	5.723	84.017
14	1.319	0.758	12.106	0.083	15.974	0.063	74.800	6.179	98.697
15	1.346	0.743	12.849	0.078	17.293	0.058	85.202	6.631	114.671
16	1.373	0.728	13.578	0.074	18.639	0.054	96.129	7.080	131.964
17	1.400	0.714	14.292	0.070	20.012	0.050	107.555	7.526	150.604
18	1.428	0.700	14.992	0.067	21.412	0.047	119.458	7.968	170.616
19	1.457	0.686	15.678	0.064	22.841	0.044	131.814	8.407	192.028
20	1.486	0.673	16.351	0.061	24.297	0.041	144.600	8.843	214.868
21	1.516	0.660	17.011	0.059	25.783	0.039	157.796	9.276	239.166
22	1.546	0.647	17.658	0.057	27.299	0.037	171.379	9.705	264.949
23	1.577	0.634	18.292	0.055	28.845	0.035	185.331	10.132	292.248
24	1.608	0.622	18.914	0.053	30.422	0.033	199.630	10.555	321.093
25	1.641	0.610	19.523	0.051	32.030	0.031	214.259	10.974	351.515
26	1.673	0.598	20.121	0.050	33.671	0.030	229.199	11.391	383.545
27	1.707	0.586	20.707	0.048	35.344	0.028	244.431	11.804	417.216
28	1.741	0.574	21.281	0.047	37.051	0.027	259.939	12.214	452.561
29	1.776	0.563	21.844	0.046	38.792	0.026	275.706	12.621	489.612
30	1.811	0.552	22.396	0.045	40.568	0.025	291.716	13.025	528.404

A9.2 Values of Conversion Factors at $i = 4\%$

n	F_{ps}	F_{sp}	F_{rp}	F_{pr}	F_{rs}	F_{sr}	F_{gp}	F_{gr}	F_{gs}
1	1.040	0.962	0.962	1.040	1.000	1.000	0.000	0.000	0.000
2	1.082	0.925	1.886	0.530	2.040	0.490	0.925	0.490	1.000
3	1.125	0.889	2.775	0.360	3.122	0.320	2.703	0.974	3.040
4	1.170	0.855	3.630	0.275	4.246	0.235	5.267	1.451	6.162
5	1.217	0.822	4.452	0.225	5.416	0.185	8.555	1.922	10.408
6	1.265	0.790	5.242	0.191	6.633	0.151	12.506	2.386	15.824
7	1.316	0.760	6.002	0.167	7.898	0.127	17.066	2.843	22.457
8	1.369	0.731	6.733	0.149	9.214	0.109	22.181	3.294	30.356
9	1.423	0.703	7.435	0.134	10.583	0.094	27.801	3.739	39.570
10	1.480	0.676	8.111	0.123	12.006	0.083	33.881	4.177	50.153
11	1.539	0.650	8.760	0.114	13.486	0.074	40.377	4.609	62.159
12	1.601	0.625	9.385	0.107	15.026	0.067	47.248	5.034	75.645
13	1.665	0.601	9.986	0.100	16.627	0.060	54.455	5.453	90.671
14	1.732	0.577	10.563	0.095	18.292	0.055	61.962	5.866	107.298
15	1.801	0.555	11.118	0.090	20.024	0.050	69.735	6.272	125.590
16	1.873	0.534	11.652	0.086	21.825	0.046	77.744	6.672	145.613
17	1.948	0.513	12.166	0.082	23.698	0.042	85.958	7.066	167.438

A9.2 (*Continued*).

n	F_{ps}	F_{sp}	F_{rp}	F_{pr}	F_{rs}	F_{sr}	F_{gp}	F_{gr}	F_{gs}
18	2.026	0.494	12.659	0.079	25.645	0.039	94.350	7.453	191.135
19	2.107	0.475	13.134	0.076	27.671	0.036	102.893	7.834	216.781
20	2.191	0.456	13.590	0.074	29.778	0.034	111.565	8.209	244.452
21	2.279	0.439	14.029	0.071	31.969	0.031	120.341	8.578	274.230
22	2.370	0.422	14.451	0.069	34.248	0.029	129.202	8.941	306.199
23	2.465	0.406	14.857	0.067	36.618	0.027	138.128	9.297	340.447
24	2.563	0.390	15.247	0.066	39.083	0.026	147.101	9.648	377.065
25	2.666	0.375	15.622	0.064	41.646	0.024	156.104	9.993	416.148
26	2.772	0.361	15.983	0.063	44.312	0.023	165.121	10.331	457.794
27	2.883	0.347	16.330	0.061	47.084	0.021	174.138	10.664	502.105
28	2.999	0.333	16.663	0.060	49.968	0.020	183.142	10.991	549.190
29	3.119	0.321	16.984	0.059	52.966	0.019	192.121	11.312	599.157
30	3.243	0.308	17.292	0.058	56.085	0.018	201.062	11.627	652.123

A9.3 Values of Conversion Factors at $i = 6\%$

n	F_{ps}	F_{sp}	F_{rp}	F_{pr}	F_{rs}	F_{sr}	F_{gp}	F_{gr}	F_{gs}
1	1.060	0.943	0.943	1.060	1.000	1.000	0.000	0.000	0.000
2	1.124	0.890	1.833	0.545	2.060	0.485	0.890	0.485	1.000
3	1.191	0.840	2.673	0.374	3.184	0.314	2.569	0.961	3.060
4	1.262	0.792	3.465	0.289	4.375	0.229	4.946	1.427	6.244
5	1.338	0.747	4.212	0.237	5.637	0.177	7.935	1.884	10.618
6	1.419	0.705	4.917	0.203	6.975	0.143	11.459	2.330	16.255
7	1.504	0.665	5.582	0.179	8.394	0.119	15.450	2.768	23.231
8	1.594	0.627	6.210	0.161	9.897	0.101	19.842	3.195	31.624
9	1.689	0.592	6.802	0.147	11.491	0.087	24.577	3.613	41.522
10	1.791	0.558	7.360	0.136	13.181	0.076	29.602	4.022	53.013
11	1.898	0.527	7.887	0.127	14.972	0.067	34.870	4.421	66.194
12	2.012	0.497	8.384	0.119	16.870	0.059	40.337	4.811	81.166
13	2.133	0.469	8.853	0.113	18.882	0.053	45.963	5.192	98.036
14	2.261	0.442	9.295	0.108	21.015	0.048	51.713	5.564	116.918
15	2.397	0.417	9.712	0.103	23.276	0.043	57.555	5.926	137.933
16	2.540	0.394	10.106	0.099	25.673	0.039	63.459	6.279	161.209
17	2.693	0.371	10.477	0.095	28.213	0.035	69.401	6.624	186.881
18	2.854	0.350	10.828	0.092	30.906	0.032	75.357	6.960	215.094
19	3.026	0.331	11.158	0.090	33.760	0.030	81.306	7.287	246.000
20	3.207	0.312	11.470	0.087	36.786	0.027	87.230	7.605	279.760
21	3.400	0.294	11.764	0.085	39.993	0.025	93.114	7.915	316.545
22	3.604	0.278	12.042	0.083	43.392	0.023	98.941	8.217	356.538
23	3.820	0.262	12.303	0.081	46.996	0.021	104.701	8.510	399.930
24	4.049	0.247	12.550	0.080	50.816	0.020	110.381	8.795	446.926
25	4.292	0.233	12.783	0.078	54.865	0.018	115.973	9.072	497.742
26	4.549	0.220	13.003	0.077	59.156	0.017	121.468	9.341	552.606
27	4.822	0.207	13.211	0.076	63.706	0.016	126.860	9.603	611.763
28	5.112	0.196	13.406	0.075	68.528	0.015	132.142	9.857	675.469

A9.3 (*Continued*).

n	F_{ps}	F_{sp}	F_{rp}	F_{pr}	F_{rs}	F_{sr}	F_{gp}	F_{gr}	F_{gs}
29	5.418	0.185	13.591	0.074	73.640	0.014	137.310	10.103	743.997
30	5.743	0.174	13.765	0.073	79.058	0.013	142.359	10.342	817.636

A9.4 Values of Conversion Factors at $i = 8\%$

n	F_{ps}	F_{sp}	F_{rp}	F_{pr}	F_{rs}	F_{sr}	F_{gp}	F_{gr}	F_{gs}
1	1.080	0.926	0.926	1.080	1.000	1.000	0.000	0.000	0.000
2	1.166	0.857	1.783	0.561	2.080	0.481	0.857	0.481	1.000
3	1.260	0.794	2.577	0.388	3.246	0.308	2.445	0.949	3.080
4	1.360	0.735	3.312	0.302	4.506	0.222	4.650	1.404	6.326
5	1.469	0.681	3.993	0.250	5.867	0.170	7.372	1.846	10.833
6	1.587	0.630	4.623	0.216	7.336	0.136	10.523	2.276	16.699
7	1.714	0.583	5.206	0.192	8.923	0.112	14.024	2.694	24.035
8	1.851	0.540	5.747	0.174	10.637	0.094	17.806	3.099	32.958
9	1.999	0.500	6.247	0.160	12.488	0.080	21.808	3.491	43.594
10	2.159	0.463	6.710	0.149	14.487	0.069	25.977	3.871	56.082
11	2.332	0.429	7.139	0.140	16.645	0.060	30.266	4.240	70.569
12	2.518	0.397	7.536	0.133	18.977	0.053	34.634	4.596	87.214
13	2.720	0.368	7.904	0.127	21.495	0.047	39.046	4.940	106.191
14	2.937	0.340	8.244	0.121	24.215	0.041	43.472	5.273	127.687
15	3.172	0.315	8.559	0.117	27.152	0.037	47.886	5.594	151.901
16	3.426	0.292	8.851	0.113	30.324	0.033	52.264	5.905	179.054
17	3.700	0.270	9.122	0.110	33.750	0.030	56.588	6.204	209.378
18	3.996	0.250	9.372	0.107	37.450	0.027	60.843	6.492	243.128
19	4.316	0.232	9.604	0.104	41.446	0.024	65.013	6.770	280.578
20	4.661	0.215	9.818	0.102	45.762	0.022	69.090	7.037	322.025
21	5.034	0.199	10.017	0.100	50.423	0.020	73.063	7.294	367.787
22	5.437	0.184	10.201	0.098	55.457	0.018	76.926	7.541	418.209
23	5.871	0.170	10.371	0.096	60.893	0.016	80.673	7.779	473.666
24	6.341	0.158	10.529	0.095	66.765	0.015	84.300	8.007	534.559
25	6.848	0.146	10.675	0.094	73.106	0.014	87.804	8.225	601.324
26	7.396	0.135	10.810	0.093	79.954	0.013	91.184	8.435	674.430
27	7.988	0.125	10.935	0.091	87.351	0.011	94.439	8.636	754.385
28	8.627	0.116	11.051	0.090	95.339	0.010	97.569	8.829	841.735
29	9.317	0.107	11.158	0.090	103.966	0.010	100.574	9.013	937.074
30	10.063	0.099	11.258	0.089	113.283	0.009	103.456	9.190	1041.040

A9.5 Values of Conversion Factors at $i = 10\%$

n	F_{ps}	F_{sp}	F_{rp}	F_{pr}	F_{rs}	F_{sr}	F_{gp}	F_{gr}	F_{gs}
1	1.100	0.909	0.909	1.100	1.000	1.000	0.000	0.000	0.000
2	1.210	0.826	1.736	0.576	2.100	0.476	0.826	0.476	1.000

A9.5 (*Continued*).

n	F_{ps}	F_{sp}	F_{rp}	F_{pr}	F_{rs}	F_{sr}	F_{gp}	F_{gr}	F_{gs}
3	1.331	0.751	2.487	0.402	3.310	0.302	2.329	0.937	3.100
4	1.464	0.683	3.170	0.315	4.641	0.215	4.378	1.381	6.410
5	1.611	0.621	3.791	0.264	6.105	0.164	6.862	1.810	11.051
6	1.772	0.564	4.355	0.230	7.716	0.130	9.684	2.224	17.156
7	1.949	0.513	4.868	0.205	9.487	0.105	12.763	2.622	24.872
8	2.144	0.467	5.335	0.187	11.436	0.087	16.029	3.004	34.359
9	2.358	0.424	5.759	0.174	13.579	0.074	19.421	3.372	45.795
10	2.594	0.386	6.145	0.163	15.937	0.063	22.891	3.725	59.374
11	2.853	0.350	6.495	0.154	18.531	0.054	26.396	4.064	75.312
12	3.138	0.319	6.814	0.147	21.384	0.047	29.901	4.388	93.843
13	3.452	0.290	7.103	0.141	24.523	0.041	33.377	4.699	115.227
14	3.797	0.263	7.367	0.136	27.975	0.036	36.800	4.996	139.750
15	4.177	0.239	7.606	0.131	31.772	0.031	40.152	5.279	167.725
16	4.595	0.218	7.824	0.128	35.950	0.028	43.416	5.549	199.497
17	5.054	0.198	8.022	0.125	40.545	0.025	46.582	5.807	235.447
18	5.560	0.180	8.201	0.122	45.599	0.022	49.640	6.053	275.992
19	6.116	0.164	8.365	0.120	51.159	0.020	52.583	6.286	321.591
20	6.727	0.149	8.514	0.117	57.275	0.017	55.407	6.508	372.750
21	7.400	0.135	8.649	0.116	64.002	0.016	58.110	6.719	430.025
22	8.140	0.123	8.772	0.114	71.403	0.014	60.689	6.919	494.027
23	8.954	0.112	8.883	0.113	79.543	0.013	63.146	7.108	565.430
24	9.850	0.102	8.985	0.111	88.497	0.011	65.481	7.288	644.973
25	10.835	0.092	9.077	0.110	98.347	0.010	67.696	7.458	733.471
26	11.918	0.084	9.161	0.109	109.182	0.009	69.794	7.619	831.818
27	13.110	0.076	9.237	0.108	121.100	0.008	71.777	7.770	940.999
28	14.421	0.069	9.307	0.107	134.210	0.007	73.650	7.914	1062.099
29	15.863	0.063	9.370	0.107	148.631	0.007	75.415	8.049	1196.309
30	17.449	0.057	9.427	0.106	164.494	0.006	77.077	8.176	1344.940

A9.6 Values of Conversion Factors at $i = 12\%$

n	F_{ps}	F_{sp}	F_{rp}	F_{pr}	F_{rs}	F_{sr}	F_{gp}	F_{gr}	F_{gs}
1	1.120	0.893	0.893	1.120	1.000	1.000	0.000	0.000	0.000
2	1.254	0.797	1.690	0.592	2.120	0.472	0.797	0.472	1.000
3	1.405	0.712	2.402	0.416	3.374	0.296	2.221	0.925	3.120
4	1.574	0.636	3.037	0.329	4.779	0.209	4.127	1.359	6.494
5	1.762	0.567	3.605	0.277	6.353	0.157	6.397	1.775	11.274
6	1.974	0.507	4.111	0.243	8.115	0.123	8.930	2.172	17.627
7	2.211	0.452	4.564	0.219	10.089	0.099	11.644	2.551	25.742
8	2.476	0.404	4.968	0.201	12.300	0.081	14.471	2.913	35.831
9	2.773	0.361	5.328	0.188	14.776	0.068	17.356	3.257	48.130
10	3.106	0.322	5.650	0.177	17.549	0.057	20.254	3.585	62.906
11	3.479	0.287	5.938	0.168	20.655	0.048	23.129	3.895	80.455
12	3.896	0.257	6.194	0.161	24.133	0.041	25.952	4.190	101.109

A9.6 (*Continued*).

n	F_{ps}	F_{sp}	F_{rp}	F_{pr}	F_{rs}	F_{sr}	F_{gp}	F_{gr}	F_{gs}
13	4.363	0.229	6.424	0.156	28.029	0.036	28.702	4.468	125.243
14	4.887	0.205	6.628	0.151	32.393	0.031	31.362	4.732	153.272
15	5.474	0.183	6.811	0.147	37.280	0.027	33.920	4.980	185.664
16	6.130	0.163	6.974	0.143	42.753	0.023	36.367	5.215	222.944
17	6.866	0.146	7.120	0.140	48.884	0.020	38.697	5.435	265.697
18	7.690	0.130	7.250	0.138	55.750	0.018	40.908	5.643	314.581
19	8.613	0.116	7.366	0.136	63.440	0.016	42.998	5.838	370.331
20	9.646	0.104	7.469	0.134	72.052	0.014	44.968	6.020	433.770
21	10.804	0.093	7.562	0.132	81.699	0.012	46.819	6.191	505.823
22	12.100	0.083	7.645	0.131	92.503	0.011	48.554	6.351	587.522
23	13.552	0.074	7.718	0.130	104.603	0.010	50.178	6.501	680.024
24	15.179	0.066	7.784	0.128	118.155	0.008	51.693	6.641	784.627
25	17.000	0.059	7.843	0.127	133.334	0.007	53.105	6.771	902.782
26	19.040	0.053	7.896	0.127	150.334	0.007	54.418	6.892	1036.116
27	21.325	0.047	7.943	0.126	169.374	0.006	55.637	7.005	1186.450
28	23.884	0.042	7.984	0.125	190.699	0.005	56.767	7.110	1355.824
29	26.750	0.037	8.022	0.125	214.583	0.005	57.814	7.207	1546.523
30	29.960	0.033	8.055	0.124	241.333	0.004	58.782	7.297	1761.106

A9.7 Values of Conversion Factors at $i = 14\%$

n	F_{ps}	F_{sp}	F_{rp}	F_{pr}	F_{rs}	F_{sr}	F_{gp}	F_{gr}	F_{gs}
1	1.140	0.877	0.877	1.140	1.000	1.000	0.000	0.000	0.000
2	1.300	0.769	1.647	0.607	2.140	0.467	0.769	0.467	1.000
3	1.482	0.675	2.322	0.431	3.440	0.291	2.119	0.913	3.140
4	1.689	0.592	2.914	0.343	4.921	0.203	3.896	1.337	6.580
5	1.925	0.519	3.433	0.291	6.610	0.151	5.973	1.740	11.501
6	2.195	0.456	3.889	0.257	8.536	0.117	8.251	2.122	18.111
7	2.502	0.400	4.288	0.233	10.730	0.093	10.649	2.483	26.646
8	2.853	0.351	4.639	0.216	13.233	0.076	13.103	2.825	37.377
9	3.252	0.308	4.946	0.202	16.085	0.062	15.563	3.146	50.610
10	3.707	0.270	5.216	0.192	19.337	0.052	17.991	3.449	66.695
11	4.226	0.237	5.453	0.183	23.045	0.043	20.357	3.733	86.032
12	4.818	0.208	5.660	0.177	27.271	0.037	22.640	4.000	109.077
13	5.492	0.182	5.842	0.171	32.089	0.031	24.825	4.249	136.348
14	6.261	0.160	6.002	0.167	37.581	0.027	26.901	4.482	168.436
15	7.138	0.140	6.142	0.163	43.842	0.023	28.862	4.699	206.017
16	8.137	0.123	6.265	0.160	50.980	0.020	30.706	4.901	249.860
17	9.276	0.108	6.373	0.157	59.118	0.017	32.430	5.089	300.840
18	10.575	0.095	6.467	0.155	68.394	0.015	34.038	5.263	359.958
19	12.056	0.083	6.550	0.153	78.969	0.013	35.531	5.424	428.352
20	13.743	0.073	6.623	0.151	91.025	0.011	36.914	5.573	507.321
21	15.668	0.064	6.687	0.150	104.768	0.010	38.190	5.711	598.346
22	17.861	0.056	6.743	0.148	120.436	0.008	39.366	5.838	703.114

A9.7 (*Continued*).

n	F_{ps}	F_{sp}	F_{rp}	F_{pr}	F_{rs}	F_{sr}	F_{gp}	F_{gr}	F_{gs}
23	20.362	0.049	6.792	0.147	138.297	0.007	40.446	5.955	823.550
24	23.212	0.043	6.835	0.146	158.659	0.006	41.437	6.062	961.847
25	26.462	0.038	6.873	0.145	181.871	0.005	42.344	6.161	1120.506
26	30.167	0.033	6.906	0.145	208.333	0.005	43.173	6.251	1302.377
27	34.390	0.029	6.935	0.144	238.499	0.004	43.929	6.334	1510.709
28	39.204	0.026	6.961	0.144	272.889	0.004	44.618	6.410	1749.209
29	44.693	0.022	6.983	0.143	312.094	0.003	45.244	6.479	2022.098
30	50.950	0.020	7.003	0.143	356.787	0.003	45.813	6.542	2334.192

A9.8 Values of Conversion Factors at $i = 16\%$

n	F_{ps}	F_{sp}	F_{rp}	F_{pr}	F_{rs}	F_{sr}	F_{gp}	F_{gr}	F_{gs}
1	1.160	0.862	0.862	1.160	1.000	1.000	0.000	0.000	0.000
2	1.346	0.743	1.605	0.623	2.160	0.463	0.743	0.463	1.000
3	1.561	0.641	2.246	0.445	3.506	0.285	2.024	0.901	3.160
4	1.811	0.552	2.798	0.357	5.066	0.197	3.681	1.316	6.666
5	2.100	0.476	3.274	0.305	6.877	0.145	5.586	1.706	11.732
6	2.436	0.410	3.685	0.271	8.977	0.111	7.638	2.073	18.609
7	2.826	0.354	4.039	0.248	11.414	0.088	9.761	2.417	27.587
8	3.278	0.305	4.344	0.230	14.240	0.070	11.896	2.739	39.001
9	3.803	0.263	4.607	0.217	17.519	0.057	14.000	3.039	53.241
10	4.411	0.227	4.833	0.207	21.321	0.047	16.040	3.319	70.759
11	5.117	0.195	5.029	0.199	25.733	0.039	17.994	3.578	92.081
12	5.936	0.168	5.197	0.192	30.850	0.032	19.847	3.819	117.814
13	6.886	0.145	5.342	0.187	36.786	0.027	21.590	4.041	148.664
14	7.988	0.125	5.468	0.183	43.672	0.023	23.217	4.246	185.450
15	9.266	0.108	5.575	0.179	51.660	0.019	24.728	4.435	229.122
16	10.748	0.093	5.668	0.176	60.925	0.016	26.124	4.609	280.781
17	12.468	0.080	5.749	0.174	71.673	0.014	27.407	4.768	341.706
18	14.463	0.069	5.818	0.172	84.141	0.012	28.583	4.913	413.379
19	16.777	0.060	5.877	0.170	98.603	0.010	29.656	5.046	497.520
20	19.461	0.051	5.929	0.169	115.380	0.009	30.632	5.167	596.123
21	22.574	0.044	5.973	0.167	134.841	0.007	31.518	5.277	711.503
22	26.186	0.038	6.011	0.166	157.415	0.006	32.320	5.377	846.344
23	30.376	0.033	6.044	0.165	183.601	0.005	33.044	5.467	1003.759
24	35.236	0.028	6.073	0.165	213.978	0.005	33.697	5.549	1187.360
25	40.874	0.024	6.097	0.164	249.214	0.004	34.284	5.623	1401.338
26	47.414	0.021	6.118	0.163	290.088	0.003	34.811	5.690	1650.552
27	55.000	0.018	6.136	0.163	337.502	0.003	35.284	5.750	1940.640
28	63.800	0.016	6.152	0.163	392.503	0.003	35.707	5.804	2278.142
29	74.009	0.014	6.166	0.162	456.303	0.002	36.086	5.853	2670.645
30	85.850	0.012	6.177	0.162	530.312	0.002	36.423	5.896	3126.948

A9.9 Values of Conversion Factors at $i = 18\%$

n	F_{ps}	F_{sp}	F_{rp}	F_{pr}	F_{rs}	F_{sr}	F_{gp}	F_{gr}	F_{gs}
1	1.180	0.847	0.847	1.180	1.000	1.000	0.000	0.000	0.000
2	1.392	0.718	1.566	0.639	2.180	0.459	0.718	0.459	1.000
3	1.643	0.609	2.174	0.460	3.572	0.280	1.935	0.890	3.180
4	1.939	0.516	2.690	0.372	5.215	0.192	3.483	1.295	6.752
5	2.288	0.437	3.127	0.320	7.154	0.140	5.231	1.673	11.968
6	2.700	0.370	3.498	0.286	9.442	0.106	7.083	2.025	19.122
7	3.185	0.314	3.812	0.262	12.142	0.082	8.967	2.353	28.564
8	3.759	0.266	4.078	0.245	15.327	0.065	10.829	2.656	40.706
9	4.435	0.225	4.303	0.232	19.086	0.052	12.633	2.936	56.033
10	5.234	0.191	4.494	0.223	23.521	0.043	14.352	3.194	75.118
11	6.176	0.162	4.656	0.215	28.755	0.035	15.972	3.430	98.640
12	7.288	0.137	4.793	0.209	34.931	0.029	17.481	3.647	127.395
13	8.599	0.116	4.910	0.204	42.219	0.024	18.877	3.845	162.326
14	10.147	0.099	5.008	0.200	50.818	0.020	20.158	4.025	204.545
15	11.974	0.084	5.092	0.196	60.965	0.016	21.327	4.189	255.363
16	14.129	0.071	5.162	0.194	72.939	0.014	22.389	4.337	316.328
17	16.672	0.060	5.222	0.191	87.068	0.011	23.348	4.471	389.267
18	19.673	0.051	5.273	0.190	103.740	0.010	24.212	4.592	476.335
19	23.214	0.043	5.316	0.188	123.414	0.008	24.988	4.700	580.075
20	27.393	0.037	5.353	0.187	146.628	0.007	25.681	4.798	703.489
21	32.324	0.031	5.384	0.186	174.021	0.006	26.300	4.885	850.117
22	38.142	0.026	5.410	0.185	206.345	0.005	26.851	4.963	1024.138
23	45.008	0.022	5.432	0.184	244.487	0.004	27.339	5.033	1230.482
24	53.109	0.019	5.451	0.183	289.494	0.003	27.772	5.095	1474.969
25	62.669	0.016	5.467	0.183	342.603	0.003	28.155	5.150	1764.464
26	73.949	0.014	5.480	0.182	405.272	0.002	28.494	5.199	2107.067
27	87.260	0.011	5.492	0.182	479.221	0.002	28.791	5.243	2512.339
28	102.967	0.010	5.502	0.182	566.481	0.002	29.054	5.281	2991.561
29	121.501	0.008	5.510	0.181	669.447	0.001	29.284	5.315	3558.041
30	143.371	0.007	5.517	0.181	790.948	0.001	29.486	5.345	4227.489

A9.10 Values of Conversion Factors at $i = 20\%$

n	F_{ps}	F_{sp}	F_{rp}	F_{pr}	F_{rs}	F_{sr}	F_{gp}	F_{gr}	F_{gs}
1	1.200	0.833	0.833	1.200	1.000	1.000	0.000	0.000	0.000
2	1.440	0.694	1.528	0.655	2.200	0.455	0.694	0.455	1.000
3	1.728	0.579	2.106	0.475	3.640	0.275	1.852	0.879	3.200
4	2.074	0.482	2.589	0.386	5.368	0.186	3.299	1.274	6.840
5	2.488	0.402	2.991	0.334	7.442	0.134	4.906	1.641	12.208
6	2.986	0.335	3.326	0.301	9.930	0.101	6.581	1.979	19.650
7	3.583	0.279	3.605	0.277	12.916	0.077	8.255	2.290	29.580
8	4.300	0.233	3.837	0.261	16.499	0.061	9.883	2.576	42.495
9	5.160	0.194	4.031	0.248	20.799	0.048	11.434	2.836	58.995

A9.10 (*Continued*).

n	F_{ps}	F_{sp}	F_{rp}	F_{pr}	F_{rs}	F_{sr}	F_{gp}	F_{gr}	F_{gs}
10	6.192	0.162	4.192	0.239	25.959	0.039	12.887	3.074	79.793
11	7.430	0.135	4.327	0.231	32.150	0.031	14.233	3.289	105.752
12	8.916	0.112	4.439	0.225	39.581	0.025	15.467	3.484	137.903
13	10.699	0.093	4.533	0.221	48.497	0.021	16.588	3.660	177.483
14	12.839	0.078	4.611	0.217	59.196	0.017	17.601	3.817	225.980
15	15.407	0.065	4.675	0.214	72.035	0.014	18.509	3.959	285.176
16	18.488	0.054	4.730	0.211	87.442	0.011	19.321	4.085	357.211
17	22.186	0.045	4.775	0.209	105.931	0.009	20.042	4.198	444.653
18	26.623	0.038	4.812	0.208	128.117	0.008	20.680	4.298	550.583
19	31.948	0.031	4.843	0.206	154.740	0.006	21.244	4.386	678.700
20	38.338	0.026	4.870	0.205	186.688	0.005	21.739	4.464	833.440
21	46.005	0.022	4.891	0.204	225.026	0.004	22.174	4.533	1020.128
22	55.206	0.018	4.909	0.204	271.031	0.004	22.555	4.594	1245.154
23	66.247	0.015	4.925	0.203	326.237	0.003	22.887	4.647	1516.184
24	79.497	0.013	4.937	0.203	392.484	0.003	23.176	4.694	1842.421
25	95.396	0.010	4.948	0.202	471.981	0.002	23.428	4.735	2234.905
26	114.475	0.009	4.956	0.202	567.377	0.002	23.646	4.771	2706.886
27	137.371	0.007	4.964	0.201	681.853	0.001	23.835	4.802	3274.264
28	164.845	0.006	4.970	0.201	819.223	0.001	23.999	4.829	3956.117
29	197.814	0.005	4.975	0.201	984.068	0.001	24.141	4.853	4775.340
30	237.376	0.004	4.979	0.201	1181.882	0.001	24.263	4.873	5759.408

A9.11 Values of Conversion Factors at $i = 22\%$

n	F_{ps}	F_{sp}	F_{rp}	F_{pr}	F_{rs}	F_{sr}	F_{gp}	F_{gr}	F_{gs}
1	1.220	0.820	0.820	1.220	1.000	1.000	0.000	0.000	0.000
2	1.488	0.672	1.492	0.670	2.220	0.450	0.672	0.450	1.000
3	1.816	0.551	2.042	0.490	3.708	0.270	1.773	0.868	3.220
4	2.215	0.451	2.494	0.401	5.524	0.181	3.127	1.254	6.928
5	2.703	0.370	2.864	0.349	7.740	0.129	4.607	1.609	12.453
6	3.297	0.303	3.167	0.316	10.442	0.096	6.124	1.934	20.192
7	4.023	0.249	3.416	0.293	13.740	0.073	7.615	2.230	30.635
8	4.908	0.204	3.619	0.276	17.762	0.056	9.042	2.498	44.374
9	5.987	0.167	3.786	0.264	22.670	0.044	10.378	2.741	62.136
10	7.305	0.137	3.923	0.255	28.657	0.035	11.610	2.959	84.806
11	8.912	0.112	4.035	0.248	35.962	0.028	12.732	3.155	113.464
12	10.872	0.092	4.127	0.242	44.874	0.022	13.744	3.330	149.426
13	13.264	0.075	4.203	0.238	55.746	0.018	14.649	3.485	194.300
14	16.182	0.062	4.265	0.234	69.010	0.014	15.452	3.623	250.046
15	19.742	0.051	4.315	0.232	85.192	0.012	16.161	3.745	319.056
16	24.086	0.042	4.357	0.230	104.935	0.010	16.784	3.852	404.248
17	29.384	0.034	4.391	0.228	129.020	0.008	17.328	3.947	509.182

A9.11 (*Continued*).

n	F_{ps}	F_{sp}	F_{rp}	F_{pr}	F_{rs}	F_{sr}	F_{gp}	F_{gr}	F_{gs}
18	35.849	0.028	4.419	0.226	158.405	0.006	17.803	4.029	638.202
19	43.736	0.023	4.442	0.225	194.254	0.005	18.214	4.101	796.607
20	53.358	0.019	4.460	0.224	237.989	0.004	18.570	4.163	990.860
21	65.096	0.015	4.476	0.223	291.347	0.003	18.877	4.218	1228.850
22	79.418	0.013	4.488	0.223	356.443	0.003	19.142	4.265	1520.197
23	96.889	0.010	4.499	0.222	435.861	0.002	19.369	4.306	1876.640
24	118.205	0.008	4.507	0.222	532.750	0.002	19.563	4.341	2312.500
25	144.210	0.007	4.514	0.222	650.955	0.002	19.730	4.371	2845.251
26	175.936	0.006	4.520	0.221	795.165	0.001	19.872	4.397	3496.206
27	214.642	0.005	4.524	0.221	971.102	0.001	19.993	4.419	4291.371
28	261.864	0.004	4.528	0.221	1185.744	0.001	20.096	4.438	5262.473
29	319.474	0.003	4.531	0.221	1447.608	0.001	20.184	4.454	6448.217
30	389.758	0.003	4.534	0.221	1767.081	0.001	20.258	4.468	7895.824

A9.12 Values of Conversion Factors at $i = 24\%$

n	F_{ps}	F_{sp}	F_{rp}	F_{pr}	F_{rs}	F_{sr}	F_{gp}	F_{gr}	F_{gs}
1	1.240	0.806	0.806	1.240	1.000	1.000	0.000	0.000	0.000
2	1.538	0.650	1.457	0.686	2.240	0.446	0.650	0.446	1.000
3	1.907	0.524	1.981	0.505	3.778	0.265	1.699	0.858	3.240
4	2.364	0.423	2.404	0.416	5.684	0.176	2.968	1.235	7.018
5	2.932	0.341	2.745	0.364	8.048	0.124	4.333	1.578	12.702
6	3.635	0.275	3.020	0.331	10.980	0.091	5.708	1.890	20.750
7	4.508	0.222	3.242	0.308	14.615	0.068	7.039	2.171	31.730
8	5.590	0.179	3.421	0.292	19.123	0.052	8.292	2.424	46.346
9	6.931	0.144	3.566	0.280	24.712	0.040	9.446	2.649	65.469
10	8.594	0.116	3.682	0.272	31.643	0.032	10.493	2.850	90.181
11	10.657	0.094	3.776	0.265	40.238	0.025	11.431	3.028	121.824
12	13.215	0.076	3.851	0.260	50.895	0.020	12.264	3.184	162.062
13	16.386	0.061	3.912	0.256	64.110	0.016	12.996	3.322	212.957
14	20.319	0.049	3.962	0.252	80.496	0.012	13.636	3.442	277.067
15	25.196	0.040	4.001	0.250	100.815	0.010	14.191	3.547	357.563
16	31.243	0.032	4.033	0.248	126.011	0.008	14.672	3.638	458.378
17	38.741	0.026	4.059	0.246	157.253	0.006	15.085	3.716	584.389
18	48.039	0.021	4.080	0.245	195.994	0.005	15.438	3.784	741.642
19	59.568	0.017	4.097	0.244	244.033	0.004	15.741	3.842	937.637
20	73.864	0.014	4.110	0.243	303.601	0.003	15.998	3.892	1181.669
21	91.592	0.011	4.121	0.243	377.465	0.003	16.216	3.935	1485.270
22	113.574	0.009	4.130	0.242	469.056	0.002	16.401	3.971	1862.735
23	140.831	0.007	4.137	0.242	582.630	0.002	16.557	4.002	2331.791
24	174.631	0.006	4.143	0.241	723.461	0.001	16.689	4.028	2914.421
25	216.542	0.005	4.147	0.241	898.092	0.001	16.800	4.051	3637.882
26	268.512	0.004	4.151	0.241	1114.634	0.001	16.893	4.069	4535.973
27	332.955	0.003	4.154	0.241	1383.146	0.001	16.971	4.085	5650.607

A9.12 (*Continued*).

n	F_{ps}	F_{sp}	F_{rp}	F_{pr}	F_{rs}	F_{sr}	F_{gp}	F_{gr}	F_{gs}
28	412.864	0.002	4.157	0.241	1716.101	0.001	17.036	4.099	7033.753
29	511.952	0.002	4.159	0.240	2128.965	0.000	17.091	4.110	8749.853
30	634.820	0.002	4.160	0.240	2640.916	0.000	17.137	4.119	10878.818

A10 Further Resources: Websites

About solar power	www.greenhouse.gov.au www.howstuffworks.com www.eere.energy.gov/solar
Australian Business Council for Sustainable Energy	www.bcse.org.au
BIPV simulation tools	www.pvsyst.com www.mauisolarsoftware.com www.valentin.de www.lapsys.co.jp/pro/pro_english.html
European Photovoltaic Industry Association (EPIA)	www.epia.org
IEA task 7 PV in the built environment	www.task7.org
International Energy Agency (IEA)	www.iea.org
PVPS programme and task reports	www.iea-pvps.org
International Solar Energy Society (ISES)	www.ises.org
Japan Photovoltaic Energy Association (JPEA)	www.ipea.gr.jp
NVPV virtual library	www.nreal.gov/ncpv/libbody.html
PV databases	www.pvdatabase.com
PV information weblinks	www.pvresources.com
PV manufacturers	www.solarbuzz.com www.pvportal.com www.pvresources.com
PV software	www.pvresources.com/en/software.php
PV standards	www.ieee.org www.iec.ch www.jrc.it
Solar Energy Industries Association (SEIA)	www.seia.org
Solar power in the built environment	www.fbe.unsw.edu.au
University of NSW PV Centre	www.pv.unsw.edu.au/solpages.html
US National Renewable Energy Laboratory (NREL)	www.nrel.gov
US NREL – National Centre for Photovoltaics (NCPV)	www.nrel.gov/ncpv

Glossary

Absorber plate	A component of the solar flat-plate collector that absorbs solar radiation and converts it into heat.
Absorptance	The ratio between the radiation absorbed by a surface (absorber) and the total amount of solar radiation striking the surface.
Air change per hour (ach)	A unit that denotes the number of times a house exchanges its entire volume of air with the outside air in an hour; also called number of air changes.
Air cleanser	A device used to remove airborne impurities.
Air mass (AM)	A measure of how far light travels through the Earth's atmosphere (see Figure 1.9).
Air permeability	The property of a building component to let air pass when it is subjected to a differential pressure.
Albedo	The ratio of the amount of light reflected by a surface to the light falling onto it.
Alternating current (ac)	An electric current that alternates direction between positive and negative cycles, usually 50 or 60 times per second. Alternating current is the current typically available from power outlets in a household.
Altitude	The Sun's angle above the horizon, as measured in a vertical plane.
Altitude angle (α)	The angle between the Sun's rays and a horizontal plane (see Figure 1.17).
Amorphous silicon (a-Si)	A thin-film PV silicon cell having no crystalline structure. Manufactured by depositing layers of doped silicon on a substrate.

RSC Energy Series No. 4
Building Integrated Photovoltaic Thermal Systems: For Sustainable Developments
By Basant Agrawal and G. N. Tiwari
© B. Agrawal and G. N. Tiwari 2011
Published by the Royal Society of Chemistry, www.rsc.org

Ampere (A)	A unit of electric current; a measure of flowing electrons.
Anemometer	Instrument used for measuring wind speed.
Angle of incidence (θ_i)	The angle between beam radiation on a surface and the normal to that surface (see Figure 1.20).
Antifreeze	Substance added to water to lower its freezing point. Solar water heaters usually use a mixture of water and propylene glycol instead of just water to prevent freezing.
Anti-reflection coating	A thin coating of a material applied to a photovoltaic cell surface, which reduces light reflection and increases light transmission.
Array	A number of solar modules connected together in a single structure.
Array current	The electrical current output of a PV array when exposed to sunlight.
Balance of system (BOS)	All the parts of a PV system excluding the solar module.
Battery	A collection of cells that store electrical energy; each cell converts chemical energy into electricity or *vice versa*, and is interconnected with other cells to form a battery for storing useful quantities of electricity.
Battery available capacity	Total maximum charge, expressed in ampere hours, that can be withdrawn from a cell or battery under a specific set of operating conditions, including discharge rate, temperature, initial state of charge, age and cutoff voltage.
Battery capacity	The maximum total electrical charge, expressed in ampere hours (Ah), that a battery can deliver to a load under a specific set of conditions.
Battery cell	The simplest operating unit in a storage battery. It consists of one or more positive electrodes or plates, an electrolyte that permits ionic conduction, one or more negative electrodes or plates, separators between plates of opposite polarity and a container for all the above.
Battery cycle life	Number of cycles, to a specified depth of discharge, that a cell or battery can undergo before failing to meet its specified capacity or efficiency performance criteria.
Battery energy capacity	Total energy available, expressed in watt hours, that can be withdrawn from a fully charged cell or battery. For a given cell it varies with temperature, rate, age and cutoff voltage.

Beam or direct radiation	Radiation coming directly from the Sun without its direction undergoing any change.
BIPV	Building integrated photovoltaics; also represented as BiPV.
BIPVT	Building integrated photovoltaics thermal; also represented as BiPVT.
Black body	A perfect absorber and emitter of radiation. Lampblack is close to a black body.
Brightness	The subjective human perception of luminance.
Bypass diode	A diode connected across one or more solar cells in a photovoltaic module such that the diode will conduct if the cell(s) become reverse biased.
Cadmium telluride (CdTe)	A polycrystalline thin-film photovoltaic material.
Calorific value	Energy content per unit mass (or volume) of a fuel, which will be released in combustion ($kWh\,kg^{-1}$, $MJ\,kg^{-1}$, $kWh\,m^{-3}$, $MJ\,m^{-3}$).
Candela (cd)	SI unit of luminous intensity. An ordinary candle has a luminous intensity of one candela.
Carbon dioxide (CO_2)	The colourless, odourless gas that is produced during normal human breathing. It is also emitted by combustion activities used to produce electricity. CO_2 is a major cause of the greenhouse effect that traps radiant energy near the Earth's surface.
Cell	A device that generates electricity, traditionally consisting of two plates or conducting surfaces placed in an electrolytic fluid.
Cell efficiency	The ratio of the electrical energy produced by a photovoltaic cell (under full Sun conditions or $1\,kW\,m^{-2}$) to the energy from sunlight falling upon the photovoltaic cell.
Celsius	The international temperature scale in which water freezes at 0 [degrees] and boils at 100 [degrees].
Charge controller	A component of a photovoltaic system that controls the flow of current to and from the battery to protect the batteries from over-charge and over-discharge.
Charge rate	The current applied to a cell or battery to restore its available capacity.
Circuit	A system of conductors (*i.e.* wires and appliances) capable of providing a closed path for an electric current
Circuit breaker	A circuit breaker acts like an automatic switch that can shut the power off when it senses too much current.

Clear sky	A sky condition with few or no clouds, usually taken as 0–2 tenths covered with clouds. Clear skies have high luminance and high radiation, and create strong shadows relative to more cloudy conditions.
Clerestory	A window that is placed vertically (or near vertical) in a wall above one's line of vision to provide natural light in a building.
Clothing factor (Clo)	A measure of the insulating value of clothing.
Collector	The name given to the device that converts incoming solar radiation to heat.
Collector efficiency	Ratio of the useful (heat) energy converted by the solar collector to the radiation incident on the device.
Collector plate	A component of the solar flat-plate collector that absorbs solar radiation and converts it into heat.
Comfort chart	A chart showing dry-bulb temperature and humidity (and sometimes air motion) by which the effects of various air conditions on human comfort may be compared.
Comfort zone	On the bioclimatic chart, the area of combined temperatures and humidities that 80% of people find comfortable.
Condensation	Process of vapour changing into the liquid state. Heat is released in the process.
Conductance	A measure of the ease with which heat flows through a specified thickness of a material by conduction. Units are $W\,m^{-2}\,°C^{-1}$.
Conduction	Process by which heat energy is transferred through materials (solids, liquids or gases) by molecular excitation of adjacent molecules.
Conductivity	Quantity of heat that will flow through one square metre of material, one metre thick, in one second, when there is a temperature difference of 1 °C between its surfaces.
Concentrating collector	A solar collector that reflects the solar radiation (direct) to an absorber plate to produce high temperatures.
Convection	The transfer of heat between a moving fluid medium (liquid or gas) and a surface, or the transfer of heat within a fluid by movements within the fluid.
Conversion efficiency	The ratio of the electrical energy generated by a solar PV cell to the solar energy impacting the cell

Copper indium gallium selenide (CIGS)	A polycrystalline thin-film photovoltaic material.
Cross ventilation	Ventilative cooling of people and spaces driven by the force of wind.
Crystalline silicon	A type of PV cell made from a single crystal or polycrystalline slices of silicon.
Current at maximum power (I_{mp})	The current at which maximum power is available from a module.
Cut-off voltage	The voltage levels at which the charge controller (regulator) disconnects the PV array from the battery, or the load from the battery.
Cycle	The discharge and re-charge of a battery; one complete charge–discharge cycle of the battery.
Czochralski process	A method of growing large-size, high-quality semiconductor crystals by slowly lifting a seed crystal from a molten bath of the material under careful cooling conditions.
Daylight	Illuminance from radiation in the visible spectrum from diffuse sky, reflected light and direct Sun that lights a room.
Declination (δ)	The angle of the Sun north or south of the equatorial plane (see Figure 1.16).
Deep cycle battery	A battery designed to regularly discharge 80% of its capacity before recharging.
Deep discharge	Discharging a battery by more than 80% of its full charge.
Dendritic web technique	A method for making sheets of polycrystalline silicon in which silicon dendrites are slowly withdrawn from a melt of silicon, whereupon a web of silicon forms between the dendrites and solidifies as it rises from the melt and cools.
Depth of discharge (DOD)	The amount of energy withdrawn from a battery or cell expressed as a percentage of its rated capacity.
Diffuse radiation	The solar radiation reaching the surface due to reflection and scattering effects.
Diffuse reflectance	Reflectance is the ratio of reflected radiation to incident radiation. Diffuse reflectance spreads the incident flux over a range of reflected angles or directions.
Direct current (dc)	The complement of ac, or alternating current, dc presents one unvarying voltage to a load. This is standard in motor vehicles.

Direct gain	The transmission of sunlight through glazing directly into the spaces to be heated, where it is converted to heat by absorption on interior mass surfaces.
Direct radiation	Radiation coming in a beam from the Sun, which can be focused.
Discharge	Withdrawal of electrical energy from a battery.
Dry bulb temperature (DBT)	The temperature of a gas or mixture of gases indicated by an accurate thermometer after correction for radiation.
Edge-defined film-fed growth (EFG)	A method for making sheets of polycrystalline silicon for photovoltaic devices in which molten silicon is drawn upward by capillary action through a mold.
Efficiency (η)	The ratio of output power (or energy) to input power (or energy) expressed as a percentage.
Electromagnetic spectrum	The entire range of wavelengths or frequencies of electromagnetic radiation extending from gamma rays to the longest radio waves and including visible light.
Embodied energy	The amount of energy required to produce an object in its final form.
Emissivity	The property of emitting heat by radiation, possessed by all materials to a varying extent.
Emittance	The numerical value for emissivity.
Energy	A measure of a system's ability to do work.
Energy audit	A survey that shows how much energy is used in a home, which helps find ways to use less energy.
Energy density	Energy required in producing unit mass of an object in its final form.
Energy intensity	The ratio of energy use in a sector to activity in that sector, *e.g.* the ratio of energy use to constant dollar production in manufacturing.
Energy payback time	The time required for any energy producing system to recover energy equal to that fed during its manufacture.
Epitaxial growth	The growth of one crystal on the surface of another crystal. The growth of the deposited crystal is oriented by the lattice structure of the original crystal.
Equinox	When the centre of the Sun is in the same plane as the Earth's equator; occurs twice a year.
Ethylene–vinyl acetate (EVA) polymer	An encapsulant used between the glass cover and the solar cells in PV modules. It is durable, transparent, resistant to corrosion and flame retardant

Evaporation	Phase change of a material from liquid to vapour at a temperature below the boiling point of the liquid. Cooling occurs during the process of evaporation.
Evaporative cooling	A heat removal process in which water vapour is added to air, increasing its humidity while lowering its temperature.
Extraterrestrial solar radiation	The solar radiation incident on the top of the terrestrial atmosphere.
Extrinsic semiconductor	The product of doping a pure semiconductor.
Fermi level	Energy level at which the probability of finding an electron is one-half. In a metal, the Fermi level is very near the top of the filled levels in the partially filled valence band. In a semiconductor, the Fermi level is in the band gap.
Fill factor (FF)	On an *I–V* (current–voltage) curve characterizing the output of a solar cell or module; the ratio of the maximum power to the product of the open-circuit voltage and the short-circuit current.
Flat-plate collector	A solar collection device for gathering the Sun's heat, consisting of a shallow metal container covered with one or more layers of transparent glass or plastic; either air or a liquid is circulated through the cavity of the container, whose interior is painted "black" and whose exterior is well insulated.
Float-zone process	A method of growing a large-size, high-quality crystal whereby coils heat a polycrystalline ingot placed on top of a single-crystal seed. As the coils are slowly raised the molten interface beneath the coils becomes a single crystal.
Fresnel lens	A stepped lens, positioned above and concave to a PV material to concentrate light on the material (see Figure 2.31).
Gallium arsenide (GaAs)	A crystalline, high-efficiency semiconductor or photovoltaic material.
Glare	The perception caused by a very bright light or a high contrast of light, making it uncomfortable or difficult to see.
Glazing	Transparent or translucent materials, usually glass or plastic, used to cover an opening without impeding (relative to opaque materials) the admission of solar radiation and light.
Global radiation	The sum of direct, diffuse and reflected radiation.
Greenhouse effect	The global warming resulting from the absorption of infrared solar radiation by carbon dioxide and other traces of gases present in the atmosphere.

Greenhouse gases	Gases which contribute to the greenhouse effect by absorbing infrared radiation in the atmosphere. These gases include carbon dioxide (CO_2), nitrous oxide (N_2O), methane (CH_4), water vapour and a variety of chlorofluorocarbons (CFCs).
Grid	An electrical utility distribution network.
Grid-connected	An energy producing system connected to the utility transmission grid.
Heat capacity	The quantity of heat required to raise one kilogram of a substance by 1 °C; product of mass and specific heat of any substance.
Heat gain	An increase in the amount of heat contained in a space, resulting from direct solar radiation and the heat given off by people, lights, equipment, machinery and other sources.
Heat loss	An unwanted decrease in the amount of heat contained in a space.
Hole	The vacancy where an electron would normally exist in a solid; behaves like a positively charged particle.
Hour angle (ω)	The angle through which the Earth must be rotated to bring the plane of the local meridian directly under the Sun.
Humidity	The amount of water vapour in the air.
HVAC	Mechanical system for heating, ventilating and air conditioning that controls temperature, humidity and air quality.
Hybrid system	A PV system that includes solar PV modules and some other electricity generating power source.
Hydrogenated amorphous silicon	Amorphous silicon with a small amount of incorporated hydrogen. The hydrogen neutralizes dangling bonds in the amorphous silicon, allowing charge carriers to flow more freely.
I–V curve	The plot of current *versus* voltage characteristics of a solar cell, module or array. *I–V* curves are used to compare various solar cell modules, and to determine their performance at various levels of insolation and temperatures (see Figure 2.9).
Illumination	Lighting of the surface by daylight or electric light.
Incident light	Light that shines on to the surface of a PV cell or module.
Incident radiation	The quantity of radiant energy striking a surface per unit time and unit area.

Infiltration	The uncontrolled movement of outdoor air into the interior of a building through cracks around windows and doors or in walls, roofs and floors. This may work by cold air leaking in during winters, or the reverse in summers.
Insolation (or incident solar radiation)	Measure of the amount of solar radiation falling on a given surface area in a given time; usually expressed in watt hours per square meter per day.
Insulation	A material that keeps energy from crossing from one place to another; on electrical wire, it is the plastic or rubber that covers the conductor; in a building, insulation makes the walls, floors and roof more resistant to the outside (ambient) temperature.
Internal gains	The energy dissipated inside the heated space by people and appliances. A portion of this energy contributes to the space heating requirement.
Intrinsic semiconductor	An undoped semiconductor.
Inverter	Electrical device that changes direct current (dc) into alternating current (ac).
Irradiance	The solar power incident on a surface, usually expressed in kilowatts per square meter.
Joule	Unit of energy or work. One joule is equal to one watt second.
Junction box	A PV junction box is a protective enclosure on a PV module where PV strings are electrically connected and where electrical protection devices such as diodes can be fitted.
Kilowatt (kW)	1000 watts; energy consumption at a rate of 1000 joules per second.
Kilowatt hour (kWh)	One kilowatt of power used for one hour. A typical house uses 750 kWh per month.
Latent heat	Change of enthalpy during a change of state.
Latitude (ϕ)	The angular position of a location north or south of the equator (see Figure 1.11).
Life cycle cost	The estimated cost of owning, operating and disposing of a system over its useful life.
Life cycle costing	A method for estimating the comparative costs of alternative energy or other systems. Life cycle costing takes into consideration such long-term costs as energy consumption, maintenance and repair.
Longitude	The angular position east or west of the Greenwich meridian (see Figure 1.12).

Low-iron glass	Glass which contains only a small percentage of iron oxide and as a result has a high transmittance.
Lux	SI unit of illuminance; it is the illuminance produced on a surface of unit area (square metre) by a luminous flux of 1 lumen uniformly distributed over that surface.
Masonry	Concrete, concrete block, brick, adobe, stone and similar building materials.
Maximum power (P_{max})	The maximum power is achieved at a point on the *I–V* characteristics where the product of current and voltage is largest.
Maximum power point (MPP)	The voltage at which a PV array produces maximum power.
Maximum power point tracker (MPPT)	A power conditioning unit that increases the power of a PV system by ensuring operation of the PV generator at its maximum power point (MPP). The ability to do this can depend on climate and the battery's state of charge.
Module	An encapsulated panel containing a number of electrically connected PV cells.
Monocrystalline silicon (sc-Si) solar cell	A form of solar cell made from a thin slice of a single large crystal of silicon; also referred as single crystalline.
Multicrystalline silicon (μc-Si)	A material that has solidified at a rate such that many small crystals (crystallites) form. The multitude of grain boundaries in the material (between the crystallites) reduces the cell efficiency. Also referred to as polycrystalline.
N-type semiconductor	A semiconductor produced by the doping of an intrinsic semiconductor with an electron-donor impurity, for example phosphorus in silicon.
Natural convection	The natural convection of heat through the fluid in a body that occurs when warm, less-dense fluid rises and cold, dense fluid sinks under the influence of gravity.
Night insulation	Movable insulation that covers a glazing at night and is removed during the day.
Night sky radiation	A reversal of the daytime insolation principle. Just as the Sun radiates energy during the day through the void of space, so also heat energy can travel unhindered at night from the Earth's surface back into space. On a clear night, any warm object can cool itself by radiating longwave heat energy to the cooler sky. On a cloudy night, the cloud cover acts as an insulator and prevents the heat from travelling to the cooler sky.

Nominal operating cell temperature (NOCT)	Estimated temperature of a PV module when operating under 800 W m^{-2} irradiance, 20 °C ambient temperature and wind speed of 1 m s^{-1}.
Nominal voltage	A rounded voltage value used to describe batteries, modules or systems based on their specification (*e.g.* a 12, 24 or 48 V battery, module or system).
Number of air changes	Denotes the number of times a house exchanges its entire volume of air with the outside air in an hour.
One-axis tracking	A PV system structure that is capable of rotating on a single axis in order to track the movement of the Sun.
Opaque	Not transparent to radiation; usually applied to a material of high absorptance.
Open circuit voltage (V_{oc})	The maximum possible voltage across a solar module or array. Open circuit voltage occurs in sunlight when no current is flowing.
Orientation	Placement of a surface, measured by its slope and azimuth angles.
Overall heat transfer coefficient	For heat transfer by conduction, convection and radiation, the heat transfer per unit area can be found by the product of the overall coefficient and the overall temperature difference; the heat transfer can be a loss, and the coefficient is then termed an overall heat loss coefficient.
Overcharge	Applying current to a fully charged battery. This can damage the battery.
P-type semiconductor	A semiconductor produced by the doping of an intrinsic semiconductor with a hole-donor impurity, for example boron or gallium in silicon.
Panel	Used interchangeably with "module".
Passive solar home	A house that utilizes part of the building as a solar collector.
Passive solar system	A system that uses non-mechanical and non-electrical means to satisfy heating, lighting or cooling loads. Purely passive systems use radiation, conduction and natural convection to distribute heat, and daylight for lighting.
Peak load	The maximum usage of electrical power occurring in a given period of time, typically a day. The electrical supply must be able to meet the peak load if it is to be reliable.
Peak Sun hours	The equivalent number of hours per day when solar irradiance averages 1000 W m^{-2}.

Peak watt (W_p)	The amount of power a solar PV cell or module will produce under standard test conditions (normally 1000 W m^{-2} and 25 °C cell temperature, AM 1.5 spectrum).
Photon	The elementary particle of electromagnetic energy; light (Greek *photos*, light)
Photovoltaic (PV)	A technology for using semiconductors to convert light directly into electricity.
Photovoltaic array	A number of PV modules connected together in a single structure.
Photovoltaic cell	The smallest discrete element in a PV module that performs the conversion of light into electrical energy to produce a dc current and voltage.
Photovoltaic conversion efficiency	The ratio of the electric power produced by a photovoltaic device to the power of the sunlight incident on the device.
Photovoltaic device	A device that converts light directly into dc electricity.
Photovoltaic effect	The effect that causes a voltage to be developed across the junction of two different materials when they are exposed to light.
Photovoltaic module	The basic building block of a photovoltaic device, which consists of a number of interconnected solar cells.
Polycrystalline cell	A wafer of silicon with a multi-grained structure.
Polycrystalline silicon	A material used to make PV cells, which consists of many crystals in contrast to single-crystal silicon.
Poly(vinyl chloride) (PVC)	A plastic used as an insulator on electrical cables.
Power	The rate at which energy is consumed or produced. The unit is the watt.
Present worth or cost	The amount of money which must be invested today at the market discount rate in order to have a specified amount at a specified future time.
Pyranometer	An instrument for measuring total hemispherical solar irradiance on a flat surface, *i.e.* global irradiance.
Pyrheliometer	Instrument used to measure the direct component of solar radiation at normal incidence.
Radiance	Power per unit solid angle per unit projected source area.
Radiant exitance/radiant emittance	Power emitted from a surface.
Radiant heat transfer	The transfer of heat by radiation. Heat radiation is a form of electromagnetic radiation. Radiant heating due to infrared radiation is commonly employed in passive systems.

Radiant temperature	The average temperature of surfaces surrounding a person or surface, with which the person or surface can exchange thermal radiation.
Radiation	Electromagnetic waves that directly transport energy through space. Sunlight is a form of radiation.
Radiosity	Emitted plus reflected power leaving a surface.
Rated battery capacity (Ah)	The maximum amount of energy that can be withdrawn from a battery at a specified discharge rate and temperature.
Reflectivity	The ratio of radiant energy reflected by a body to that falling upon it.
Relative humidity	The ratio of the partial pressure of water vapour (in a gaseous mixture of air and water vapour) to the saturated vapour pressure of water at a prescribed temperature.
Renewable energy	An energy source that renews itself without effort; solar energy is renewable because the light from the Sun that we harvest today has no effect on the light that we can harvest tomorrow.
Resistor	Any electronic component that restricts the flow of electrical current in circuits.
Reverse bias	Condition where the current producing capability of a PV cell is significantly less than that of other cells in its series string. This can occur when a cell is shaded, cracked or otherwise degraded or when it is electrically poorly matched with other cells in its string.
Semiconductor	A material such as silicon, which has a crystalline structure that will allow current to flow under certain conditions. Semiconductors are usually less conductive than metals, but not insulators like rubber.
Sensible heat	Heat that results in a change in air temperature.
Short circuit current (I_{sc})	Current across the terminals when a solar cell or module in strong sunlight is not connected to a load (measured with an ammeter).
Silicon (Si)	A semiconductor material commonly used to make PV cells.
Single-crystal structure	A material having a crystalline structure such that a repeatable or periodic molecular pattern exists in all three dimensions
Solair temperature	An equivalent temperature which will produce the same heating effect as the incident radiation in conjunction with the actual external air temperature.

Solar altitude	The Sun's angle above the horizon, as measured in a vertical plane.
Solar azimuth angle (γ_s)	The angle in a horizontal plane between the line due south and the projection of beam radiation on the horizontal plane (see Figure 1.19).
Solar cell	A device that converts light energy or solar radiation (photons) directly into dc electricity.
Solar cell module	Groups of encapsulated solar cells framed in a glass or plastic unit, usually the smallest unit of solar electric equipment available to the consumer.
Solar collector	A device that gathers and accumulates solar radiation to produce heat.
Solar concentrator	A device that uses reflective surfaces in a planar, parabolic trough or parabolic bowl configuration to concentrate solar radiation onto a smaller surface.
Solar constant	The amount of radiation arriving from the Sun at the edge of the Earth's atmosphere. The accepted value is about $1367\,W\,m^{-2}$.
Solar cooling	The use of solar thermal energy or solar electricity to power a cooling appliance.
Solar declination	The angle of the Sun north or south of the equatorial plane.
Solar energy	The electromagnetic radiation generated by the Sun.
Solar gain	Heat transferred to a space by solar radiation through glazing.
Solar incident angle (θ_i)	The angle at which an incoming solar beam strikes a surface.
Solar module	A device used to convert light from the sun directly into dc electricity by using the photovoltaic effect. Usually made of multiple solar cells bonded between glass and a backing material.
Solar noon	The mid-point between sunrise and sunset; the time when the Sun reaches its highest point in its daily arc across the sky.
Solar radiation	The radiant energy received from the Sun, from both direct and diffuse or reflected sunlight.
Solar spectrum	The total distribution of electromagnetic radiation emitted from the Sun.
Solar time	The time measured with respect to the Sun.
Solstice	This happens twice each year, when the tilt of the Earth's axis is most inclined toward or away from the Sun, causing the Sun's apparent position in the sky to reach its northernmost or southernmost extreme.

Specific heat	A measure of the ability of a material to store heat. Specifically, the quantity of heat required to raise the temperature of unit mass of a substance by one degree ($J\,kg^{-1}\,K^{-1}$).
Specific humidity	The ratio of water vapour to air (including water vapour and dry air) in a particular mass.
Staebler–Wronski effect	The tendency of the sunlight to electricity conversion efficiency of amorphous silicon photovoltaic devices to degrade (drop) upon initial exposure to light.
Stand alone PV (SAPV) system	A solar PV system that operates without connection to a grid a supply of electricity.
Standard test conditions (STC)	Conditions under which a module is typically tested in a laboratory (irradiance intensity of $1000\,W\,m^{-2}$, AM1.5 solar reference spectrum and a cell/module temperature of $25\pm2\,°C$).
Standard time	The time reckoned with respect to the standard meridian of a specific time zone. It is the everyday clock time used within each one hour zone.
String	A number of cells, modules or panels interconnected electrically in series to produce the required operating voltage.
Substrate	The physical material upon which a photovoltaic cell is made.
Sunshine recorder	An instrument used to measure the duration (in hours) of bright sunshine during the course of the day.
Superstrate	The covering on the Sun side of a PV module, providing protection for the PV materials from impact and environmental degradation while allowing maximum transmission of the appropriate wavelengths of the solar spectrum.
Surface azimuth angle (γ)	The angle in the horizontal plane between the line due south and the projection of the normal to the inclined surface on the horizontal plane (see Figure 1.18).
Surge	An excessive amount of power drawn by an appliance when it is first switched on.
Temperature	Degree of hotness or coldness measured on one of several arbitrary scales based on some observable phenomenon.
Temperature compensation	Adjustment *via* the use of electronic circuitry to change the charge controller activation points, depending on battery temperature.
Temperature swing	The range of indoor temperatures in the building between day and night.

Terrestrial radiation	The longwave electromagnetic energy emitted by the Earth's surface and by the gases, aerosols and clouds of the atmosphere; it is also partly absorbed within the atmosphere.
Thermal conductivity	The ability of a material to conduct heat.
Thermal mass	A material used to store heat, thereby slowing the temperature variation within a space.
Thin-film silicon	Usually amorphous silicon and non-crystalline material used to make PV cells.
Tilt angle	The angle at which a solar collector is tilted upwards from the horizontal surface for maximum heat collection.
Tracking array	A PV array that is moved to follow the path of the Sun in order to maintain the maximum incident solar radiation on its surface.
Transmittance	The ratio of the solar radiation transmitted through glass to the total radiant energy falling on its surface.
Trombe wall	A vertical glass wall facing south (northern hemisphere) with a solid masonry wall for heat storage behind it; natural circulation carries the energy to the space to be heated.
Two-axis tracking	A system capable of rotating independently about two axes (*e.g.* vertical and horizontal) and following the Sun for maximum efficiency of the solar array.
U-value	The amount of heat that flows in or out of a system at steady state in one hour, when there is a one degree difference in temperature between the fluid inside and outside.
Ultraviolet radiation	A portion of the electromagnetic radiation in the wavelength range 4–400 nm.
Universe time	The time ascertain with respect to the zero meridian. It is the same as Greenwich meantime.
Visible radiation	Electromagnetic radiation in the wavelength band 0.38–0.78 μm.
Volt	The unit of electric potential and potential difference. The amount of work done per unit charge in moving a charge from one place to another.
Voltage at maximum power (V_{mp})	The voltage at which a PV device is operating at maximum power.
Voltage drop	The voltage lost along a length of wire or conductor due to the resistance of that conductor.
Voltage regulator	A device that controls the operating voltage of a photovoltaic array.

Wafer	Raw material for a solar cell; a thin sheet of crystalline semiconductor material is made by mechanically sawing it from a single-crystal boule or by casting it.
Watt (W)	The unit of electrical power commonly used to define the electricity consumption of an appliance.
Watt hour (Wh)	A common energy measure arrived at by multiplying the power by the hours of use. Grid power is ordinarily sold and measured in kilowatt hours.
Watt peak (W_p)	The number of watts output when a PV module is illuminated under standard conditions.
Wavelength	The distance between two similar points of a given wave.
Wet-bulb temperature (WBT)	The air temperature measured using a thermometer with a wetted bulb moved rapidly through the air to promote evaporation.
Zenith	The top of the sky dome. A point directly overhead, 90° in altitude angle above the horizon.
Zenith angle	The angle between the Sun's rays and a perpendicular line to the horizontal plane (see Figure 1.17).

Bibliography

M. K. Selçuk, Glossary of terms used in solar energy, *Sol. Energy*, 1984, **33**, 97–114.

Subject Index

Page numbers in *italics* refer to Figures and Tables